Springer Handbook of Auditory Research

For further volumes:
http://www.springer.com/series/2506

Sunil Puria • Richard R. Fay • Arthur N. Popper
Editors

The Middle Ear

Science, Otosurgery, and Technology

With 60 Illustrations

Springer

Editors
Sunil Puria
Department of Mechanical Engineering
and Otolaryngology–Head
and Neck Surgery
Stanford University
Stanford, CA, USA

Richard R. Fay
Marine Biological Laboratory
Woods Hole, MA, USA

Arthur N. Popper
Department of Biology
University of Maryland
College Park, MD, USA

ISSN 0947-2657
ISBN 978-1-4614-6590-4 ISBN 978-1-4614-6591-1 (eBook)
DOI 10.1007/978-1-4614-6591-1
Springer New York Heidelberg Dordrecht London

Library of Congress Control Number: 2013932708

Printed on acid-free paper

Springer is part of Springer Science+Business Media (www.springer.com)

*This volume is dedicated by
Sunil to his Love, Neshie.*

Series Preface

The following preface is the one that we published in Volume 1 of the Springer Handbook of Auditory Research (SHAR) back in 1992. Thus, 2012 marks the 20th year of SHAR. As anyone reading the original preface, or the many users of the series, will note, we have far exceeded our original expectation of eight volumes. Indeed, with books published to date and those in the pipeline, we are now set for more than 50 volumes in SHAR, and we are still open to new and exciting ideas for additional books.

We are very proud that there seems to be consensus, at least among our friends and colleagues, that SHAR has become an important and influential part of the auditory literature. While we have worked hard to develop and maintain the quality and value of SHAR, the real value of the books is very much because of the numerous authors who have given their time to write outstanding chapters and to our many coeditors who have provided the intellectual leadership to the individual volumes. We have worked with a remarkable and wonderful group of people, many of whom have become great personal friends of both of us. We also continue to work with a spectacular group of editors at Springer, currently Ann Avouris. Indeed, several of our past editors have moved on in the publishing world to become senior executives. To our delight, this includes the current president of Springer US, Dr. William Curtis.

But the truth is that the series would and could not be possible without the support of our families, and we want to take this opportunity to dedicate all of the SHAR books, past and future, to them. Our wives, Catherine Fay and Helen Popper, and our children, Michelle Popper Levit, Melissa Popper Levinsohn, Christian Fay, and Amanda Fay, have been immensely patient as we developed and worked on this series. We thank them and state, without doubt, that this series could not have happened without them.

Preface 1992

The Springer Handbook of Auditory Research presents a series of comprehensive and synthetic reviews of the fundamental topics in modern auditory research. The volumes are aimed at all individuals with interests in hearing research including advanced graduate students, postdoctoral researchers, and clinical investigators. The volumes are intended to introduce new investigators to important aspects of hearing science and to help established investigators to better understand the fundamental theories and data in fields of hearing that they may not normally follow closely.

Each volume presents a particular topic comprehensively, and each serves as a synthetic overview and guide to the literature. As such, the chapters present neither exhaustive data reviews nor original research that has not yet appeared in peer-reviewed journals. The volumes focus on topics that have developed a solid data and conceptual foundation rather than on those for which a literature is only beginning to develop. New research areas will be covered on a timely basis in the series as they begin to mature.

Each volume in the series consists of a few substantial chapters on a particular topic. In some cases, the topics will be ones of traditional interest for which there is a substantial body of data and theory, such as auditory neuroanatomy (Vol. 1) and neurophysiology (Vol. 2). Other volumes in the series deal with topics that have begun to mature more recently, such as development, plasticity, and computational models of neural processing. In many cases, the series editors are joined by a coeditor having special expertise in the topic of the volume.

Richard R. Fay, Falmouth, MA
Arthur N. Popper, College Park, MD

Volume Preface

To date, the middle ear has not been the focus of any single SHAR volume despite its importance in auditory function. In this volume, however, we take a broad look at this structure from a wide range of interdisciplinary perspectives, starting with basic science and evolutionary approaches and ending at clinical issues.

In Chap. 2, Manley and Sienknecht discuss the evolution and embryonic development of the middle ear, while in Chap. 3 Rosowski compares the middle ears across diverse vertebrate species.

In Chap. 4, Voss, Nakajima, Huber, and Shera review the overall physiological functioning of normal and diseased middle ears. In Chap. 5, Dirckx, Marcusohn, and Gaihede focus on mechanisms by which the balance of pressure is maintained between the middle ear and the atmosphere, while Stenfelt, in Chap. 6, focuses on mechanisms of bone conduction. In Chap. 7, Funnell, Maftoon, and Decraemer describe the role of computational approaches in helping to further our understanding of middle ear structure and function.

Clinical issues are more specifically discussed starting in Chap. 8, where Popelka and Hunter describe the clinical techniques for measuring and diagnosing the human middle ear. In Chap. 9, Merchant and Rosowski follow this with a description and discussion of the various middle ear pathologies that clinicians are able to repair as well as the surgical procedures they use. Finally, in Chap. 10, Puria describes various types of hearing devices that operate by mechanically vibrating the middle ear.

While the middle ear has not been the focus of past volumes, it has been discussed in chapters throughout the series. These include a chapter on the outer and middle ears by Rosowski in Volume 4, *Comparative Hearing: Mammals* (edited by Fay and Popper, 1993) and a chapter in Volume 6, *Auditory Computation* (edited by Hawkins, McMullen, Popper, and Fay, 1996) on models, also by Rosowski. The middle ear in birds and mammals was discussed in a chapter by Saunders et al. in Volume 13, *Comparative Hearing: Birds and Reptiles* (edited by Dooling, Fay, and

Popper, 2000), and the middle ear in amphibians was covered in a chapter by Mason and Narins in Volume 28, *Hearing and Sound Communication in Amphibians* (edited by Narins, Feng, Fay, and Popper, 2007).

Sunil Puria, Stanford, CA
Richard R. Fay, Falmouth, MA
Arthur N. Popper, College Park, MD

Editors' Note

We are saddened by the announcement that co-author of Chapter 9 Saumil N. Merchant, MD passed away on June 27th 2012. He was one of the finest clinicians, researcher scientist, scholar and teacher. To his patients he generously gave the gift of time and used his surgical skills to enable them to hear. We will forever miss his warm and infectious smile and guiding presence in our lives.

Contents

Contributors

Willem F. Decraemer Laboratory of BioMedical Physics, University of Antwerp, Antwerp, Belgium

Joris J.J. Dirckx Laboratory of Biomedical Physics, University of Antwerp, Antwerp, Belgium

W. Robert J. Funnell Departments of BioMedical Engineering and Otolaryngology – Head & Neck Surgery, McGill University, Montréal, QC, Canada

Michael L. Gaihede Department of Otolaryngology, Head and Neck Surgery, Aalborg University Hospital, DK-9000 Aalborg, Denmark

Alexander M. Huber Department of Otorhinolaryngology, University Hospital of Zurich, Zurich, Switzerland

Lisa L. Hunter Department of Otolaryngology and Division of Audiology, University of Cincinnati, Cincinnati, OH, USA

Nima Maftoon Department of BioMedical Engineering, McGill University, Montréal, QC, Canada

Geoffrey A. Manley Cochlear and Auditory Brainstem Physiology, IBU, Faculty V, Carl von Ossietzky University Oldenburg, Oldenburg, Germany

Yael Marcusohn Zichron Ya'acov, Israel

Saumil N. Merchant† Massachusetts Eye and Ear Infirmary, Boston, USA

Hideko Heidi Nakajima Department of Otology and Laryngology, Harvard Medical School and Eaton-Peabody Laboratories, Massachusetts Eye and Ear Infirmary, Boston, MA, USA

†Deceased

Gerald R. Popelka Department of Otolaryngology–Head and Neck Surgery, Stanford University, Stanford, CA, USA

Sunil Puria Departments of Mechanical Engineering and Otolaryngology–Head and Neck Surgery, Stanford University, Stanford, CA, USA

EarLens Corporation, Redwood City, CA, USA

John J. Rosowski Eaton-Peabody Laboratories, Massachusetts Eye and Ear Infirmary, Boston, MA, USA

Christopher A. Shera Eaton-Peabody Laboratories, Massachusetts Eye and Ear Infirmary, Boston, MA, USA

Ulrike J. Sienknecht Cochlear and Auditory Brainstem Physiology, IBU, Faculty V, Carl von Ossietzky University Oldenburg, Oldenburg, Germany

Stefan Stenfelt Department of Clinical and Experimental Medicine/Technical Audiology, Linköping University, Linköping, Sweden

Susan E. Voss Picker Engineering Program, Smith College, Ford Hall Northampton, MA, USA

Chapter 1
The Middle Ear: Science and Applications

Sunil Puria

Keywords Cochlea • Conductive hearing impairment • Hearing aids • Hearing devices • Middle ear cavity • Middle ear development • Middle ear evolution • Middle ear muscles • Ossicles • Sensorineural hearing impairment • Tympanic membrane

The clinical and scientific study of the middle ear attracts professionals from disciplines as diverse as evolutionary and developmental biology, biophysics, engineering, otology, and audiology; however, because each of these professions works with its own set of journals and societies, it can be difficult to find a single resource that provides comprehensive overviews of the corresponding wide-ranging literature. This volume aims to provide just such a resource, for newcomers and specialists alike, by compiling knowledge bases and gateways to the literature for the major subfields of middle ear study.

Regardless of whether they concern themselves primarily with promoting scientific, surgical, or technological advancements, each discipline of middle ear study is intimately concerned with the functional implications of middle ear structure. In the chapters of this volume, one can appreciate the evolutionary wonder of the mammalian middle ear and its unique structural suitability for high-frequency hearing, as well as the various avenues by which researchers and engineers continue to leverage their understanding of middle ear structure–function relationships to deliver the practical results of new and improved diagnostic methods, surgical procedures, passive prostheses for repairing the middle ear, and devices for sound amplification through the direct stimulation of the middle ear and cochlea.

S. Puria (✉)
Departments of Mechanical Engineering and Otolaryngology–Head and Neck Surgery, Stanford University, 496 Lomita Mall, Stanford, CA 94305, USA

EarLens Corporation, 200 Chesapeake Drive, Redwood City, CA 94063, USA
e-mail: puria@stanford.edu

S. Puria et al. (eds.), *The Middle Ear: Science, Otosurgery, and Technology*,
Springer Handbook of Auditory Research 46, DOI 10.1007/978-1-4614-6591-1_1,

In Chap. 2, Geoffrey A. Manley and Ulrike J. Sienknecht discuss the development of the middle ear, in terms of both its evolutionary history and its growth within a developing embryo. They argue against the idea that the middle ear evolved as a direct result of vertebrates transitioning from water to land, arguing instead that the development of the tympano-ossicular system did not occur until more than 100 million years later. They also argue that each of the three middle ear bones of mammals evolved independently, rather than through the addition of two more bones to the less complex one-bone system of amphibians, birds, and other nonmammals. High-frequency hearing appears to have arisen in small mammals over a very long period of time, and very possibly to the detriment of low-frequency hearing. Sensitivity to low frequencies in larger mammals and more specialized small mammals likely evolved later.

The results of developmental studies are also discussed, which suggest that the primary jaw joint of nonmammals, as well as the columella/stapes and the malleus and incus, all arise from a common developmental foundation that can transform into these diverse structures through processes controlled by gene patterning and cellular interactions. Changes in the number of genes and in their temporal and spatial expression during development can then lead, in turn, to the kinds of morphological transformations that are observed over evolutionary time.

The significant variability in the middle ear anatomies of nonmammals such as amphibians, reptiles, and birds, as well as land and marine mammals, is well known. In Chap. 3, John J. Rosowski summarizes the results from a number of studies comparing the middle ears of these different vertebrates, and argues that the wide variations in hearing capabilities among different mammalian species correlate with the form and size of their middle ear structures, with these in turn correlating with body size. He further argues that the different parts of the middle ear and the inner ear coadapted in parallel with one another to meet the demands required for survival, rather than developing independently of one another.

The approach taken by Susan E. Voss, Hideko Heidi Nakajima, Alex M. Huber, and Chris A. Shera in Chap. 4 is to review the functional differences between a normal middle ear and middle ears with alterations due to the effects of disease or other structural changes (e.g., tympanic-membrane perforations, stapes fixation, stapes disarticulation, middle ear fluid). They also describe techniques for performing accurate in vivo and in vitro physiological measurements (e.g., impedance, eardrum motions, 3D ossicular vibrations, cochlear fluid pressure) of both human and animal ears, which vary relatively smoothly with frequency, and use simple mathematical models to provide a theoretical framework for drawing conclusions from such measurements.

In Chap. 5, Joris J. J. Dirckx, Yael Marcusohn, and Michael L. Gaihede describe the different physiological mechanisms for actively controlling the volume of gas in the middle ear cleft, which is the combined airspace of the middle ear cavity and the mastoid, and thus maintaining a pressure balance between the middle ear cleft and the atmosphere. Changes to the volume of the cleft (e.g., due to a thickening of the mucosa) can alter the pressure balance. Gas can enter or leave the middle ear cavity either through the Eustachian tube during the action of swallowing, or by gas exchange through the mucosa either by diffusion or by perfusion. The overall regulation of middle ear pressure includes active neural feedback control based possibly on peripheral mechanoreceptors. Various pathologies can occur when the pressure balance is not maintained (particularly in children), causing a disruption of the normal function of the tympano-ossicular system.

Although hearing depends primarily on sound reaching the cochlea after passing through the tympano-ossicular system via air conduction, in Chap. 6, Stefan Stenfelt explains some of the known ways that sound can alternately reach the cochlea via bone conduction. While the measurement of bone-conducted hearing is clinically important, the mechanisms of bone conduction—and their relationship to air-conducted hearing—are still not well understood. Changes to the mass and stiffness of the middle ear can affect bone-conducted hearing, although the middle ear generally has less of an effect on the bone-conduction route of hearing than it does on the air-conduction route. For this and other reasons, a number of implantable (percutaneous and transcutaneous) and nonimplantable hearing devices (such as cross-aids and dental transducers) have been developed that make use of the bone-conduction route to stimulate the cochlea.

In Chap. 7, W. Robert J. Funnell, Nima Maftoon, and Willem F. Decraemer describe how various computational approaches have helped to codify our modern understanding of middle ear structure and function. Finite-element models, which offer realistic representations of anatomic features and material properties but can be computationally expensive, are contrasted with simpler two-port and circuit modeling approaches. Finite-element models have been formulated for the simulation of middle ear prostheses and implants, perforations and pathologies of the eardrum, ventilation tubes, fluid in the middle ear cavity, and bone conduction, as well as reverse transmission to determine the effects of the middle ear on ear-canal measurements of otoacoustic emissions. Finite-element models can also be quite useful for studies of the effects of anatomic variability and changes in material properties. Although a majority of the finite-element models have been formulated for the human ear, others have been formulated and validated for cat and gerbil ears. Most of these models have not incorporated the effects of active control of the middle ear muscles. Another area of research interest discussed in this chapter is the measurement and modeling of the high-frequency behavior of the middle ear.

In Chap. 8, Gerald R. Popelka and Lisa L. Hunter delve into the existing conventional, and upcoming, clinical measures and technologies used to quantify the physical and functional status of the human middle ear in a minimally invasive manner. These include measures based on behavioral voluntary responses, physical attributes, and physiological responses, as well as the use of recent imaging techniques. From the individual and combined results of these methods, assessments of the different components making up the middle ear, including the two middle ear muscles, can be formed. Significant challenges still lie ahead with regard to quantifying conductive and sensorineural hearing impairments at frequencies above 4–6 kHz, which would make it possible to assess and potentially improve surgical outcomes with respect to high-frequency hearing. This requires the development of clinical techniques for producing and calibrating high-frequency acoustic stimuli, as well as the development of suitable bone-conduction transducers.

When a conductive hearing impairment is diagnosed, an ENT (ear, nose, and throat) physician or a more specialized otologist might recommend surgical repair or reconstruction of the middle ear. In Chap. 9, Saumil N. Merchant and John J. Rosowski outline the different kinds of pathologies that clinicians can repair, as well as the approaches they use. Well-covered surgical procedures include the repair of the tympanic membrane, as well as the reconstruction of eroded or missing middle ear bones resulting from, for example, chronic otitis media. In other cases, a stapedectomy is performed to remove a fixation of the stapes to the surrounding bone. Various passive prostheses can be used to repair the ossicles to alleviate conductive impairment. Outlined areas for future development include better understanding the structure–function correlations for reconstructed eardrums, improving methods of coupling passive prostheses to the stapes, and finding better methods of assessing the effects of tension in ossicular reconstruction.

Although acoustic hearing aids are currently the standard of care in cases of sensorineural hearing impairment, in Chap. 10, Sunil Puria describes various types of middle ear hearing devices (MEHDs) that mechanically vibrate the middle ear. The basic configurations of acoustic hearing aids and MEHDs are very similar, with the primary difference being in the output transducer: acoustic hearing aids use a tiny loudspeaker whereas MEHDs typically use a tiny mechanical actuator. The latter has the potential to overcome limitations of acoustic hearing aids by offering a broad-spectrum output, an increased gain margin due to reduced feedback, and better sound quality. A new classification system for transducers is introduced, which is based on the number of required anatomical connection points and how those points are linked to one another. Surgical devices are discussed, including totally and partially implanted systems, as well as nonsurgical, nonimplanted devices that contact and mechanically vibrate the eardrum. Technologies such as these have the potential to change the standard of care by providing a variety of new treatment options for individuals who are not well served by acoustic hearing aids.

A recurring point of discussion in many of the chapters in this volume is the ability of the middle ear to transmit high frequencies from the outer ear to the inner ear. The range of mammalian hearing varies from up to 10 kHz in elephants, 20 kHz in humans, 100 kHz in mice, and even higher in some marine mammals and bats. At frequencies below a few hundred Hz, the middle ear is reasonably straightforward to characterize because at those frequencies the eardrum surface moves more uniformly and the ossicles are minimally constrained by mass inertia. As the frequency increases, however, the eardrum begins to exhibit more and more complex modes of vibration, and the motions of the ossicles might become more and more constrained due to their mass inertia. A challenge has been to understand how the middle ear is still able to transmit sound to the cochlea smoothly over such a wide range of frequencies in spite of these higher-frequency effects. An exciting development for the investigation of structure–function topics such as these is the recent availability of genetically engineered varieties of mice that exhibit well-characterized alterations to middle ear structures.

People with vision impairment currently have multiple treatment options, such as eyeglasses, contact lenses, and surgical methods of vision correction. In the treatment of hearing impairment, on the other hand, acoustic hearing aids have long been the standard of care and the only option available in most cases. This appears to be changing, however, as alternate nonsurgical treatments, akin to the contact lens, and surgical treatments are either becoming available now or are well on their way to becoming available in the near future. Owing to their potential for broad-spectrum amplification, these new options could help to overcome limitations of acoustic hearing aids related, for example, to hearing in noise and sound quality.

The ongoing, highly cross-disciplinary efforts to better understand the structural and functional interrelationships of the middle ear will no doubt continue to bring forth compelling scientific insights, while at the same time leading to improved care and treatment options for individuals with hearing impairment.

Acknowledgments This work was supported in part by grant R01 DC 005960 from the National Institute on Deafness and Other Communication Disorders (NIDCD) of the National Institutes of Health.

Chapter 2
The Evolution and Development of Middle Ears in Land Vertebrates

Geoffrey A. Manley and Ulrike J. Sienknecht

Keywords Embryology of middle ear • Middle ear development • Middle ear evolution • Three-ossicle • Tympanic

2.1 Introduction

This chapter is an attempt to reconcile interpretations of the structures of fossil mammalian middle ears with what is known about the development, anatomy, and physiology of modern mammalian and nonmammalian ears. As Bennett and Ruben (1986) wrote: "It is obviously difficult to ascertain physiological characters from dead animals. It is even more difficult to infer those characters from fossilized animals" (p. 207). In spite of these truisms, it is possible, when taking all known paleontological, developmental, anatomical, and physiological data into account and observing the traditional rules pertaining to the interpretations of each set of data, to come to a consistent view of the changes in structure and function of the hearing of mammals over geological time. Detailed overviews of the structure and physiology of amniote middle ears already exist (see, e.g., Rosowski, Chap. 3 and Rosowski 1994).

The term *middle ear* applies to any structure that improves the transmission of sound energy between a conductive medium outside the body and the inner ear. Strictly speaking, the term could be applied where water or air is the conductive medium, thus also in certain kind of fishes—even though they swim in a medium whose acoustic impedance is essentially the same as that of the inner-ear fluids. In those animals, the presence of a gas-filled swim bladder creates an interface

G.A. Manley (✉) • U.J. Sienknecht
Cochlear and Auditory Brainstem Physiology, IBU, Faculty V, Carl von Ossietzky University Oldenburg, 26111 Oldenburg, Germany
e-mail: geoffrey.manley@uni-oldenburg.de; ulrike.sienknecht@uni-oldenburg.de

S. Puria et al. (eds.), *The Middle Ear: Science, Otosurgery, and Technology*,
Springer Handbook of Auditory Research 46, DOI 10.1007/978-1-4614-6591-1_2,

within the body where there is a large change in acoustic impedance, and stronger acoustic vibrations occur at that interface. Connecting the inner ear to this interface, as with the Weberian ossicles in certain fish groups, greatly improves sensitivity to water-borne sound (Ladich and Popper 2004) and fulfills the definition of a middle ear. In the present discourse, however, coverage is restricted to the middle ears of land vertebrates.

The emergence of vertebrate animals onto the land was, without doubt, one of the most far-reaching events in evolution. As so often in science, early concepts of this "event" have had to be strongly modified in the face of newer evidence. For example, examination of the first fossils of this period led early to a number of dogmata that have since been shown to be false. One example is the idea that the earliest vertebrates transitional to the amphibians were at least partially land-living and possessed pentadactile, or five-toed, appendages. It has since been shown that limbs, as opposed to fins, in fact developed in water-living animals, limbs that were presumably used to move around more easily among water plants and that these animals possessed more than five toes on their appendages (Coates and Clack 1990; Clack 2009). Another dogma, which is very relevant to our understanding of middle ears, is that vertebrates developed a tympanic (or eardrum-bearing) middle ear at the time of the water-to-land transition and that all subsequent vertebrates inherited this kind of middle ear and modified it accordingly. In fact, the history of hearing in land vertebrates is, at least for the first half of their evolutionary story, much more varied than expected. As described later, most lacked a tympanic middle ear and were presumably "hard-of-hearing."

A second "auditory" dogma has also fallen victim to the clarity that has emerged from newer fossils. The mammalian middle ear did not emerge by the addition of two more ossicles to an existing, one-ossicle middle ear, for the simple reason that mammalian ancestors, like all other vertebrate lineages of those late Permian-early Triassic times, lacked a tympanic middle ear. These and other issues are the topics briefly discussed in the text that follows.

2.2 The Water–Land Transition and Early Attempts at Middle Ears

It is not the intention of this chapter to go deeply into paleontological issues, but of course the history of land vertebrate middle ears is being discussed and—besides comparisons between modern lineages—fossils are the main source of information. Older textbooks reiterate the story that developed from the early descriptions in Paleozoic amphibians of a deep notch in the back of the skull that, among the various changes to sensory organs that were necessary when vertebrates emerged on to land, was assumed to be the start of the evolution of a tympanic, impedance-matching middle ear. Air-borne sound reflects strongly from a surface with a higher impedance and this development would have improved hearing sensitivity by at least 40 dB compared to the absence of such a middle ear (Manley 2011; Puria and Steele 2008). As it turns out, however, although there is evidence of some highly interesting innovations for hearing in air and water in early fish (e.g., Clack et al. 2003; Clack and Allin 2004; Brazeau and Ahlberg 2006), none of these innovations survived very long or they were found only in lineages that themselves died out. Reinterpretation of some early fossils led to the conclusion that at least some of the skull notches interpreted as tympana instead housed a spiracle, an open passage for water between the buccal cavity and the outside world (e.g., Clack 2002). For the best part of 100 million years (Ma) after vertebrates emerged onto land, fossil indications of a tympanic middle ear are scattered and provide no evidence for the early development of a middle ear that was inherited by all later forms.

2.3 Middle Ears Developed Late in Evolution and Many Times Independently

Over the course of land vertebrate evolution, several kinds of tympanic middle ears developed, only to be lost again or in lineages that died out. Some forms in the late Carboniferous (310 Ma; e.g., Clack 2002) and late Permian (265 Ma; Müller and Tsuji 2007) show evidence of possessing a middle ear, but died out during, for example, the great extinction event of the Permian-Triassic, at the transition from the Paleozoic to the Mesozoic. Until the beginning of the Triassic (~250 Ma ago) the majority of land vertebrate lineages showed no history of a tympanic middle ear (Clack and Allin 2004). During the Triassic period, probably over a period of tens of millions of years, however, all lineages of tetrapods that survive until today developed a tympanic middle ear—and all independently of each other (Clack and Allin 2004; Manley and Clack 2004). Although the skeletal elements that were used to create these middle ears were common to all groups, the formation of these elements into a functional tympanic middle ear was independent in all cases, as it has been shown that their respective ancestors did not have a middle ear and presumably heard only louder, lower frequencies (e.g., Kemp 2007).

The aforementioned conclusions mean that the middle ear of amphibians, of archosaurs (birds and their crocodilian relatives), of lepidosaurs (tuataras, lizards, and snakes), and of mammals do not have a common ancestry, although their individual components do. The independent emergence of middle ears and the scattered attempts at middle ears in earlier vertebrate history was possible thanks to an amazing flexibility in development provided by a cell type unique to vertebrate animals, the neural crest cells (see Sect. 2.6). A close look at the middle ear of amphibians shows clearly that, among middle ears, it is unusual (Smotherman and Narins 2004). Among other interesting features, there is a unique linkage in the columellar system such that—in contrast to all other middle ear systems—when the eardrum is pushed inwards, the columellar footplate is pulled outwards. In spite of their independent origins, the middle ears of mammals and nonmammals share important features in individual development or ontogeny (see Sect. 2.6). The mammalian middle ear is, of course, the only one that uses three ossicles to connect the eardrum to the inner ear, and the above discussion makes clear that it developed de novo and was not an "improvement" on a preexisting, single-ossicle middle ear (Manley 2010). In fact, as shown later, it also arose multiply and independently within several related groups of early mammals, some of which did not survive until modern times.

2.4 The Single-Ossicle Middle Ear of Archosaurs and Lepidosaurs

In these two groups, as also perhaps in the others, a change in jaw-movement patterns during evolution led to adjustments in the structures bracing the jaws against the rest of the skull. For our purposes, the most important change was that the columella ("stapes") bone lost its most important function. At that time, it was a substantial skeletal element that had until Triassic times braced the rear part of the outer skull (specifically the quadrate bone, later to become the incus in mammals) against the braincase. The columella thinned greatly and changed its orientation, the outer end migrating dorsally, where an eardrum evolved and connected to the columella via a new extension, the extracolumella. This apparatus lay directly behind the skull, above and behind the jaw joint. Thus in these lineages, the changes in skull and head structure necessary to evolve a tympanic middle ear were not very great, as the columella-stapes had always connected on its inner end to the bones surrounding the inner ear at a location that later became the oval window. It has been suggested that the relatively massive columella-stapes bones of the amniote ancestors might have worked as an inertial system (Manley 1973, based on Hotton 1959). Thus head vibration caused by low-frequency sound or ground vibrations might have been accompanied by a delay in the movement of the (large) stapes, which would have vibrated out-of-phase with the rest of the head and thus provided a stimulus to the inner ear.

There has, in the past, been considerable confusion in the literature with regard to the performance of the ears of mammals and nonmammals, also with regard to their middle ears. Earlier, the multiple-ossicle middle ear was considered to be responsible for the fact that mammals heard "better" than nonmammals, "better," however, generally not being clearly defined (Masterton et al. 1969; Taylor 1969). The middle ear of nonmammals was supposed to be inferior to that of mammals, and this idea was based partly on the belief that (supposedly) mammals added two ossicles to a preexisting middle ear and this presumably would not have happened if it had not led to an improvement in performance. We now know that in fact the mammalian three-ossicle middle ear evolved de novo (see later) and thus the relationship between the two types of middle ear must be discussed quite independently of any assumptions of "improvement." All three mechanisms that are used by the three-ossicle middle ear to match impedances (area ratio between the eardrum and the footplate, lever ratio between the malleus and incus "arms," and the curved-membrane lever system) are also all found in single-ossicle middle ears (Manley 1972; Fig. 2.1). The only difference is that, in contrast to the primary lever system of mammals, the single-ossicle system uses a secondary lever along the extracolumella–columella system (Fig. 2.1a). The "performance" at the level of the eardrum is equivalent (Fig. 2.1c), but above about 4 kHz, the secondary lever system is less efficient at passing along the stimulus, resulting in an increasingly large loss at the footplate for the higher frequencies. This is, however, at least partly due to an increase in inner-ear impedance at higher frequencies (Manley 1972). In the guinea pig, there is also a dramatic decrease in middle ear performance at frequencies exceeding those processed by the inner ear (Manley and Johnstone 1974).

Manley (1973), comparing the inner and middle ears of mammals and nonmammals, came to the conclusion that in general, the mammalian ear was superior to that of nonmammals only with respect to its frequency-hearing range. Generally, but not always, the upper frequency range of hearing in mammals is higher or much higher—leaving aside new evidence for ultrasonic hearing in frogs (Feng et al. 2006) and an upper frequency limit in lizards of 14 kHz (Manley and Kraus 2010). The upper frequency limits of inner and middle ears in all species have apparently coevolved and, despite earlier concepts to the contrary, the upper frequency limit of the middle ear does not alone determine the upper limit of hearing. Instead, middle ear performance also depends on the frequency range "accepted" by the inner ear. Above the highest frequencies of the inner-ear receptor, the impedance of the inner ear rises and this influences the upper limit of the middle ear (Manley 1972). The discussion concerning the relative importance of inner and middle ears regarding the shape of the audiogram has more recently been extended and strengthened by Hemilä et al. (1995) and Ruggero and Temchin (2002). A discussion of the evolution of the mammalian middle and inner ears must be carried out fully free of preconceptions of "better" or "poorer" and concentrated on the status of inner and middle ears during the fascinating evolutionary innovations of the Triassic period.

2.5 The Origins of Mammalian Middle Ears

The title of this section is couched in the plural to emphasize that the mammalian type of three-ossicle middle ear originated several times, perhaps indeed many times. Modern (extant) mammals are divided into three groups: the placental (eutherian), marsupial (metatherian), and egg-laying monotreme mammals. Placentals and marsupials together are termed therian mammals. Before the origin of true mammals in the late Triassic (Lucas and Luo 1993), the ancestral synapsid "reptiles" had already developed some features that are considered uniquely mammalian. Indeed, the features that today are considered as mammalian (some of which were present in now-extinct nonmammals) arose over a very long period of time: there was no "big

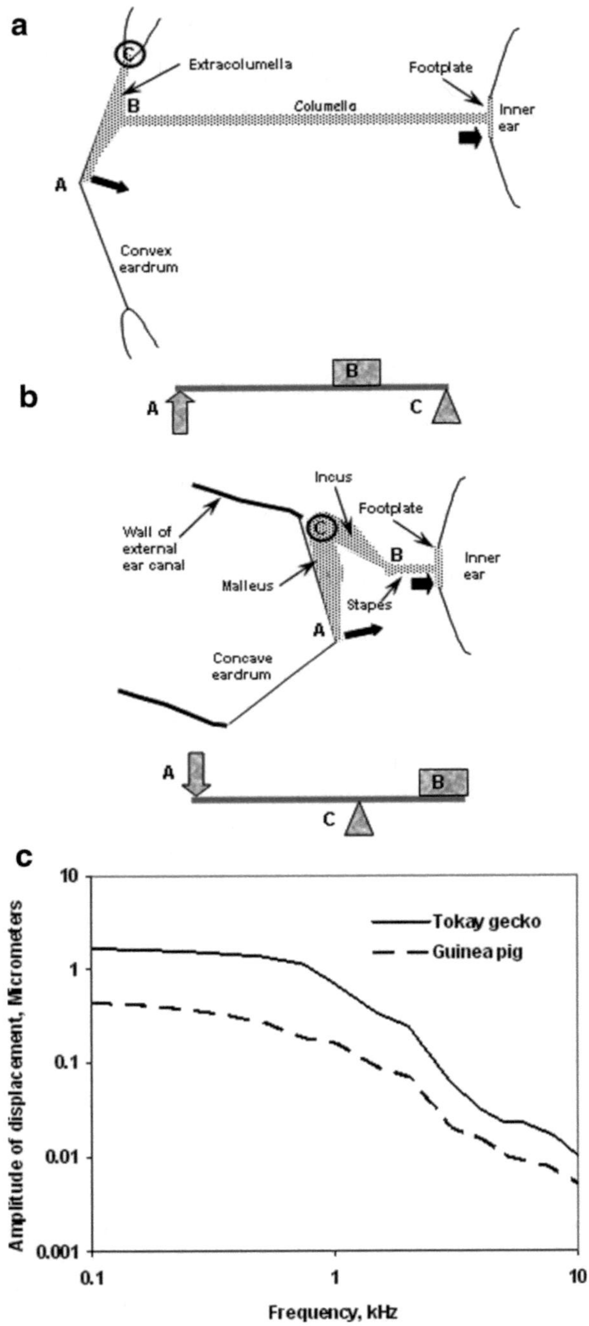

Fig. 2.1 Schematic representation of middle ear function, comparing (**a**) nonmammalian amniote and (**b**) mammalian middle ears, both in (**c**). In both cases, a diagram of the lever system involved is shown, with the capital letters corresponding to the positions of force application (*A*, idealized to

bang" origin for mammals. One of the first of the features typical of mammals (but that had its origin in the lineage well before true mammals arose) is a heterodont set of teeth, which indicated a substantial change in diet. This change in diet was accompanied by a coordinated series of changes in the muscles that moved the jaws and the bones that made up the lower jaw. The lower jaw progressively became simplified, from originally seven bones to one single bone, the dentary, which was later part of a new, secondary jaw joint. All of the jaw muscles thus became attached to the dentary, a process that involved migration of the muscle–tendon attachments. The final stage brought forth a jaw suitable for chewing, correlated with the processing of food in the mouth cavity, rather than the typical nonmammalian bite-and-swallow technique. Detailed, comparative examination of individual development in nonmammals and mammals strongly supports the ideas generated from paleaontological evidence and indicates that changes in the genetic control of the ontogenetic processes that led to the jaw-joint and middle ear components could gradually re-mold this region of the head (see Sect. 2.6).

A further, parallel, development was the growth of a bony plate, the secondary palate, separating the mouth from the nasal cavity. This structural feature is also—with the exception of its independent evolution in crocodilians—uniquely mammalian and arose more or less parallel to the loss of the primary jaw joint (Carroll, 1988). The secondary palate prevented food particles entering the nasal cavity and thus permitted uninterrupted breathing during chewing. This innovation permitted mammals to begin the masticatory and enzymatic digestive processes in the mouth itself. It has been suggested that this palate—and other changes—would also have played an important role in separating the middle ears of mammals from each other and from the mouth cavity, thus leading to the loss of a previously existing pressure-gradient received system (Christensen-Dalsgaard 2010; Manley 2010). A reinterpretation of the evidence indicates, however, that

Fig. 2.1 (continued) the middle of the eardrum), load (*B*), and fulcrum (*C*). The axis of rotation is shown as a circle around the fulcrum. The necessity for transforming a rotation of the extracolumella in the nonmammalian middle ear into a piston-like movement of the columella is enabled by a flexible joint between the extracolumella and the columella. The amplitude and force at the eardrum (*longer black arrow*) is changed by the lever into a smaller amplitude and greater force at the footplate of the columella/stapes (*shorter but wider black arrow*). (**c**) Comparison of the displacement amplitudes of the middle of the eardrum in (*continuous line*) the Tokay gecko and (*dashed line*) the guinea pig over the same frequency range and using the same apparatus for stimuli at 100 dB SPL. In both cases, the outer ear was driven by a closed sound system. Although these are similar measurement conditions, the relative amplitudes may be influenced by the different impedance conditions on the inside of the eardrum (opened mouth floor in the gecko, open bulla condition in the guinea pig) (Partially after Manley 2011; Tokay gecko data from Manley 1972; guinea pig data from Manley and Johnstone 1974)

the immediate ancestors of mammals did not in fact have a tympanic middle ear, and thus had no pressure-gradient receiver that they could lose.

Thus the immediate ancestors of true mammals had changed their jaw construction and eliminated six bones from the lower jaw, making it more stable. During the transition period from a primary to a secondary jaw joint (the latter between the squamosal in the upper jaw and the dentary), species with a double jaw joint existed. The primary jaw joint was gradually eliminated because its lower-jaw component, the articular bone, which connected to the upper-jaw quadrate, was moved medial to and out of the lower jaw. The secondary jaw joint evolved lateral to the primary joint, and contemporary species such as *Diarthrognathus* used both joints simultaneously (Allin and Hopson 1992). With time, the old joint moved deeper and entered the middle ear while retaining a connection to the lower jaw over a long period of time. There is a general consensus that the mammalian middle ear, including its eardrum, evolved at a completely different location from that of the single-ossicle middle ear (e.g., Allin 1986). Instead of directly behind the head, the tympanum originated near the rear end of the lower jaw, over those bones that were in transition out of the jaw and into the middle ear. The angular bone of the lower jaw became known as the ectotympanic, and grew into a circular support for the eardrum; the articular became the incus. The malleus originated from the upper-jaw quadrate. This series of events were, in basic form, elucidated very many years ago, of course, by Reichert (1837) and later Gaupp (1912) and provided an early and very convincing case of evolutionary transformation of function. Since then, this research area has been enormously enriched by new fossil material but has not been free of controversy. Some authors suggested, for example, that early mammals had a double middle ear, with two tympana, or that the early tympana were perhaps also sound-producing, rather than only sound-absorbing organs (see, e.g. Allin 1986). Maier (1990), however, considered it unlikely that early mammals had anything other than a single tympanum behind the lower jaw.

The three ossicles of the mammalian middle ear evolved independently at least three times. In monotremes, for example, the jaw depressor muscles and thus the relative placements of middle ear structures, differ from the therian situation, indicating independent evolutionary acquisition (Rich et al. 2005). In therian mammals, the three ossicles of the middle ear did not suddenly detach from the lower jaw and become freely suspended in a middle ear space. Although middle ear spaces are difficult to find in early mammals, it is obvious that the malleus, in particular, remained attached to the inside of the lower jaw via an ossified Meckel's cartilage (a remnant of the embryonic lower jaw of vertebrates). This condition is considered as an intermediate stage in the evolution of freely suspended ossicles and persisted for a remarkably long time (transitional mammalian middle ear [TMME]; Allin and Hopson 1992; Figs. 2.2 and 2.3). This morphological stage can be seen in a very similar form today in embryonic monotreme (egg-laying) mammals, as the ossicles in modern monotreme mammals separate fully from the lower jaw only around the time of hatching (Luo 2007) but remain very stiff throughout life (Aitkin and Johnstone 1972).

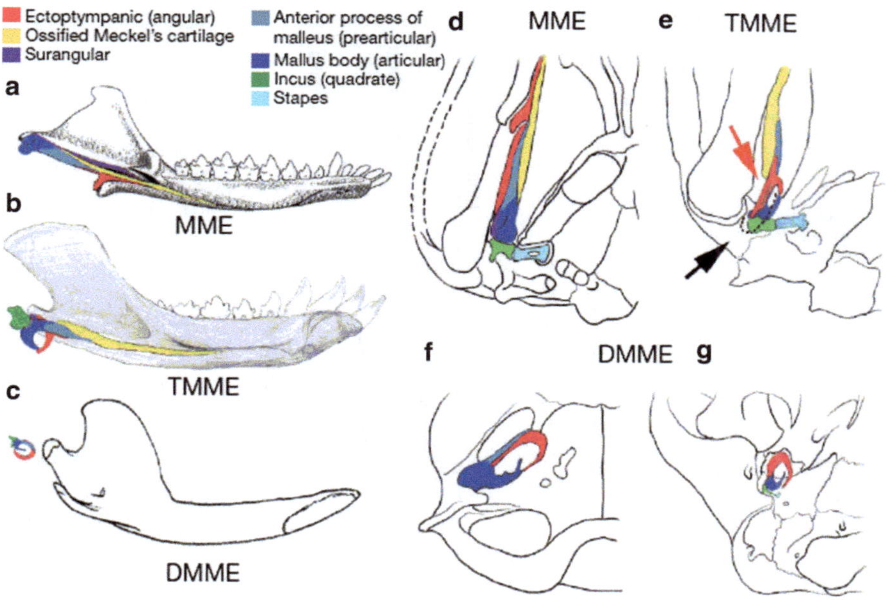

Fig. 2.2 Different morphological states of mammalian middle ears, illustrating the transitions from a mandibular middle ear (MME, **a**, **d**) to the transitional mammalian middle ear (TMME, **b**, **e**) and finally to the definite mammalian middle ear (DMME, **c**, **f**, **g**). (**a–c**) Medial views of the mandibles of the fossil species *Morganucodon* (early Jurassic) and *Liaoconodon* (early Cretaceous) and a generalized modern therian mammal, showing the relationship with the ossified Meckel's cartilage (in *yellow*, absent in modern adult mammals) and the ear ossicles (see color coding). (**d–g**) Ventral views of the ear regions in *Morganucodon*, *Liaoconodon*, *Ornithorhynchus* (modern Platypus), and *Didelphis* (modern marsupial), illustrating the relationship of the ossified Meckel's cartilage, ear ossicles, the dentary bone, and the nearby cranium. The *black arrow* in (**e**) points to the external auditory meatus, the *red arrow* to the gap between the ossicles and the inside of the dentary (From Meng et al. 2011. Reprinted by permission from Macmillan Publishers Ltd. *Nature* 472, 181–185, copyright 2011.)

In some very early mammalian groups, such as the genus *Morganucodon* of the early Jurassic, this condition prevailed. In a study of the *Morganucodon* middle ear, Rosowski and Graybeal (1991) came to the conclusion that it was so stiff that it very likely best transmitted higher frequencies to the inner ear. This correlated with the analysis by Masterton et al. (1969) of mammalian hearing, in which they speculated that the earliest mammals perhaps heard only high frequencies. However, since 1991, new fossil finds of *Morganucodon* indicate that Rosowski and Graybeal's specimens were distorted and in fact the ossicles were not so confined as they thought (Hurum 1998), which influences any functional interpretation. In *Morganucodon*, the cochlear canal was straight and less than 3 mm in length (Graybeal et al. 1989). Aitkin and Johnstone (1972) studied the middle ear of the Echidna or "spiny anteater" *Tachyglossus aculeatus* of Australia and showed that, although its best frequency was at 6 kHz, it had a very low upper frequency limit

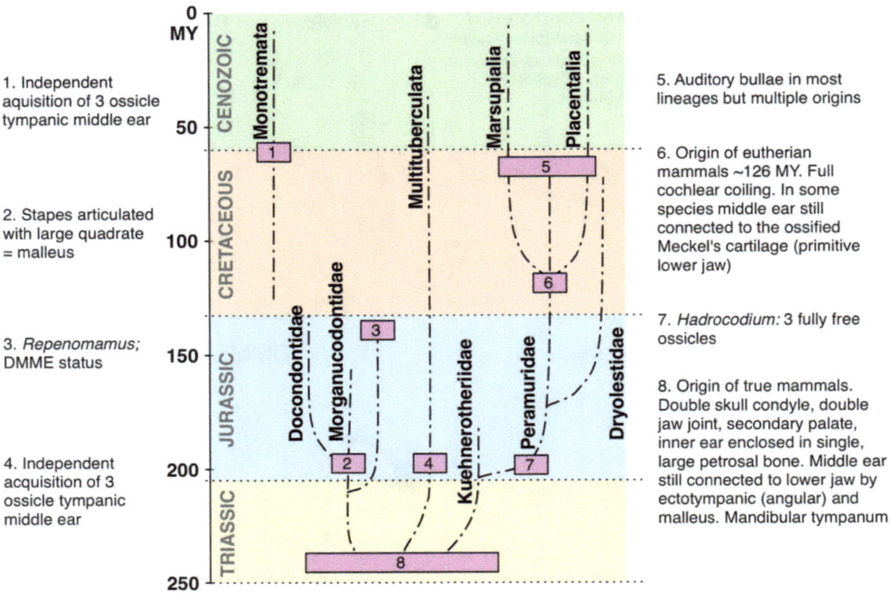

Fig. 2.3 A schematic diagram of the events occurring in the middle ears of mammalian lineages over 230 million years of evolutionary time. The time scale is to the *left*, in millions of years before the present. The four main geological periods shown—the Triassic, Jurassic, Cretaceous (all Mesozoic eras), and Cenozoic—are color-coded. *Dashed lines* indicate the approximate times of origin and extinction of the various lineages. Only placental, marsupial, and monotreme lineages survived to modern times. *Small boxes* enclose time blocks during which major events occurred or important fossil finds indicate the acquisition of new features, as shown in the appropriate labels

near 14 kHz and was about 20 dB less sensitive compared to the middle ears of other mammals—and, indeed when compared to those of lizards. Hearing in the related Platypus *Ornithorhnychus anatinus* is very similar (Gates et al. 1974). Rosowski (1992) suggested that *Morganucodon* had an audiogram similar to that of modern monotreme mammals. If anything, *Morganucodon* is more related to modern monotremes than to therians (Fig. 2.3).

Studies of fossil middle ears with a view to understanding their frequency response is, however, bound to be a very difficult and inconclusive enterprise because the frequency response of the middle ear is strongly influenced by what the inner ear can process (Manley 1973; Ruggero and Temchin 2002). What is very clear, however, is the fact that from the early beginnings of mammalian middle ears, during which the malleus was still connected to the lower jaw, it took something like 100 Ma before the ancestors of placental and marsupial mammals had ossicles that were freely suspended in the middle ear. There is evidence that this free suspension did occur in some other early forms (*Hadrocodium*; Luo et al. 2001; Martin and Luo 2005; Fig. 2.3) but either it never happened, or it was not sustained, in the ancestors of the later-evolving therian mammals. Meng et al. (2011), following Allin and Hopson (1992), define a "transitional" middle ear seen, for example in *Laioconodon*, an early Cretaceous (~120 Ma) mammal. We cite their definition of the TMME to illustrate the state of these structures almost 100 Ma after the origin of mammals. Meng et al. (2011) wrote: "The TMME can be characterized by several features: the articular, prearticular and angular lose their direct contact with the dentary (thus called the malleus and ectotympanic) and are supported anteriorly by a persistent Meckel's cartilage, but not by cranial structures, in adult; the malleo-incudal articulation is hinge-like and lost its primary function for jaw suspension; all ear ossicles are primarily auditory structures but are not completely free from the feeding effect; the tympanic membrane is not fully suspended by the ectotympanic, and the manubrium of the malleus has not developed" (p. 184). In contrast to this situation, the direct ancestors of placental and marsupial mammals, which split from each other about 100 Ma after the actual origin of the group Mammalia, possessed what has been termed the "definitive mammalian middle ear," or DMME. As seen in modern representatives, in which all ossicles are freely suspended, the malleus had developed a manubrium and other middle ear structures as per the aforementioned TMME definition. Apart from its evolution in the lineage leading to placental and marsupial mammals, this DMME evolved independently (i.e., is homoplastic; Martin and Luo 2005; Rich et al. 2005) in monotreme mammals (as indicated by the structural relationships; Rich et al. 2005) and in their earlier relatives, the multituberculates (Fig. 2.3), and perhaps also independently in other related lineages that did not survive until today. Even late Cretaceous multituberculates had cochleae that were only approximately 6 mm long, with some evidence of low-frequency hearing (Luo and Ketten 1991).

2.6 Middle Ear Development in the Ontogeny of Mammals and Nonmammals

The historical evidence from fossils has been extended and corroborated by comparative developmental studies, more recently using general genetic and gene misexpression techniques. The homologies between bones of the ancestral jaw and of the mammalian middle ear have been confirmed through these studies, which traced the cellular origin of each structure and the genetic control of its development.

The development of the middle ear during embryogenesis involves complex morphogenetic processes and reciprocal interactions between mesenchymal (mesodermal) and epithelial (both ectodermal and endodermal) cells (Fig. 2.4). During middle ear evolution, the process of homeosis, that is, the transformation of one body part into another due to mutation(s) in, or altered expression of, specific developmentally critical genes, has been centrally important. In the course of middle ear development, both the origin of the cells and the local signaling environment mutually specify cell fates and thus the final identity of the forming structures. All

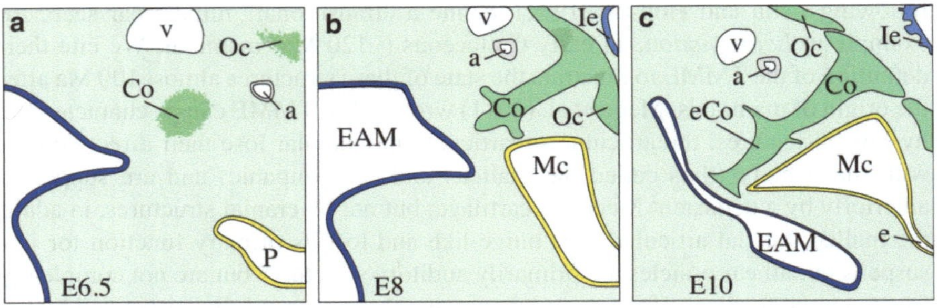

Fig. 2.4 Embryonic development of the middle ear as exemplified by a single-ossicle avian ear (chicken). (**a–c**) Schematics of transversal sections through the middle ear region at different stages of development between embryonic day (*E*) 6.5 and E10. (**a**) At E6,5, middle ear ossicles such as the columella (*Co*) and the forming otic capsule (*Oc*) are visible as condensing mesenchymal cells (*green clusters*). (**b**) Morphogenesis appears as chondrogenesis of the otic capsule (*Oc*) and primordial columella (*Co*) takes place. From the outside, the embryo's ectoderm invaginates to form the external auditory meatus (*EAM*). At the same time the pharyngeal pouch (*P*) extends and establishes the middle ear cavity (*Mc*). (**c**) At E10, before endochondral ossification takes place, the columella morphology differentiates, and the processes of the distal extracolumella (*eCo*) extend into the mesenchyme. Toward the inner ear (*Ie*) the columella footplate inserts into the oval window, which has formed by separation of the cartilage of columella (*Co*) and otic capsule (*Oc*). After narrowing, the descending connection of the middle ear cavity (*Mc*) communicates with the pharynx via the remaining Eustachian tube (*e*). *a* artery, *Co* columella, *E* embryonic day, *e* Eustachian tube, *EAM* external auditory meatus, *eCo* extracolumella, *Ie* inner ear, *Mc* middle ear cavity, *Oc* otic capsule, *P* pharyngeal pouch, *v* vein

three germ layers contribute to the formation of middle ear elements, and the foundation for these elements is laid by hindbrain-derived neural crest cells that migrate into the embryo's branchial ("gill") arches (Fig. 2.5a). Neural crest cells are unique to vertebrates and originate along the lateral edges of the developing central neural tube. Branchial cleft ("gill slit") ectoderm, pharyngeal pouch (mouth-throat cavity) endoderm, and branchial arch ("gill support") mesenchyme together give rise to the middle ear. The "gill" structures are truly homologous to fish gill arches and slits and the arches form the ventral component of the vertebrate skull known as the viscerocranium.

The development of the ear drum (tympanic membrane) clearly reveals this assemblage of different tissues, as it is composed of epithelia of the branchial arch and of the pouch, with mesenchymal cells sandwiched between (Chin et al. 1997; Mallo et al. 2000). From the outside, invagination of the first pharyngeal cleft surface ectoderm creates the external auditory meatus, whereas the medially lying tympanic cavity results from expansion of the pharyngeal pouch. Thus the entire epithelium of the mature middle ear cavity is of endodermal origin. The middle ear cavity later communicates with the pharynx via the auditory (Eustachian) tube (Fig. 2.4), the latter being a narrow extension of the pharyngeal pouch (Jaskoll and Maderson 1978) that permits air-pressure equalization within the middle ear space.

The segmental structure of the embryonic head region—as reflected in serial hindbrain sections known as rhombomers and in embryonic branchial arches—was

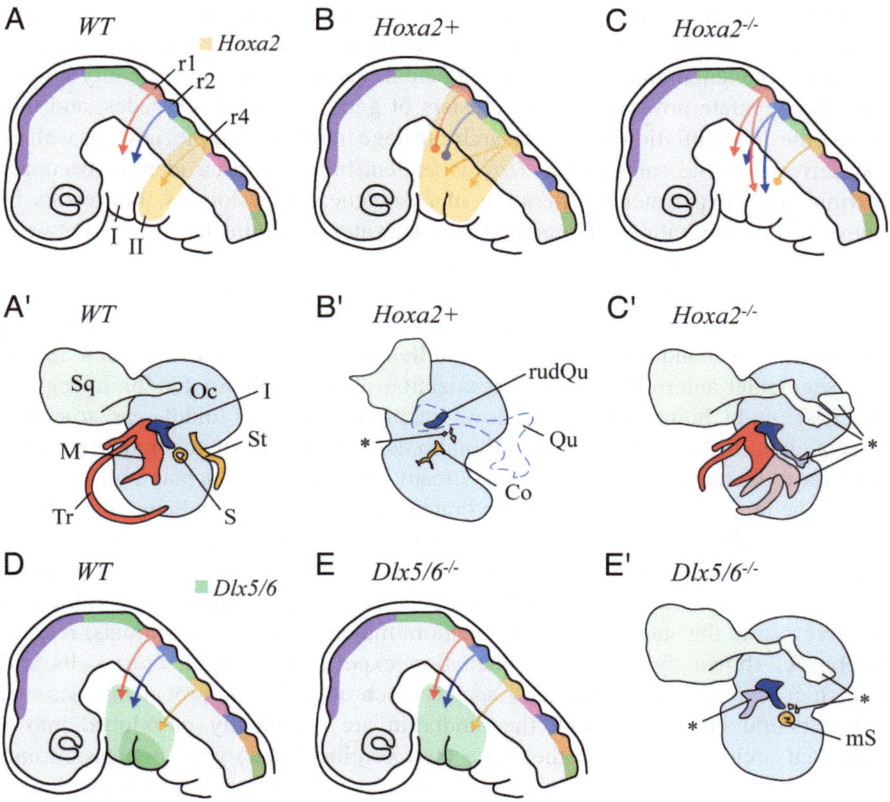

Fig. 2.5 Developmental origin of vertebrate middle ear structures. (**a–e**) Vertebrate embryo head region, lateral view. (**a**) Neural crest streams (*arrows*) originating from hindbrain rhombomers (r) migrate into the branchial arches (*I, II*), where they give rise to middle ear structures. Expression domains of *Hox* (*Hoxa2*) (**a**) and *Dlx* (such as *Dlx5/6)* (**d**) genes influence the migrating neural crest cells and thus pattern the branchial arches. (**b**) Broad aberrant rostral expansion of the *Hoxa2* domain causing misguided r4 derived neural crest cells to invade first branchial arch. (**c**) Missing *Hoxa2* expression domain leading to neural crest cells from *r1* and *r2* entering and populating the second branchial arch. (**d**) Nested expression pattern (different densities of *green color*) of vertebrate *Dlx* gene pairs (*Dlx1/2, Dlx5/6, Dlx3/7*). (**e**) Missing *Dlx5/6* expression domain in *Dlx5/6*$^{-/-}$ double mutant mice. (**a'–c', e'**) Schematics of middle ear structures resulting from expression patterns **a–c**, **e**. (**a'**) Wild type (*WT*) mouse (Modified after Mallo 2001). (**b'**) Chicken. *Hoxa2* overexpression causes reduction of first branchial arch structures such as the quadrate (*Qu*) leaving an unarticulated rudiment (*rudQu*) and supernumery second arch structures (*asterisk*) (After data from Grammatopoulos et al. 2000). (**c'**) *Hoxa2* mutant mice *(Hoxa2*$^{-/-}$*)* display homeotic and mirror image transformation of first arch derived middle ear structures (Gendron-Maguire et al. 1993; Rijli et al. 1993; modified after Mallo 2001). (**e'**) *Dlx5/6*$^{-/-}$ double mutant mice lack distal branchial arch elements but duplicate structures (*asterisk*) derived from the proximal branchial arch (maxillary), such as the incus. The hyomandubular stapes, lacking a foramen (*mS*), is associated with ectopic cartilages (After data from Depew et al. 2002). *Co* columella, *I, II* branchial arches, *I* incus, *M* malleus, *mS* malformed stapes, *Oc* otic capsule, *Qu* quadrate, *r1, r2, r4* hindbrain rhombomers, *rudQu* rudimental quadrate, *S* stapes, *Sq* squamosal, *St* styloid process, *Tr* tympanic ring, *WT* wild type, *arrows* neural crest streams, * supernumery structures resulting from gene misexpression

fundamental to the evolution of the vertebrate viscerocranium. During development, distinct homeobox (*Hox*) gene expression domains in the hindbrain rhombomeres and in the branchial arches establish segmental identity. Segmental identity in turn provides separate novel regulatory clusters of gene interaction cascades, and this allows the differentiation of visceral arch cartilage into, for example, jaws, as well as the derived middle ear ossicles. *Hox*-dependent branchial arch identity becomes obvious after experimental alteration of *Hox* gene expression, as this results in homeotic transformations of branchial arch derivates, swapping tissue fates between first and second arches (see later for details).

Early in development, neural crest cells originating from dorsal neural tube rhombomeres migrate in so-called streams laterally and ventrally, populate the branchial arches and give rise to the middle ear (Koentges and Lumsden 1996). The segmental anterior–posterior organization of the midbrain–hindbrain regions that is set up by homeobox genes allows tracing the origin of middle ear structures to sequential neural crest streams. Rhombomeres r1 and r2 (numbered from anterior) form the mandibular neural crest stream into the first branchial arch and give rise to the incus (maxillary part of the branchial arch) and the malleus (mandibular part of the arch) of mammals. In nonmammals, these same neural crest streams give rise to the articular and quadrate bones of the skull—the primary jaw joint. Derivatives from the second branchial arch stream, originating from rhombomere r4, give rise to the stapes or the columella in mammals and nonmammals, respectively. As shown by tissue transplantation experiments, neural crest cells and branchial arch endoderm mutually specify each other by reciprocal interactions. When neural crest cells from the hindbrain are ectopically introduced into a branchial arch environment, they lose their original (*Hox*) gene expression and take on the fate dictated by their new environment. Conversely, transplantation of larger portions of neural crest results in the cells keeping their original identity (Trainor and Krumlauf 2000; Schilling et al. 2001). These neural crest cells instruct gene expression in the surrounding cells and influence patterning and growth of the branchial arch (Noden 1991; Schneider and Helms 2003).

Pluripotent migrating neural crest cells are thus a prerequisite for the evolution of the middle ear. In birds, the application of retinoic acid to the hindbrain can induce the loss of the columella and ectopic formation of a retroarticular process cartilage in the first branchial arch (equivalent to the lateral process of the mammalian malleus [O'Gorman 2005]). Rhombomere r4–derived neural crest cells are, in this situation, misguided into the first branchial arch, concomitantly with the expression of *Hoxa2* extending into the first arch (which normally has no endogenous *Hox* gene expression; Fig. 2.5b). Thus, after migrating cells encounter an altered *Hox* gene expression and—subsequently—changed local signals, the first branchial arch cartilage transforms into second arch derivates (Plant et al. 2000). Broad misexpression of *Hoxa2* in the hindbrain and in the branchial arch leads to duplication of tongue skeleton (also viscerocranium) elements as a result of transformation of the Meckel's cartilage (forerunner of the lower jaw) and of the quadrate (Grammatopoulos et al. 2000) (Fig. 2.5b'). Loss of *Hoxa2* in mutant mice mimics first branchial arch identity and misguides neural crest cells ectopically into the second branchial arch (Fig. 2.5c), resulting in a mirror-image duplication of first arch derivates such as malleus and incus in the second branchial arch (Fig. 2.5c').

The evolution of a different set of homeobox genes—*Dlx*—was crucial for neural crest patterning. Whereas vertebrates exhibit expression of three pairs of *Dlx* genes (*Dlx1/2, Dlx5/6, Dlx3/7)*, responsible for the elaborate proximodistal patterning of branchial arches, the ancient cephalochordate relative of vertebrates, *Branchiostoma*, has only a single copy of *Dlx* expressed in the epidermis and nervous system. Thus, probably via gene duplication, vertebrates extended and diversified the *Dlx* expression domain. *Dlx2*, for example, is expressed in the surface ectoderm as well as in the neural crest–derived mesenchyme and plays a key role in ectoderm-mesenchyme interaction. Mutation of this gene leads to malformation of the incus and stapes (Qiu et al. 1995). Further candidate genes for locally-operating signals that mediate epithelial–mesenchymal interactions include *Endothelin1* and *Fgf8*, as both, when mutated, result in the absence of first arch-derived malleus and incus (reviewed in Mallo 2001). Growing evidence supports the interpretation that the multiplication of *Dlx* genes was also crucial for the evolution of jaw diversification, as an independent regulatory system for upper and lower jaw structures became available, allowing the exploitation of new feeding niches (Koentges and Matsuoka 2002).

Early in development, initial cues from the branchial endoderm impose a first proximodistal patterning axis onto the arriving neural crest cells, leading to a nested expression pattern of *Dlx* genes. This nested *Dlx* expression (Fig. 2.5d), however, appears to be a novelty of jawed vertebrates, as lampreys (jawless relatives) exhibit the full range of *Dlx* genes, but not in a nested expression pattern (Shigetani et al. 2002). The gene pair *Dlx5/6* controls distal branchial arch identities and is implicated in the elaboration of the lower jaws, as a double mutation of *Dlx5* and *Dlx6* (and thus failure to form this expression domain, Fig. 2.5e) leads to a transformation of the lower jaw into an upper jaw. In the forming middle ear, loss of distal branchial arch elements due to *Dlx5/6* knock-down is concomitant with a duplication of structures originally derived from the proximal branchial arch, such as the incus (Depew et al. 2002) (Fig. 2.5e').

Middle ear ossicles undergo endochondral ossification; thus their development is visible as foci of mesenchymal condensation within the embryonic branchial arches, the sites where cartilage is generated (Fig. 2.4). A single condensation gives rise to both the malleus and incus (Hall and Miyake 1995). Separation of these ossicle primordia from each other appears to be influenced by the gene *Eya1*; hence mice mutant for this gene display different forms of malleus–incus fusions (Xu et al. 1999). First and second branchial arch mesenchyme, respectively, is characterized by expression of specific genes such as *Ptx1, Six2, Lhx6, Alx, Bapx1* in the first arch and *Msx1* in the second (reviewed in Chapman 2011). Loss of distal identity also leads to subsequent down-regulation of second arch expressors such as *Wnt5a* (Depew et al. 2002). Although the expression of a plethora of *Wnt*-related genes, most prominently *Wnt11, Fzd9, Frzb,* and *SFRP2,* in the developing middle ear has been described (Sienknecht and Fekete 2008), their functional interactions still remain to be elucidated and are currently under investigation.

Developing middle ear ossicles influence the positioning of the outer ear meatus and tympanic membrane, as well as, at the other end, the positioning of the oval window on the inner ear (reviewed in Mallo 2001). Vice versa, the otic capsule contributes to the columellar footplate (Jaskoll and Maderson 1978). Thus, evolution and development of the vertebrate middle ear appears to be a playground of various gene-interaction scenarios that determine structure–function relationships.

Comparisons of the embryonic development of different extant mammalian groups (monotremes, marsupials, and placentals) present sufficient variation to exemplify the phylogenetic separation of the middle ear from the jaw. In all mammalian groups, similar morphogenetic processes contribute to this separation of the middle ear, but to a varying extent. These processes are (1) embryonic displacement of middle ear anlagen (in the medial and posterior direction); (2) negative allometry of middle ear structures (i.e., early ossification of middle ear components, thus limiting their relative increase in size); and (3) developmental resorption of Meckel's cartilage, leading to separation of the middle ear and achievement of the DMME status (reviewed in Luo 2011).

In summary:

1. There is a common developmental origin for the primary jaw joint and the middle ear elements in all amniotes.
2. These elements arise through processes controlled by common gene patterning and developmental pathways.
3. Modifications in the number of genes and in their temporal and spatial expression during development and over evolutionary time are sufficient to lead to the observed morphological transformations.

2.7 Function of the Early Mammalian Middle and Inner Ear

One of the most important questions relating to the functional aspects of the earliest three-ossicle middle ears is: What was the hearing organ like in the earliest mammals? If we can assume that in these animals, as in all modern amniotes, the middle and inner ear performances were well matched (Ruggero and Temchin 2002), the configuration of the inner ear should help us decide whether early mammalian middle ears transmitted high frequencies or not. The main evolutionary events are summed up on the appropriate time scale in Fig. 2.3. For the present purposes, it will suffice to note the following conclusions about the early mammalian inner ear:

1. A cladistic outgroup analysis of hearing in amniotes clearly reaches the conclusion that the ancestral condition of hearing in stem amniotes, before the several lineages

split from each other, was low-frequency hearing only (below ~1 kHz; Manley and Köppl 1998; Manley 2000).

2. The cochlea of the earliest mammals (defined by those animals that had three ossicles and thus a secondary jaw joint) was very short indeed (Fig. 2.3; *Hadrocodium, Dryolestes, Henkelotherium,* 2–3 mm; Luo et al. 2001, 2010; Ruf et al. 2009) and not coiled. Although in some cases a short cochlea (e.g., mouse, 7 mm) can be correlated (in modern mammals) with high-frequency hearing (Rosowski 1992), the kind of cochlear structure and the cochlear ancestry (see 4) tell a different story in the earliest mammals.
3. The space available in the earliest mammalian cochlea was restricted further by the presence of a lagenar macula (which is still found in modern monotreme mammals), reducing the length of the putative basilar membrane to approximately 1.5 mm.
4. A short hearing organ (~1–1.5 mm) is compatible with the kind of structure that must have been the ancestral hearing organ of stem amniotes (Kemp 2007) and, also because of the absence of a tympanic middle ear in those animals, could have responded to only loud, low-frequency sounds.
5. Any middle ear cavity present was surrounded only by membranous tissues. Auditory bullae partly or wholly surrounded by cartilage and/or bone(s) evolved only late in therian mammal history, vary greatly in their structural components, and probably arose many times independently (Novacek 1977).

All of the foregoing indicate that, contrary to the conclusions reached via a comparison of modern mammals (Masterton et al. 1969), the hearing of the earliest mammals (~220 Ma ago) was low frequency only, even though these were small animals (Kemp 2007). Smallness does not necessarily correlate with high-frequency hearing. Indeed, almost all modern nonmammals are small and all survive only hearing frequencies well below 10–15 kHz. Among modern mammals, size is not a reliable indicator of frequency responses (Heffner et al. 2001), and modern mammals have of course a quite different inner ear from that of their earliest forebears.

The fossil evidence indicates that it took more than 50 Ma for mammals to achieve full coiling of the cochlea, before a peak in mammalian diversity approximately 90 Ma ago (Bininda-Emonds et al. 2007; Fig. 2.3). At that time the lengths of the hearing organ were approximately 5 mm, which is shorter than the cochlea of any modern mammal. Nonetheless, almost all Cretaceous mammals were small to very small and because of this, their middle ear structures were also small and not well suited for the transmission of low frequencies. Achievement of a coiled cochlea in such small animals likely led to a gradual improvement in higher-frequency hearing in parallel in placentals and marsupials. In more recent evolutionary history, following the demise of the dinosaurs at the end of the Cretaceous, mammalian radiations led in many parallel lines to increases in body size. Even very large changes in body size were possible in 10–20 Ma (Evans et al. 2012). Such increases in body size almost certainly led to an improvement in low-frequency hearing, in some cases (e.g., elephants) accompanied by a reduction in the upper frequency limit. This phenomenon has been clearly demonstrated during the history of primates, including humans (Coleman and Boyer 2012), but was seen in many lineages, the results for the audiograms of each group differing as a result of their independent evolutionary trajectories. Thus the evolution of hearing ranges in the mammalian cochlea had two major phases: (1) an early phase of greater than 100 Ma, before the cochlea coiled, a phase marked by an extension of the initial low-frequency hearing range toward higher frequencies; and (2) after coiling, a major differentiation during the next 100 Ma between many diverse lineages, resulting also in the extremes of ultrasonic hearing in some (bats, some rodents, whales) and subsonic hearing in others (large mammals). The origin and evolution of the mammalian cochlea and the consequences of coiling will be discussed in a further publication (Manley 2012).

2.8 Mammals: Correlations of Middle and Inner Ear and Brain Evolution

Although the evolution of the middle and inner ear of mammals correlates with changes in brain size, so do many other features of early mammals. Thus, for example, the evolution of hair and vibrissae, together with the accompanying homeothermy and the olfactory expansion associated with the nocturnal life style of early mammals, also correlate with—and were likely largely responsible for—an early expansion of the brains of mammals, including the olfactory lobes (Rowe et al. 2011). A correlation has also been described between the changes in middle ears and forebrain expansion in early mammals (Rowe 1996). Rowe (1996) suggested that the expansion of the forebrain, pushing the middle ear toward the rear of the head (the middle ear completes its individual development before the brain does) was ultimately responsible for the separation of the middle ear from the lower jaw, forming the DMME. However, as Takechi and Kuratani (2010) point out, "Although it was suggested that brain expansion increased the distance between the postdentary elements and dentary to separate them during mammalian evolution (Rowe 1996; Luo et al. 2001), this view was refuted by a fossil with a small brain size and with postdentary elements separate from the dentary" (Wang et al. 2001, p. 421).

In spite of these correlations, Rowe et al. (2011) imply that, in the first phases of brain expansion following the origin of the first mammals, the auditory system played no obvious role. Thus features that later take up large volumes of the brain, such as well developed auditory nuclei, did not undergo significant expansion following the origin of the mammalian middle ear. For the first phase of mammalian evolution, the auditory system was far less important than, for example, olfaction and somatosensation. This supports the idea that the earliest mammalian inner ears were essentially the same as those of their immediate therapsid ancestors, processing lower-frequency sounds for which the requisite mental capacities were essentially already present. During the first 100 Ma of mammalian evolution, the brain expanded slowly until, in the mid-Cretaceous, it was two to three times more voluminous than in the first mammals. The brain capacity necessary for processing extensive auditory information was apparently a very late development in mammalian evolution, most of the early evolution of the mammalian brain being dominated by olfaction (Rowe et al. 2011).

These ideas make it difficult to interpret the significance of correlations between brain size and auditory structural characteristics (e.g., Rowe 1996) because recent fossil data suggest that there was little or no causality between these developments. It appears likely that high-frequency hearing in small mammals, which first made it possible for them to localize sounds efficiently and rapidly and was processed by an appropriate expansion of auditory brain centers, probably did not occur until the late Cretaceous. Remarkably, this was 30–40 Ma after the cochlea had achieved full coiling (Fig. 2.3; Manley 2012) and the middle ear had reached the DMME status. Because mammals developed their middle ear de novo and it was connected to the upper buccal cavity only by narrow Eustachian tubes to allow for periodic ventilation, they never had a pressure-gradient system such as that of lizards that provides sound lateralization information at the level of the auditory nerve (Christensen-Dalsgaard and Manley 2008). The mammalian brain thus had to develop methods of neural-computation processing of sound level and sound arrival-time information derived from the two ears and aided by the (newly evolved) pinnae.

2.9 Pinnae

One other major difference exists between mammalian and nonmammalian hearing systems, in that mammals generally have pinnae. Because pinnae are a soft-tissue feature, we do not know when—and how many times—they evolved. They are absent in monotreme mammals, which probably indicates that pinnae evolved after the early major split of lineages in mammalian evolution. Pinnae of course act as sound collectors, increasing the sound pressure at the eardrum in some frequency ranges by up to 20 dB or more, but they also enlarge the distance between the ears, which increases the time-of-arrival difference of sounds at the two eardrums. Thus features of pinnae, once evolved, undoubtedly played an important role in the subsequent evolution of mammalian hearing. Their origin is likely to have been well after the evolution of cochlear coiling, accompanying an increase in the upper frequency limit of hearing.

Lacking a pressure-gradient hearing system, mammals were mainly dependent on interaural cues and neural computation for sound localization. In bats, which evolved relatively late in mammalian evolution and in which we might expect the greatest specializations for sound localization, there are indeed correlations between pinnae performance and emitted echolocation sounds, at least among the constant-frequency bats (e.g., Obrist et al. 1993). The largest increase in sound pressure due to the pinna, however, is found in bats that do not echolocate, but capture their prey using passive sound localization. Such bats fly very slowly in the vicinity of their prey and their very large pinnae are thus less of an aerodynamic problem than in bats that capture prey during fast flight. Evolutionary pressures on the size and shape of pinnae have been diverse and were certainly not confined to optimal sound reception. In general, pinnae tend to be smaller in species living in colder regions of the world, *reducing* heat loss. In a few large species, such as elephants, the pinnae play an important role in *increasing* heat loss. Among birds, the owls evolved facial disks that function as sound collectors that are, in some cases such as the barn owl, as efficient as mammalian pinnae.

2.10 Summary

All amniote vertebrates of the Mesozoic era inherited a small, low-frequency (upper limit estimated at ~1 kHz) hearing organ from their Paleozoic ancestors. Mammals and nonmammals evolved tympanic middle ears independently during late Triassic times. Despite the polyphyletic relationship of amniote middle ears, deep homology of the constituting structures is supported by both the ontogenetic development and the fossil record. Developmental studies clearly demonstrate a common origin for the middle ear ossicles of all land vertebrates and thus for the columella/stapes, the bones of the primary jaw joint of nonmammals and the malleus and incus of mammals. The development of these structures is controlled by processes of common gene patterning and cellular interactions. Over evolutionary time, changes in the number and the temporal and spatial expression of genes led to the observed morphological transformations. In mammals, a three-ossicle middle ear evolved at least three times. After these events, all groups independently enlarged and specialized their inner ears toward higher frequencies and better parallel processing of signals. The primary lever system of the therian mammalian three-ossicle middle ear was fortuitously preadapted to transmit higher frequencies more easily. In spite of this, it took almost 100 Ma after the origin of mammals for the mammalian middle ear ossicular chain to free itself fully from the lower jaw and for the cochlea to fully coil and achieve hearing organ lengths of more than 5–7 mm. This coincided roughly with the time of the origin of eutherian (placental) mammals. The (independent) evolution of the monotreme mammal middle ear and cochlea substantially lags that of therian mammals. The major period of cochlear differentiation in therian mammals, including the evolution of echolocation capabilities and of middle ear bullae, postdates the split between placental and marsupial lineages in the early Cretaceous. Brain size increases associated with hearing specializations occurred first during the Cretaceous. Thus, contrary to expectations, high-frequency hearing arose in mammals over a very long period of time, perhaps, especially in small mammals, to the detriment of low-frequency hearing. An improved sensitivity to low frequencies in large mammals (as a correlate of large size) and in specialized small mammals evolved later and independently in the various lineages.

References

Aitkin, L. M., & Johnstone, B. M. (1972). Middle-ear function in a monotreme: The echidna (*Tachyglossus aculeatus*). *Journal of Experimental Zoology,* 180, 245–250.

Allin, E. F. (1986). The auditory apparatus of advanced mammal-like reptiles and early mammals. In H. Hotton, P. D. MacLean, J. J. Roth, & E. C. Roth (Eds.), *The ecology and biology of mammal-like reptiles* (pp. 283–294). Washington, DC: Smithsonian Institution Press.

Allin, E. F., & Hopson, J. A. (1992). Evolution of the auditory system in Synapsida ("Mammal-like reptiles" and primitive mammals) as seen in the fossil record. In D. B. Webster, R. R. Fay, & A. N. Popper (Eds.), *The evolutionary biology of hearing* (pp. 587–614). New York: Springer-Verlag.

Bennett, A. F., & Ruben, J. A. (1986). The metabolic and thermoregulatory status of therapods. In H. Hotton, P. D. MacLean, J. J. Roth, & E. C. Roth (Eds.), *The ecology and biology of mammal-like reptiles* (pp. 207–218). Washington, DC: Smithsonian Institution Press.

Bininda-Emonds, O. R. P., Cardillo, M., Jones, K. E., MacPhee, R. D. E., Beck, R. M. D., Grenyer, R., Price, S. A., Vos, R. A., Gittleman, J. L., & Purvis, A. (2007). The delayed rise of present-day mammals. *Nature,* 446, 507–512.

Brazeau, M. D., & Ahlberg, P. E. (2006). Tetrapod-like middle ear architecture in a Devonian fish. *Nature,* 439, 318–321.

Carroll, R.L. (1988) Vertebrate paleontology and evolution. New York: Freeman.

Chapman, S. C. (2011). Can you hear me now? Understanding vertebrate middle ear development. *Frontiers of Bioscience,* 16, 1675–1692.

Chin, K., Kurian, R., & Saunders, J. C. (1997). Maturation of tympanic membrane layers and collagen in the embryonic and post-hatch chick (*Gallus domesticus*). *Journal of Morphology,* 233, 257–266.

Christensen-Dalsgaard, J. (2010). Vertebrate pressure-gradient receivers. *Hearing Research,* 273, 37–45.

Christensen-Dalsgaard, J., & Manley, G. A. (2008). Acoustical coupling of lizard eardrums. *Journal of the Association for Research in Otolaryngology,* 9, 407–416.

Clack, J. A. (2002). Patterns and processes in the early evolution of the tetrapod ear. *Journal of Neurobiology,* 53, 251–264.

Clack, J. A. (2009). The fin to limb transition: New data, interpretations, and hypotheses from paleontology and developmental biology. *Annual Review of Earth and Planetary Sciences,* 37, 163–179.

Clack, J. A., & Allin, E. (2004). The evolution of single- and multiple-ossicle ears in fishes and tetrapods. In G. A. Manley, A. N. Popper, & R. R. Fay (Eds.), *Evolution of the vertebrate auditory system* (pp. 128–163). New York: Springer.

Clack, J. A., Ahlberg, P. E., Finney, S. M., Dominguez Alonso, P., Robinson, J., & Ketcham, R. A. (2003). A uniquely specialised ear in a very early tetrapod. *Nature,* 425, 65–69.

Coates, M. I., & Clack, J. A. (1990). Polydactyly in the earliest known tetrapod limbs. *Nature,* 347, 66–69.

Coleman, M. N., & Boyer, D. M. (2012). Inner ear evolution in primates through the Cenozoic: Implications for the evolution of hearing. *The Anatomical Record*, DOI 10.1002/ar.22422.

Depew, M. J., Lufkin, T., & Rubenstein, J. L. R. (2002). Specification of jaw subdivisions by *Dlx* genes. *Science,* 298, 381–385.

Evans, A. R., Jones, D., Boyer, A. G., Brown, J. H., Costa, D. P., Morgan Ernest, S. K., Fitzgerald, E,. M. G.,, Fortelius, M., Gittleman, J. L., Hamilton, M. J., Harding, L. E., Lintulaakso, K., Lyons, S.K., Okie, J. G., Saarinen, J. J., Siblyo, R. M., Smith, F. A., Stephens, P. R., Theodor, J. M., & Uhen, M. D.. (2012) The maximal rate of mammal evolution. *Proceedings of the National Academy of Sciences of the USA*,doi/10.1073/pnas.1120774109.

Feng, A. S., Narins, P. M., Xu, C. H., Lin, W. Y., Yu, Z. L., Qiu, Q., Xu, Z. M., Shen, J. X. (2006). Ultrasonic communication in frogs. *Nature,* 440, 333–336.

Gates, G. R., Saunders, J., & Bock, G. R. (1974). Peripheral auditory function in the platypus, Ornithorhynchus anatinus. *Journal of the Acoustical Society of America,* 56, 152–156.

Gaupp, E. (1912). Die Reichertsche Theorie. *Archives of Anatomy and Physiology Supplement,* 1–416.

Gendron-Maguire, M., Mallo, M., Zhang, M., & Gridley, T. (1993). *Hoxa-2* mutant mice exhibit homeotic transformation of skeletal elements derived from cranial neural crest. *Cell,* 75, 1317–1331.

Grammatopoulos, G. A., Bell, E., Toole, L., Lumsden, A., & Tucker, A. S. (2000). Homeotic transformation of branchial arch identity after *Hoxa2* overexpression. *Development,* 127, 5355–5365.

Graybeal, A., Rosowski, J. J., Ketten, D. R., & Crompton, A. W. (1989). Inner-ear structure in *Morganucodon,* an early Jurassic mammal. *Zoological Journal of the Linnean Society,* 96, 107–117.

Hall, B. K., & Miyake, T. (1995). Divide, accumulate, differentiate: Cell condensation in skeletal development revisited. *International Journal of Developmental Biology,* 39, 881–893.

Heffner, R. S., Koay, G., & Heffner, H. E. (2001). Audiograms of five species of rodents: Implications for the evolution of hearing and the perception of pitch. *Hearing Research,* 157, 138–152.

Hemilä, S., Nummela, S., & Reuter, T. (1995). What middle ear parameters tell about impedance matching and high frequency hearing. *Hearing Research,* 85, 31–44.

Hotton, N. (1959). The pelycosaur tympanum and early evolution of the middle ear. *Evolution,* 13, 99–121.

Hurum, J. H. (1998). The inner ear of two late Cretaceous multituberculate mammals, and its implications for multituberculate hearing. *Journal of Mammalian Evolution,* 5, 65–93.

Jaskoll, T., & Maderson, P. (1978). A histological study of the development of the avian middle ear and tympanum. *Anatomical Record,* 190, 177–200.

Kemp, T. S. (2007). Acoustic transformer function of the postdentary bones and quadrate of a nonmammalian cynodont. *Journal of Vertebrate Paleontology,* 27, 431–441.

Koentges, G., & Lumsden, A. (1996). Rhombencephalic neural crest segmentation is preserved throughout craniofacial ontogeny. *Development,* 122, 3229–3242.

Koentges, G., & Matsuoka, T. (2002). Evolution: Jaws of the fates. *Science*, 298, 371–373.

Ladich, F., & Popper, A. N. (2004). Parallel evolution in fish hearing organs. In G. A. Manley, A. N. Popper, & R. R. Fay (Eds.), *Evolution of the vertebrate auditory system.* (pp. 95–127). New York: Springer.

Lucas, S. G., & Luo, Z. (1993). *Adelobasileus* from the upper Triassic of west Texas: The oldest mammal. *Journal of Vertebrate Paleontology,* 13, 309–334.

Luo, Z.-X. (2007). Transformation and diversification in early mammal evolution. *Nature,* 450, 1011–1019.

Luo, Z.-X. (2011). Developmental patterns in Mesozoic evolution of mammal ears. *Annual Review of Ecology and Evolution Systematics,* 42, 355–380.

Luo, Z., & Ketten, D. R. (1991). CT scanning and computerized reconstructions of the inner ear of multituberculate mammals. *Journal of Vertebrate Paleontology,* 11, 220–228.

Luo, Z-X., Crompton, A. W., & Sun, A. L. (2001). A new mammaliaform from the early Jurassic and evolution of mammalian characteristics. *Science,* 292, 1535–1540.

Luo, Z-X., Rif, I., Schultz, J. A., & Martin, T. (2010). Fossil evidence on evolution of inner ear cochlea in Jurassic mammals. *Proceedings of the Royal Society B: Biological Sciences,* 278, 28–34.

Maier, W. (1990). Phylogeny and ontogeny of mammalian middle ear structures. *Netherlands Journal of Zoology,* 40, 55–74.
Mallo, M. (2001). Formation of the middle ear: Recent progress on the developmental and molecular mechanisms. *Developmental Biology,* 231, 410–419.
Mallo, M., Schrewe, H., Martin, J. F., Olson, E. N., & Ohnemus, S. (2000). Assembling a functional tympanic membrane: Signals from the external acoustic meatus coordinate development of the malleal manubrium. *Development,* 127, 4127–4136.
Manley, G. A. (1972) Frequency response of the middle ear of geckos. *Journal of Comparative Physiology A: Neuroethology, Sensory, Neural, and Behavioral Physiology,* 81, 251–258.
Manley, G. A. (1973). A review of some current concepts of the functional evolution of the ear in terrestrial vertebrates. *Evolution,* 26, 608–621.
Manley, G. A. (2000). Cochlear mechanisms from a phylogenetic viewpoint. *Proceedings of the National Academy of Sciences of the USA,* 97, 11736–11743.
Manley, G. A. (2010). An evolutionary perspective on middle ears. *Hearing Research,* 263, 3–8.
Manley, G. A. (2011). Vertebrate hearing: Origin, evolution and functions. In F. G. Barth, H.-D. Klein, & P. Giampieri-Deutsch (Eds.), *Sensory perception: Mind and matter* (pp. 23–40). Vienna, New York: Springer.
Manley, G.A. (2012) Evolutionary paths to mammalian cochleae. JARO 13, 733–743.
Manley, G. A., & Johnstone, B. M. (1974). Middle-ear function in the guinea pig. *Journal of the Acoustical Society of America,* 56, 571–576.
Manley, G. A., & Köppl, C. (1998). Phylogenetic development of the cochlea and its innervation. *Current Opinion in Neurobiology,* 8, 468–474.
Manley, G. A., & Clack, J. A. (2004). An outline of the evolution of vertebrate hearing organs. In G. A. Manley, A. N. Popper, & R. R. Fay (Eds.), *Evolution of the vertebrate auditory system* (pp. 1–26). New York: Springer.
Manley, G. A., & Kraus, J. E. M. (2010). Exceptional high-frequency hearing and matched vocalizations in Australian pygopod geckos. *Journal of Experimental Biology,* 213, 1876–1885.
Martin, T., & Luo, Z-X. (2005). Homoplasy in the mammalian ear. *Science,* 307, 861–862.
Masterton, B., Heffner, H., & Ravizza, R. (1969). The evolution of mammalian hearing. *Journal of the Acoustical Society of America,* 45, 966–985.
Meng, J., Wang, Y., & Li, C. (2011). Transitional mammalian middle ear from a new Cretaceous Jehol eutriconodont. *Nature,* 472, 181–185.
Müller, J., & Tsuji, L. A. (2007). Impedance-matching hearing in paleozoic reptiles: Evidence of advanced sensory perception at an early stage of amniote evolution. *PLoS ONE,* 2, e889.
Noden, D. M. (1991). Cell movements and control of patterned tissue assembly during craniofacial development. *Journal of Craniofacial Genetics and Developmental Biology,* 11, 192–213.
Novacek, M. J. (1977). Aspects of the problem of variation, origin and evolution of the eutherian auditory bulla. *Mammal Reviews,* **7**, 131–149.
Obrist, M. K., Fenton, M. B., Eger, J. L., & Schlegel, P. A. (1993). What ears do for bats: A comparative study of pinna sound pressure transformation in Chiroptera. *Journal of Experimental Biology,* 180, 119–152.
O'Gorman, S. (2005). Second branchial arch lineages of the middle ear of wild-type and *Hoxa2* mutant mice. *Developmental Dynamics,* 234, 124–131.
Plant, M. R., MacDonald, M. E., Grad, L. I., Ritchie, S. J., & Richman, J. M. (2000). Locally released retinoic acid repatterns the first branchial arch cartilages in vivo. *Developmental Biology,* 222, 12–26.
Puria, S., & Steele, C. R. (2008) Mechano-acoustical transformations. In R. R. Hoy, G. M. Shepherd, A. I. Basbaum, A. Kaneko, & G. Westheimer (Eds.), *The senses: A comprehensive reference* (pp. 165–201). Amsterdam: Elsevier.
Qiu, M., Bulfone, A., Martinez, S., Meneses, J. J., Shimamura, K., Pedersen, R. A., & Rubenstein, J. L. (1995). Null mutation of *Dlx-2* results in abnormal morphogenesis of proximal first and

second branchial arch derivatives and abnormal differentiation in the forebrain. *Genes and Development,* 9, 2523–2538.

Reichert, K. B. (1837). Über die Visceralbogen der Wirbelthiere im Allgemeinen und deren Metamorphosen bei den Vögeln und Säugethieren. *Archiv der Anatomie, Physiologie und Wissenschaftliche Medizin,* 1837, 120–220.

Rich, T. H., Hopson, J. A., Musser, A. M., Flannery, T. F., & Vickers-Rich, P. (2005). Independent origins of middle ear bones in monotremes and therians. *Science,* 307, 910–914.

Rijli, F. M., Mark, M., Lakkaraju, S., Dierich, A., Dollé, P., & Chambon, P. (1993). A homeotic transformation is generated in the rostral branchial region of the head by disruption of *Hoxa-2,* which acts as a selector gene. *Cell,* 75, 1333–1349.

Rosowski, J. J. (1992). Hearing in transitional mammals: Predictions from the middle ear anatomy and hearing capabilities of extant mammals. In D. B. Webster, R. R. Fay, & A. N. Popper (Eds.), *The evolutionary biology of hearing* (pp. 615–632). New York: Springer.

Rosowski, J. J. (1994). Outer and middle ears. In R. R. Fay & A. N. Popper (Eds.), *Comparative hearing: Mammals* (pp. 172–247). New York: Springer.

Rosowski, J. J., & Graybeal, A. (1991). What did *Morganucodon* hear? *Zoological Journal of the Linnean Society,* 101, 131–168.

Rowe, T. (1996). Coevolution of the mammalian middle ear and neocortex. *Science,* 273, 651–654.

Rowe, T. B., Macrini, T. E., & Luo, Z-X. (2011). Fossil evidence on origin of the mammalian brain. *Science,* 332, 955–957.

Ruf, I., Luo, Z-X., Wible, J. R., & Martin, T. (2009). Petrosal anatomy and inner ear structures of the Late Jurassic Henkelotherium (Mammalia, Cladotheria, Dryolestoidea): insight into the early evolution of the ear region in cladotherian mammals. *Journal of Anatomy,* 214, 679–693.

Ruggero, M. A., & Temchin, A. N. (2002). The roles of the external, middle, and inner ears in determining the bandwidth of hearing. *Proceedings of the National Academy of Sciences of the USA,* 99, 13206–13210.

Schilling, T. F., Prince, V., & Ingham, P.W. (2001). Plasticity in zebrafish *hox* expression in the hindbrain and cranial neural crest. *Developmental Biology,* 231, 201–216.

Schneider, R. A., & Helms, J. A. (2003). The cellular and molecular origins of beak morphology. *Science,* 299, 565–568.

Shigetani, Y., Sugahara, F., Kawakami, Y., Murakami, Y., Hirano, S., & Kuratani, S. (2002). Heterotopic shift of epithelial-mesenchymal interactions in vertebrate jaw evolution. *Science,* 296, 1316–1319.

Sienknecht, U. J., & Fekete, D. M. (2008). Comprehensive *Wnt*-related gene expression during cochlear duct development in chicken. *Journal of Comparative Neurology,* 510, 378–395.

Smotherman, M., & Narins, P. (2004). Evolution of the amphibian ear. In G. A. Manley, A. N. Popper, & Fay, R. R. (Eds.), *Evolution of the vertebrate auditory system.* (pp. 164–199). New York: Springer.

Takechi, M., & Kuratani, S. (2010). History of studies on mammalian middle ear evolution: A comparative morphological and developmental biology perspective. *Journal of Experimental Zoology B: Molecular and Developmental Evolution,* 314B, 417–433.

Taylor, G. D. (1969). Evolution of the ear. *Laryngoscope,* 79, 638–651.

Trainor, P., & Krumlauf, R. (2000). Plasticity in mouse neural crest cells reveals a new patterning role for cranial mesoderm. *Nature Cell Biology,* 2, 96–102.

Wang, Y., Hu Y., Meng J., & Li, C. (2001). An ossified Meckel's cartilage in two cretaceous mammals and origin of the mammalian middle ear. *Science,* 294, 357–361.

Xu, P. X., Adams, J., Peters, H., Brown, M. C., Heaney, S., & Maas, R. (1999). *Eya1*–deficient mice lack ears and kidneys and show abnormal apoptosis of organ primordia. *Nature Genetics,* 23, 113–117.

Chapter 3
Comparative Middle Ear Structure and Function in Vertebrates

John J. Rosowski

Keywords Auditory ossicles • Inner ear • Periotic system • Middle ear • Middle ear air spaces • Middle ear muscles • Sound conduction in fish • Swim bladder • Sound conduction in amphibians • Sound conduction in reptiles and birds • Sound conduction in mammals • Tympanic membrane

3.1 Introduction

This chapter is an attempt to provide a brief review comparing middle ear structure and function in the vertebrate line. A vast amount of information on this topic is readily available in the literature. Hundreds of relevant scientific papers, book chapters, and books on comparative hearing that have been published within the last 170 years, and a significant fraction of the earliest papers contain information on structure as well as data and intuitions on function. There are also extensive reviews of this topic that, more often than not, concentrate on selected vertebrate classes (e.g., Rosowski 1994; Lewis and Narins 1999; Saunders et al. 2000; Mason 2007). This chapter heavily references those earlier reviews, while touching on a fraction of the information contained in those more detailed works. The interested reader should use this chapter as a gateway to those more fundamental chapters and research papers.

The chapter starts with a discussion of the unique issues associated with the coupling of environmental sound to the inner-ear sensory organs and the different adaptations used to help overcome these issues in different vertebrate classes. Following in the footsteps of some of the extensive reviews of this topic, notably

J.J. Rosowski (✉)
Eaton-Peabody Laboratories, Massachusetts Eye and Ear Infirmary, 243 Charles Street, Boston, MA 02114, USA
e-mail: John_Rosowski@meei.harvard.edu

S. Puria et al. (eds.), *The Middle Ear: Science, Otosurgery, and Technology*, Springer Handbook of Auditory Research 46, DOI 10.1007/978-1-4614-6591-1_3,

Henson (1974), a broad definition of "middle ear" is used that includes any anatomical mechanisms that assist the coupling of sound from the external environment to the sensory mechanisms within the inner ear.

3.2 Sound Stimulation of the Inner Ears of Vertebrates

3.2.1 *What Is Sound?*

Sound is a time-varying physical disturbance in pressure, velocity, density, and temperature within a medium that propagates in space, where the medium can be a fluid (such as air or water) or a solid (Wever and Lawrence 1954; Beranek 1993). Owing to differences in compressibility, these physical disturbances travel faster in water than in air, and faster still in materials with even lower compressibility such as compacted earth and rock and metal (Kinsler et al. 1982).

3.2.2 *Sound Reception Within Vertebrate Inner Ears*

The basic anatomical specialization used by vertebrates for sound reception is the hair cell (Wever 1974; Coffin et al. 2004). Sound stimulation produces a motion of the ciliary bundle at the apex of these cells relative to their cell body. Collections of hair cells are grouped in multiple sensory organs within the inner ear of vertebrates, including the ampullae of the semicircular canals, the utricle, the saccule, the lagena (all of which may also be sensitive to more constant inertial and rotary stimulation), as well as papillary organs that appear to be more specialized for sensitivity to sound (Wever 1974; Lewis and Narins 1999; Gleich and Manley 2000).

The hair cell organs that are most receptive to sound vary within the different vertebrate orders. In individual species within the orders of fish, the auditory organs include the utricle, saccule, and lagena (Popper and Fay 1999). In Amphibia the hearing organs include several specialized papillary organs (Feng et al. 1975; Wever 1985) and the saccule, which responds to vibratory stimuli at low sound frequencies (Narins and Lewis 1984; Christensen-Dalsgaard and Narins 1993). The hearing organ in reptiles and birds is the basilar papilla (Wever 1978; Miller 1980; Smith 1985), whereas in mammals the hearing organ is the cochlea (Slepecky 1996). Associated with these different organs are different specializations for producing the relative motions of the ciliary bundles that stimulate the hair cells. These specializations vary greatly between the different vertebrate orders and suborders. The next section of this chapter describes these specializations more or less along phylogenetic lines, starting with fish and ending with the ears of mammals.

3.3 Specializations for Conducting Sound to the Inner Ear

3.3.1 Fish Without a Specialized Sound-Conduction Apparatus

The multiple vertebrate classes that include the cartilaginous and bony fish exhibit a great complexity of hearing specializations including significant differences in the structure of the inner ear and the inner-ear hair cell organs that are most sensitive to sound. The least specialized of this collection of species have relatively poor hearing (Popper and Fay 1999, 2011). In these nonspecialized ears, the mechanisms for sensing sound are differences in inertia that induce relative motions between the hair cell body and those structures that determine the motions of the ciliary bundles as the entire body of the fish moves back and forth within the surrounding water that has been set into vibration by a sound source. Generally the inertial structure that determines the motion of the ciliary bundle is a macular membrane that sits on the tips of the bundle (Popper and Fay 1999, 2011). As the body of the fish is set into motion, the hair cells and the macula move with a different magnitude and phase, producing the relative motion of the ciliary tips and cell bodies necessary for sensory stimulation of the hair cells underneath the macula. Because the sound-induced displacements of fluid particles fall off more quickly with distance from a sound source than the sound pressures associated with these motions (Kinsler et al. 1982; Kalmijn 1988, 1989), this dependence on the motion of the surrounding fluid results in an ear whose sensitivity falls off faster with distance than an ear that is directly sensitive to sound pressure.

One outcome of this relatively simple mechanism for stimulating auditory hair cells is that populations of hair cells with specific orientations of their hair bundles relative to the body axes will respond in a direction selective manner to sound-induced displacement of these macula organs (Fay 1984; Fay and Edds-Walton 1997). The introduction of directional cues by the mechanisms that couple sound to the inner ear is a common theme throughout vertebrates.

3.3.2 Fish: Ears with Close Connections to a Swim Bladder

While fish with unspecialized sound-conductive systems are sensitive to sound, the range of sound frequencies and levels that elicit an auditory response are generally restricted to relatively intense sounds of frequencies below 1,000 Hz (Popper and Fay 1999). Multiple lines of bony fish have independently developed various specializations that enhance the frequency and level ranges of the sounds that can stimulate the auditory sensors. The basic anatomical feature that is common across these specializations is an association of a gas-filled swim bladder with the ear (Popper and Fay 1999, 2011), where the primary function of the swim bladder is to help fishes maintain the proper buoyancy within their aquatic environment.

A swim bladder is essentially a bubble of gas held within a sack within the fish's body. When subjected to an alternating sound pressure, the gas in the bladder is alternately squeezed and rarified resulting in time-varying displacements of the walls of the bladder. The displacements of the bladder walls produce displacements of the surrounding fluids and body tissues that are larger than the sound-induced displacements that would have occurred if there were no compressible bladder (Popper et al. 2003; Popper and Schilt 2008). Additional specializations, such as the development of chains of bone to connect better the displacements of the swim bladder with the fluid-filled inner ears, or the close approximation of the bladder or air pockets within the bladder to the inner ear, are apparent in varied fish species (Henson 1974; Popper and Fay 1999).

3.3.3 Amphibians: The Transition to Land

As the early vertebrates first moved to land, their ears were poorly designed for the reception of airborne sound. Although the fish with specialized conductive apparatuses used their swim bladders to increase the inner-ear fluid displacements associated with small sound pressures, these mechanisms were less effective when the stimulus was airborne sound, which is not efficiently conducted through the body walls. The basis of the problem is the impedance mismatch between air and the body tissues. Only a small fraction of the sound energy available in air borne sound enters the body's tissues, while the rest is reflected away. The loss in sound energy in these circumstances is on the order of 30 dB (Wever and Lawrence 1954; Dallos 1973). Several different improvements in inner-ear sound conduction occur in amphibians to increase the displacements of inner-ear fluid produced by sound. The various specializations in structure in the amphibian ear are well described in other reviews (Wever 1985; Lewis and Narins 1999; Mason 2007).

One improvement is the change in the location of the air pocket from inside the body to a position between the surface of the skull and the bony inner ear. This change permitted more efficient compression of the pocket by airborne sound as well as improved coupling to the inner ear. While apparent in several of the suborders in Amphibia, this structural alteration may have already occurred in the some of the lobe-finned fishes that are thought to be the evolutionary precursors of terrestrial vertebrates. For example, in Latimeria (e.g., *Latimeria chalumnae*), a modern example of the lobe-finned fishes, there is a spiracular pouch associated with the gills between the surface of the head and the bone covering the inner ear that may be air filled (Fritzsch 1992).

A second improvement was the development of a tympanic membrane on the lateral surface of the air space that is coupled by an ossicular chain to the inner ear. An important part in the development of such a "tympano-ossicular" middle ear is the presence of a window in the bone surrounding the inner ear that allows direct coupling of the inner-ear fluids and the mobile ossicular element bound in the window by a flexible ligament. (Following the common practice this window will be referred to as the "oval window," a name that comes from human anatomy. In reality the window is more round than oval in many nonhuman terrestrial vertebrates.) The combination of a large tympanic membrane and a small oval window is the basis for a hydromechanical transformer that better couples airborne sound energy to the inner-ear fluids. Such a transformation brings a new problem. In fish ears with conductive specializations, the swim bladder acts to increase the sound-induced displacements of fluid near the inner ear structures so as to produce significant relative displacements between the ciliary bundles and the hair-cell bodies. The tympano-ossicular system, however, actually decreases the sound-induced displacements presented to the inner ear while increasing the sound pressure at the oval window.

To take advantage of the increased sound pressures produced at the entrance of the inner ear by the tympano-ossicular system, the third improvement was the development of inner-ear specializations in the lymph-filled periotic (around the ear) pathways that connect motions of the ossicle to the sense organs (Lewis and Narins 1999). Three distinct and separate specializations can be seen in amphibians. The common effect of these specializations is to increase the displacement of fluid within the inner ear by taking advantage of the increased sound pressures produced by the tympano-ossicular system (Wever 1985; Mason 2007).

3.3.3.1 Specializations in the Periotic Lymphatic Spaces of Amphibians

For an increase in sound pressure at the oval window interface of the tympano-ossicular system and the inner ear to induce significant displacements within the relatively incompressible lymphs of the inner ear, there must be some compliant or compressible "window" on the other side of the hair-cell sensors where the sound pressure is smaller in magnitude. In such a system, the displacement of the inner-ear lymphs is directly related to the sound pressure difference between the two windows. Three different types of pressure-difference mechanism are observed in amphibians.

Caecilians (limbless borrowing amphibians) have a "re-entrant" inner-ear fluid system (Wever 1985), in which the ossicular "footplate" (the flat end of the ossicle in the oval window) is sealed within a perilymph filled space that is connected to both the proximal input to the inner ear (on the medial surface of the footplate) and the distal "output" of the system via a fluid path that ends at the lateral surface of the footplate. The hearing sense organs of the amphibian inner ear are located between the input and the output pathways to and from the inner ear. Inward motions of the footplate simultaneously produce a positive pressure on the medial surface of the footplate and a negative pressure on the lateral surface. This pressure difference between the two sides of the footplate allows a wave of fluid displacement to progress from the medial surface of the footplate to stimulate the receptors. The wave then circles through the "re-entrant" lymphatic pathway to the lateral surface of the moving footplate.

In some urodeles (salamanders and newts) the second window for lymphatic displacements is a connection between the inner-ear lymphatic spaces and the fluid spaces with the brain case (Wever 1985; Mason 2007). The relatively large size of the brain case (compared to the small volume of the inner ear) and the small but nonzero compressibility of body fluids and tissues within the brain, allows a fluid displacement wave produced by motion of the footplate to be relieved by small compressions of the fluid and tissues within the brain. (A similar path for fluid displacement has been suggested in some pathologic human ears due to abnormal fluid pathways between the inner ear and the brain [Songer and Rosowski 2007; Merchant and Rosowski 2008]). It has also been suggested that some fraction of the sound-related displacements is conducted through the brain to the contralateral ear, where the displacements induce opposite motions of the contralateral footplate (Wever 1985). This idea has not been tested extensively, and any compressions within the brain would act to reduce such coupling.

The anurans (frogs and toads) display a third mechanism for increased lymphatic displacements that result from sound pressure at the entrance to the inner ear, which is identical to the specialization seen in most reptiles and in birds and mammals. In these species a second membrane-covered window exists between the inner ear and the air-filled middle ear (Wever 1985; Mason 2007). This "round window" allows ready motion of the cochlear lymphs in response to a difference between the relatively high sound pressure within the lymphs at the oval window and the low sound pressure within the middle ear air space around the round window. Placement of the inner-ear sensing organs astride the fluid path provides efficient stimulation of the sensors (Lewis and Narins 1999; Purgue and Narins 2000a, b).

3.3.3.2 Variations in the Structure of the Middle Ear in Amphibia

The middle ear system of frogs and toads is similar to that in many amniotes. The distinguishing features of such a system include (1) a relatively thin sound receptive membrane (the tympanic membrane [TM]) placed laterally on the skull, possibly at the medial termination of an ear canal; (2) a compressible air-filled cavity behind the TM that allows the TM to move in and out when there is a difference between the sound pressure on the TM's outer surface and the sound pressure within the cavity; and (3) an ossicular system consisting of bony and cartilaginous rods and connecting levers that couple the motions of the TM to the inner ear. The footplate is the distal end of the ossicular system and is sealed in the oval window by a flexible ligament. The area of the footplate is small relative to the TM, and the combination of the hydraulic advantage of the difference in area between the TM and the footplate, and any mechanical advantages derived from potential ossicular-lever mechanisms are the source of the transformer action of the tympano-ossicular middle ear (Wever and Lawrence 1954; Dallos 1973; Rosowski 2010).

Although many anurans exhibit the tympano-ossicular middle ear described in the preceding text, individual anuran species and the non-anuran amphibians show significant variations in middle ear anatomy, including the lack of a specialized tympanic membrane, or the lack of an air-filled middle ear cavity (Wever 1985; Mason 2007). On the other hand, most amphibian middle ears contain an operculum, a bony structure not found in amniote (reptile, bird, and mammal) middle ears. This small rounded bone sits in the oval window alongside the ossicular footplate in many amphibians. The operculum often interdigitates with the footplate, and appears to be loosely held in place by muscles. The function of the operculum is controversial. Wever (1979, 1985) ascribed an auditory feedback control mechanism to the anuran operculum, where contractions and relaxations of opercular supporting muscles and a muscle attached to the footplate either inhibited or allowed large sound-induced motions of the footplate. Hetherington (1994) argued against such a protective mechanism and instead suggests the operculum has a role in the conduction of extratympanic sounds and vibrations to the inner ear.

3.3.3.3 Extratympanic Sound Conduction in Amphibians

It has been well established that sound can cause significant vibration of the skin of frogs, especially when the skin covers the respiratory air spaces (e.g., the lungs and the buccal cavity) (Narins et al. 1988; Ehret et al. 1990). Although the sound-induced displacements of these regions tend to be smaller than the vibration of the TM, the vibration of these larger surfaces can lead to significant volume displacements of the air cavities in the thorax and pharynx. It has been demonstrated that some fraction of these volume displacements is coupled to the middle ear via the normally open bony "Eustachian-tube" orifice that couples the middle ear air spaces to the pharynx (Narins et al. 1988). Further, the coupling of the bilateral middle ears to the pharynx allows sounds that displace the TM on one side of the head to propagate across the head to the contralateral middle ear (Aertsen et al. 1986). How all of these air paths for sound conduction interact when the whole body is immersed in a sound field is not completely understood. However, there is considerable evidence that these multiple sound pathways to the middle ear impart directionality to the ear that is used in the localization of sound sources (Feng and Shofner 1981; Gerhardt and Rheinlaender 1982; Vlaming et al. 1984).

In some anurans (Hetherington 1988, 1992), sound acts on the body wall and sets skeletal components into vibration (especially the broad scapula of the shoulder girdle). These vibrations can be conducted to the inner ear through displacements of the operculum, which is coupled to the scapula by the opercular muscle. The data of Hetherington (1992) suggest that the opercular path of sound conduction is most significant in small frogs. This scapular to opercular path has also been implicated in the conduction of substrate vibrations to the saccule within the inner ear (Hetherington 1985, 1987; Lewis and Narins 1999).

3.3.3.4 Feedback Control of Middle Ear Function

A common feature of the ears of terrestrial vertebrates is the feedback control of middle ear sound conduction via the efferent control of muscles whose actions modify the transduction process (Borg et al. 1984). Two separate control systems have been hypothesized in anurans. As noted previously, Wever (1979, 1985) has presented evidence that muscles attached to the operculum and footplate control the magnitude of sound-induced displacement of these structures, where variations in muscle tension alter the coupling between the footplate and the operculum. Such variations could either increase the motion of the ossicular footplate–operculum complex (Lombard and Straughan 1974) or decrease footplate motion (Wever 1979). Evidence against the reduction in motion hypothesis has been presented by Hetherington (1994), who argues that Wever misidentified the different muscle groups involved and that variations in the tension of the opercular muscle do not modulate sound transmission by the tympano-ossicular system.

Although the use of muscles to modulate the stiffness of the ossicular system is a common theme throughout amniotes, a unique form of control of middle ear sound transmission has been described in a frog that appears specialized for the reception of sound frequencies well above the normal range of amphibians (Gridi-Papp et al. 2008). The torrent frog (*Amolpos tormotus*) lives in rapidly moving streams in China, where the highly turbulent water flow produces significantly loud sounds in the frequency range of hearing of many amphibians. These frogs appear to have the ability to use muscular contractions to seal the normally open bony orifice that couples the middle ear cavity to the pharynx. Closing the tubal orifice increases the stiffness of the air within the middle ear, which reduces the sound-induced motion of the TM and ossicular system at low frequencies, but also moves the middle ear resonance to a higher frequency. The shift in resonance leads to an increase in sensitivity near the new resonance.

3.3.4 The Columellar Ear of Reptiles and Birds

3.3.4.1 The Inner Ears of Reptiles and Birds

The inner ears of reptiles and birds contain a basilar papilla specialized for the sensation of sound energy. Unlike the specialized hearing organs in fish and Amphibia, this organ sits on a basilar membrane, which vibrates when sound is presented to the inner ear (Wever 1978; Miller 1980; Gleich and Manley 2000). The direct vibration of the basilar membrane that supports the hair cells represents a significant adaptation that takes advantage of the trans-inner-ear pressure differences produced by the development of the two-window inner ear. Such a membrane produces an increase in the displacements of the hair cells and their ciliary bundles in response to a sound pressure difference between the two cochlear windows.

There are many publications describing different schema for turning these displacements of the hair cell body into relative motions of the cell and its ciliary bundle (e.g., Wever 1971; Weiss and Leong 1985; Manley et al. 1988).

Even though all reptilian and bird inner ears exhibit a specialized papilla placed on a mobile membrane (Schwartzkopf 1973; Wever 1978; Miller 1980), there is significant variability in the size and form of the papilla, especially in lizards, as well as great variation in the accessory "tectorial" structures that enhance the motion of the ciliary bundles (Wever 1978; Manley 2000). The presence of the round window is another point of variance. The inner ears of lizards, crocodylids, and birds contain a true round window membrane that releases the sound pressure produced in the inner ear by the motion of the ossicular footplate in the oval window, and thereby increase the motions of the lymphs within the inner ear. Snakes, turtles, and the rare amphisbaenids and sphenedon have no round window and seem to rely on the "re-entrant" model of inner-ear lymph displacement described previously (Wever 1978).

3.3.4.2 The Tympano-ossicular Middle Ear in Reptiles and Birds

The middle ears of most reptiles and birds use a single bony ossicle and associated cartilaginous connecting elements to couple sound-induced motions of the tympanic membrane to the oval window of the middle ear (Schwartzkopf 1973; Wever 1978; Saunders et al. 2000). As in anurans, the distal end of the single column-like ossicle (here called the columella) widens out into a flat bony plate that fits into the oval window. (The term *columella* describes the shape of this long and thin ossicle; however, because this bone is believed homologous to the mammalian stirrup-shaped stapes, many authors equate the columella and the stapes.) The area of the columellar footplate is anywhere from 10 to 60 times smaller (Kirikae 1960; Schwartzkopf 1973; Rosowski and Graybeal 1991) than the area of the tympanic membrane, and this ratio of areas is the primary determinant of the transformer function of the reptilian/avian middle ear (Wever 1978; Gummer et al. 1989a; Saunders et al. 2000).

There is a secondary transformer within the reptilian and avian middle ear: The shape of the cartilaginous "extra-columella" (or extra-stapes), which couples the columella to the TM, allows a lever action between the motion of the center of the TM and the motion of the columella that provides a mechanical lever advantage of a factor of 2–3 (Gaudin 1968; Wever 1978; Gummer et al. 1989a). The combination of the mechanical lever and the hydroacoustic transformation of the TM-to-footplate area ratio provide an increase in sound pressure and a decrease in volume displacements at the entrance to the inner ear, relative to the sound stimulus and volume displacement of air lateral to the tympanic membrane (Rosowski et al. 1985; Gummer et al. 1989a, b). This "transformation" of the sound between the TM and inner ear, together with the two inner-ear windows and the basilar membrane, allows more efficient transfer of ear canal sound pressure to the inner ear and increased motions of the stereocilliary bundle (Wever 1978; Manley et al. 1988).

The introduction of a mechanical lever into the middle ear to help transform the impedance of the air in the ear canal and the fluids in the cochlea introduces a problem into the design of the middle ear. The basis for such levers is rotation about a fulcrum, where rotation can lead to significant three-dimensional motions at the ends of the lever. Specifically, in a one-ossicle ear, the lever action produced by the rotation of the extracolumellar cartilage embedded in the TM evokes a combination of a side-to-side and in-and-out motion of the proximal end of the bony columella (e.g., Kirikae 1960; Gaudin 1968). However, the annular ligament constrains the footplate within the oval window and reduces the side-to-side motions of the distal bony columella. This reduction, relative to the side-to-side motions at the proximal end, is made possible by a bending within the cartilaginous extracolumellar. Such bending leads to a reduction in signal transfer through the middle ear that is most prominent with higher frequency sounds (Manley 1972b; Gummer et al. 1989a; see also Manley and Sienknecht, Chap. 2).

There is significant variation in the middle ear air spaces in reptiles and birds (Henson 1974; Wever 1978). Individual turtle and snake species have very small or nonexistent middle ear air spaces (Kirikae 1960; Wever 1978). Most lizards have middle ear air spaces that are not constrained by bony walls; the air spaces are really pockets of the larger pharynx (Wever 1978). The middle ear air spaces of birds and crocodylids are generally surrounded by bone (except of course for the TM); however, a bony-walled air-filled channel connects the binaural air spaces, allowing sound conduction between the middle ears (Wever and Vernon 1957; Hill et al. 1980; Rosowski and Saunders 1980). The potential for each TM to respond to a difference in sound pressures between sounds entering the middle ear from the two sides of the head could again be useful in providing directional cues for the localization of sound sources (Coles and Guppy 1988; Christensen-Dalsgaard and Manley 2005; cf. Moiseff and Konishi 1981).

3.3.4.3 The Columellar Muscle

A few reptiles—the varanids, geckos, and crocodylids (Wever 1978)—and the birds (Henson 1974; Counter and Borg 1979, 1982) exhibit a muscle attached to the bony columella. This muscle appears to be innervated by the facial nerve (cranial nerve VII), and is thought to be homologous to the mammalian stapedius muscle. There has been little work describing the function of this muscle in reptiles, but it has been noted that the reptilian species that possess a columellar muscle are those that phonate. Evidence exists that contraction of this muscle in birds reduces the transfer of low-frequency sound energy to the inner ear (Counter and Borg 1982); however, the evidence that the columellar muscle is activated by external sound is mixed. Sound-induced contractions of the muscle have been looked for but not found in pigeons and chickens, but have been described in a species of owl (Borg et al. 1984). It is known, though, that the avian muscle contracts during phonation and may function to reduce the transfer of self-generated sounds to the inner ear (Counter and Borg 1979).

3.3.5 *The Three-Ossicle Ear of Mammals*

The ears of mammals have many features in common with birds and reptiles, but there are also significant differences compared to other vertebrates. Variations in mammalian middle ear structure and function are the subjects of many earlier reports and reviews (e.g., Doran 1879; Keen and Grobbelaar 1941; Fleischer 1973).

3.3.5.1 The Inner Ear of Mammals and Its Sensitivity to Sound

The auditory inner ear, or cochlea, of mammals is a highly organized collection of hair cells, and their accessory structures supported on a flexible basilar membrane (Slepecky 1996). As in most reptiles and birds, there are two windows (the oval and the round) in the bone surrounding the inner ear, both of which open to the middle ear. Also as in most reptiles and birds, the footplate at the termination of the mammalian ossicular apparatus is bound in the oval window by a ligament (Guinan and Peake 1967; Bolz and Lim 1972), while the round window is covered by a thin membrane (Wever and Lawrence 1954; Békésy 1960; Paparella et al. 1983).

Some of the best evidence for the sensitivity of the two-windowed inner ear to the difference in sound pressure acting on the windows has been gathered in mammals (Wever and Lawrence 1950, 1954; Voss et al. 1996). These data demonstrate that the cochlear response to simultaneous and direct sound stimulation of the two windows is largest when the sound pressures applied to the windows are of equal magnitude and opposite phase, a condition that leads to the largest fluid displacements within the inner ear. The data also demonstrate a greatly reduced response of the inner ear when the stimuli applied to the two windows are identical in magnitude and phase. This latter demonstration is a strong argument against the possibility that the sensory apparatus within the inner ear responds directly to applied pressure, although some small response to such a stimulus might result if the contents of the cochlea were compressible (Shera and Zweig 1992), or if there were a normal "third window" allowing fluid motion within the inner ear when there was no pressure difference between the oval and round window (Ranke 1953; Békésy 1960; Tonndorf and Tabor 1962).

3.3.5.2 The Three Mammalian Ossicles

The mammalian ossicular chain (Figs. 3.1 and 3.2) is made up of three interconnected bones: the malleus, incus, and stapes (Dahmann 1929; Wever and Lawrence 1954; Henson 1974). These ossicles are coupled by joints and supported by ligaments in a manner that allows rotation of the malleus and incus about their combined center of gravity (Puria and Steele 2010; Lavender et al. 2011). Differences in the lengths of the malleus and incus lever arms provide a transformation of sound energy consistent with an increase in sound pressure inside the

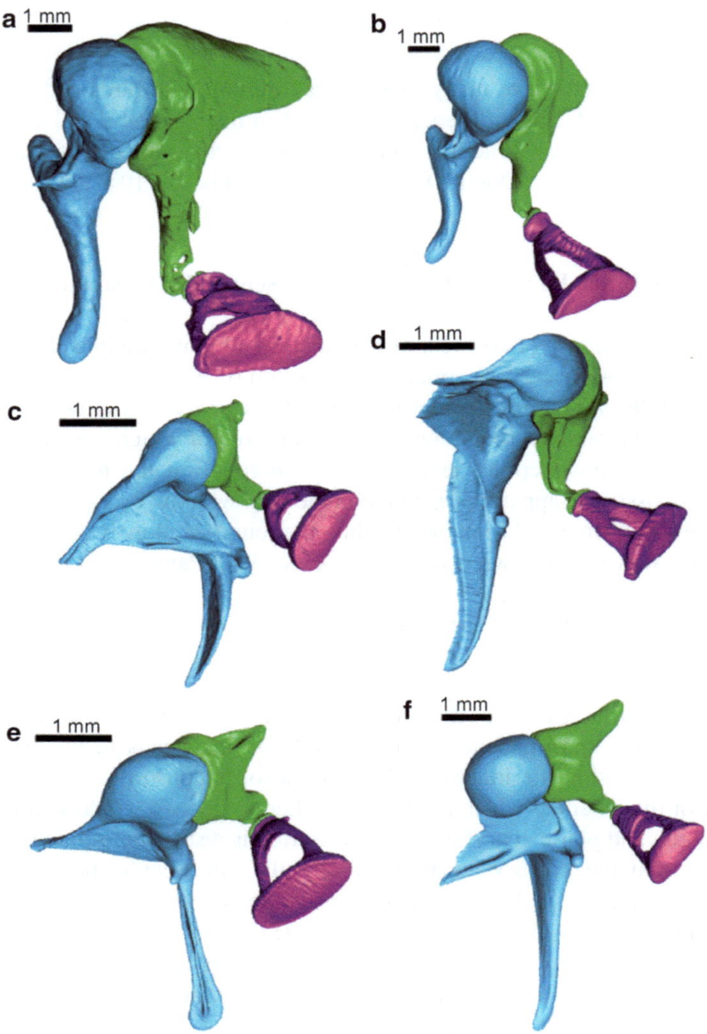

Fig. 3.1 The ossicular chains of five different mammalian species (Modified from Salih et al. 2012). The mallei are in *blue*, the inci in *green*, and the stapes in *magenta*. Each of the chains is viewed from and antero-medial direction. Ossicles from (**a**, **b**) two humans of different age and body size; (**c**) a rat; (**d**) a rabbit *(Oryctolagus cuniculus*); (**e**) a gerbil (*Meriones unguiculatus*); and (**f**) a cat. Each of the scales bars is 1 mm in length; the largest ossicles are those from the two humans; the smallest are from the gerbil and rat. Note the large variation in (1) the absolute sizes of the ossicles, (2) the relative sizes of the malleus and incus, (3) the orientation of the malleus and incus lever arms, and (4) the shape of the malleus and incus. The shape of the stapes also varies, but less dramatically

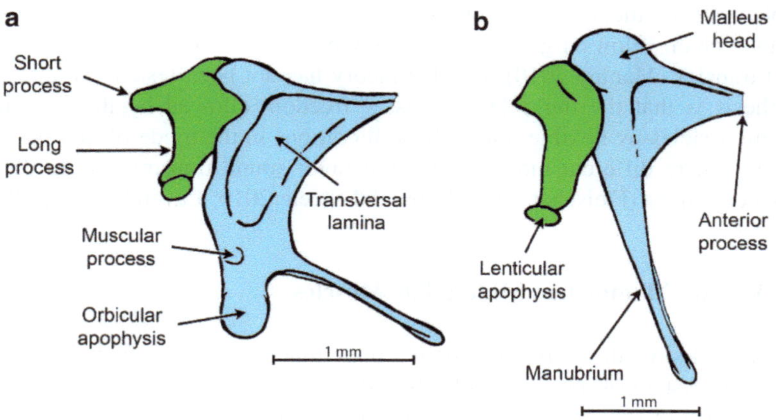

Fig. 3.2 Large variations in mammalian ossicular form (Modified from Lavender et al. 2011). The malleus and incus from two small mammals. The views are of the medial surface of a left ear, with anterior to the right and dorsal to the top. (**a**) A mouse: Note the elongated and widened shape of the malleus neck (including the transversal lamina), with the long process of the incus nearly perpendicular to the manubrium (the handle) of the malleus. This arrangement is consistent with Fleischer's (1973, 1978) categorization of a "microtype" ossicular system. (**b**) A golden hamster (*Mesocricitus auratus*): this ossicular system approaches Fleischer's (1973, 1978) "free-standing" ossicular chain, with the incus long process more parallel to the manubrium—as in human (Fig. 3.1)

cochlea and a decrease in the volume displacement of the stapes (Helmholtz 1868; Wever and Lawrence 1954; Dallos 1973).

Historically, ideas regarding the evolutionary development of the three ossicles have been varied, but it is now generally assumed that the malleus and incus developed from the bones of the jaw of early reptiles in a manner that was independent of the development of the reptilian and avian tympano-ossicular system (Allin and Hopson 1992; Bolt and Lombard 1992; Manley 2010; see also Manley and Sienknecht, Chap. 2). Although many have suggested that the development of the three-ossicle middle ear allows mammals to hear sounds of higher frequency than other terrestrial animals (e.g., Masterton et al. 1969; Manley 2010), the precise mechanism that determines this hypothesized superior ossicular function has not been clearly elucidated. It has been suggested that the rotary lever action inherent in the mammalian three-ossicle chain is responsible for its superior conduction of high-frequency sounds; however, lever actions have been described in the ossicular systems of amphibians (Wever 1985), reptiles (Manley 1972a; Wever 1978), and birds (Gaudin 1968; Saunders and Johnstone 1972; Gummer et al. 1989a). It has also been suggested that flexibility within the cartilaginous components of the reptile and avian ossicular connection is the root of the lower frequency limits of hearing in reptiles and birds (Manley 1972b, 1990), but similar flexibility exists within the joints that connect the mammalian ossicles, and this flexibility can reduce the transfer of high-frequency sound through the middle ear (Zwislocki 1962; Guinan and Peake 1967; Willi et al. 2002). It may be that the presence of the flexible joints closer to the rotary axis of the mammalian ossicular chain allows tighter ossicular coupling and improved high-frequency sound transfer (Manley 2010), but this theory has not been tested. An alternative hypothesis is that the multiple degrees of freedom allowed by the coupling of separate ossicles by flexible joints allow alterations in the mode of motions of the ossicles, where different modes of motion favor sound transmission in different frequency ranges (Fleischer 1978; Puria and Steele 2010; Lavender et al. 2011).

3.3.5.3 Two Mammalian Middle Ear Muscles

A significant structural difference between the mammalian ossicular chain and the columellar system of reptiles and birds is the presence of two separate middle ear muscles. One of the mammalian middle ear muscles, the stapedius muscle, has a tendinous insertion on the stapes, is innervated by the facial nerve (cranial nerve VII), and is thought homologous to the columellar muscles in birds and some reptiles. Contractions of the stapedius muscle stretch the ligament holding the stapes in the oval window, resulting in a stiffening of the ligament (Bennett 1984; Pang and Peake 1986) and a reduction in the transfer of lower-frequency sounds through the middle ear (Wever and Lawrence 1954; Møller 1964, 1974). The lower frequencies are those at which the transfer of sound through the middle ear is governed by stiffness (Zwislocki 1962; Møller 1965); the muscular contraction increases the stiffness and leads to the reduction in sound transfer. Sounds of frequencies above the region of stiffness control are generally unaffected, but may show small improvements in sound conduction depending on the interaction of the increased ossicular stiffness with the other impedances that govern middle ear sound transmission (Wever and Lawrence 1954; Henson 1970; Borg and Zakrisson 1975).

The second mammalian middle ear muscle is the tensor tympani, which is innervated by the motor branch of the trigeminal nerve (cranial nerve V). The tendon of the tensor tympani is attached to the malleus neck or handle (the manubrium of the malleus), and contraction of the tensor tympani pulls the TM and malleus into the middle ear air space. This contraction stiffens the middle ear by increasing the stiffness of the tympanic membrane (Nuttall 1974). Contraction of the tensor also increases the static pressure within the middle ear by reducing the volume of the middle ear spaces; this result is thought to aid in opening the Eustachian tube (Ingelstedt and Jonson 1966) as reflex contraction of the tensor occurs during swallowing.

Both the stapedius and tensor tympani are known to contract in response to loud sounds, but the stapedius generally responds at lower stimulus levels (Borg et al. 1984; Silman 1984). This sound-induced contraction is part of an acoustic reflex arc that has been suggested to reduce the sensation of loud sounds (Wever and Lawrence 1954; Wever and Vernon 1961; Henson 1965); however, contractions of the middle ear muscles take time to develop, and these muscles would be of little use in protecting the ear from rapidly developing impulsive sounds (see Popelka and Hunter, Chap. 8). Nonetheless, data suggest that an intact functioning stapedius muscle can reduce the noise exposure of the ears of workers exposed to continuous loud noise (Zakrisson et al. 1980; Borg and Nilsson 1984). Another hypothesized function is that the middle ear muscles may help reduce the masking effects of low-frequency sounds in noisy environments (Borg and Zakrisson 1974; Pang and Guinan 1997). Both muscles also contract before phonation and other activities (Carmel and Starr 1963; Borg and Zakrisson 1975; Borg et al. 1984).

A commonly suggested function of the tensor tympani is that it applies a pretension to the tympano-ossicular system that helps in the conduction of sounds to the inner ear (see Borg et al. 1984). However, there is little evidence for such a function. Measurements of middle ear function have been made in multiple animal species, and the tensor is generally found to be relaxed, except under circumstance of periodic contraction (e.g., Wiggers 1937; Rosowski et al. 2006). Also, stimulation of the tensor tympani in anesthetized animals where the muscle is relaxed produces a stiffening of the TM and a reduction of sound transfer (e.g., Wever and Lawrence 1954; Nuttall 1974). It has also been demonstrated that the sound-induced displacements of the TM and malleus in human cadaveric ears, in which the tensor is relaxed, are very similar to such measurements in live human ears (Rosowski et al. 1990; Goode et al. 1996). Finally, if such a pretension was necessary to better couple the TM and the ossicular chain, then one might expect that the application of a static pressure gradient across the TM might counterbalance the tensor effect and lead to a more compliant TM and poorer sound conduction to the ear. Multiple studies of the effect of static pressure on TM mechanics and middle ear sound conduction (Wever et al. 1948; McPherson et al. 1976; Erlandsson et al. 1980) demonstrate that both TM compliance and middle ear sound transfer are at or near maximum when there is no static pressure difference across the TM. The maximum in TM mobility with zero static pressure is one of the basic tenets of clinical tympanometry (Jerger 1970; Lidén et al. 1970; Margolis and Shanks 1985).

The presence of two middle ear muscles in the mammalian ear suggests a separation of function, which is generally borne out by the higher sensitivity of the stapes muscle contractions to sound, and the contribution of tensor contractions in the Eustachian tube reflex. Such a separation of function is possible in the mammalian ear because of the existence of multiple ossicles separated by ossicular joints (Marquet 1981). In such an arrangement, contractions of the stapes can alter the mechanics of the stapes without inducing large TM motions (Pang and Peake 1986), and contraction of the tensor tympani can pull the TM and malleus into the middle ear cavity without inducing large inward motions of the stapes (Hüttenbrink 1988; Marquet 1981). Of course such a separation of function can be achieved by ossicular systems that have the two muscles on either side of a single flexible ossicular joint, as is functionally the case in many mammals where the joint between the malleus and incus is either very tight or the two ossicles have fused (e.g., Fleischer 1973; Rosowski et al. 1999; Puria and Steele 2010).

3.3.5.4 The Mammalian Tympanic Membrane

Like the TMs of most other vertebrates, the mammalian TM is composed of three distinct layers: an external dermal layer that is continuous with the skin layer of the ear canal, a middle fibrous layer, and an internal layer of mucosal epithelium that is continuous with the epithelial layer that lines the middle ear air spaces (Lim 1968a, b; Funnell and Laszlo 1982; Decraemer and Funnell 2008). The TMs of reptiles and birds have a similar lamellar structure, but the fibrous layer appears less organized in nonmammals (Funnell and Laszlo 1982). The mammalian TM has a tentlike shape with the spine of the tent defined by the manubrium of the malleus, which is more medially placed than the tympanic bone that surrounds much of the TM. This arrangement produces a TM that appears to protrude into the middle ear air spaces. The TMs of birds and reptiles appear more flattened, and the central spine of the tent (defined by the connection of the cartilaginous extracolumellar with the TM) appears to protrude slightly into the ear canal.

Unlike other vertebrates the mammalian TM has two components, the central pars tensa and the more posterior-dorsal pars flaccida (Shrapnell 1832; Funnell and Laszlo 1982; Kohllöffel 1984). As the names imply, the pars tensa is generally stiffer and less deformable than the pars flaccida, whose shape is easily altered by small static pressure differences on either side of the membrane (Decraemer and Dirckx 1998; Dirckx et al. 1998). Indeed the pars flaccida is often assumed to play a role in maintaining equal static pressures on either side of the TM: It is thought to buffer small changes in middle ear volume produced by the absorption or generation of middle ear gas by and from the blood (Hellstrom and Stenfors 1983; Sadé et al. 1996; Decraemer and Dirckx 1998). Further, although only the pars tensa is directly coupled to the ossicular chain, it has been suggested that the pars flaccida, whose motions appear independent of the pars tensa, plays a role in equalizing low-frequency sound pressures across the pars tensa, thereby indirectly reducing the motion of the pars tensa in response to low-frequency sounds (Kohllöffel 1984; Teoh et al. 1997).

The mammalian TM is generally supported by the tympanic bone, which also can form parts of the bony ear canal and the bony wall of the middle ear air spaces. There is a wide variation in the size and extent of the tympanic bone and its support for the TM, which varies from "U-shaped" to nearly circular (van der Klaauw 1931; Keen and Grobbelaar 1941; Henson 1974). In general, the ring of tympanic bone around the TM is more complete in carnivores and primates but also in selected species of other families, including the chinchilla (*Chinchilla laniger*) and the guinea pig (*Cavia procellus*). The ring appears more "U-shaped" in nonplacental mammals and in many small mammals, for example, rats, mice, and shrews (Fleischer 1973, 1978; Henson 1974). The pars flaccida occurs at the edge of the pars tensa that is not bound by the tympanic bone (Henson 1974). The absolute area of the pars tensa varies widely among mammals (Hunt and Korth 1980; Nummela 1995; Coleman and Ross 2004). There is also significant variation in the absolute and relative size of the pars flaccida (Kohllöffel 1984; Vrettakos et al. 1988;

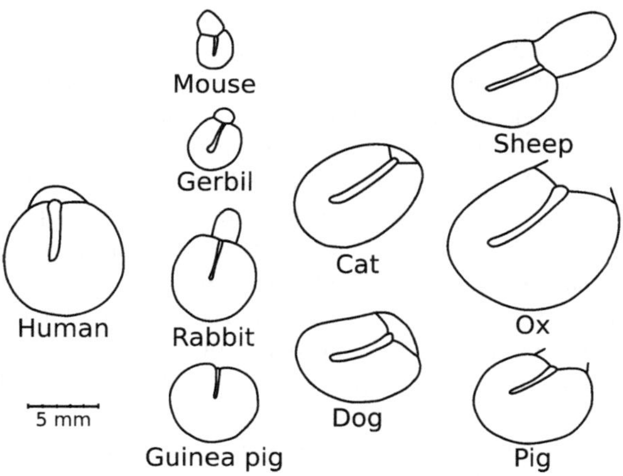

Fig. 3.3 Outlines of the TM from ten mammalian species, including human, cat, dog, mouse, gerbil (*Meriones unguiculatus*), rabbit (*Oryctolagus cuniculus*), guinea pig (*Cavia procellus*), sheep, ox (*Bos taurus*), and pig (*Sus scrofa*). The *pars tensa* of the TM is attached to the arm of the manubrium (the long thin structure). The *pars flaccida* is not apparent or complete in all of the specimens (Redrawn from Decraemer and Funnell 2008. The mouse TM is after Reijnen and Kuijpers 1971; the cat after Khanna 1970; the sheep after Lim 1968b; the gerbil after Decraemer and Funnell 2008; others after Fumagalli 1949.)

Rosowski et al. 1997); the variation in the relative size of flaccida and tensa has been associated with variation in the low-frequency hearing capabilities of different mammalian species (Kohllöffel 1984; Rosowski et al. 1997) (Fig. 3.3).

3.3.5.5 The Middle Ear Air Spaces

The middle ear air spaces behind the TM of terrestrial vertebrates act as a compressible cushion that allows large sound-induced motions of the TM. The motion of the TM is related to the difference in sound pressures acting on its middle ear and external-ear surfaces, where, in mammals, the sound pressure in the middle ear air spaces is generally related to the acoustic impedance of the air spaces and the inward and outward displacements of the entire TM (Møller 1965; Huang et al. 1997; Ravicz and Rosowski 1997). (As discussed previously, in nonmammalian terrestrial vertebrates, e.g. amphibians, reptiles and birds, there are often other pathways for sound to enter the middle ear, and these other pathways complicate the computation of sound pressure within the middle ear air spaces just medial to the TM.) Motions of both the pars flaccida and tensa contribute to the volume displacement of the TM with sound, and therefore both contribute to the sound pressure within the mammalian middle ear air spaces. Therefore, when the middle ear air spaces are intact, as noted earlier, motions of the pars flaccida indirectly affect the motion of the pars tensa (Kohllöffel 1984; Teoh et al. 1997).

The contribution of the middle ear air spaces to sound-induced motions of the pars tensa and the coupled ossicles depends on the relative impedances of the air spaces and the rest of the middle ear. There is significant variation in the relative and absolute magnitudes of these impedances in different mammalian species (Dallos 1970; Rosowski 1994). In some species (mouse [*Mus musculus*], chinchilla, guinea pig), the impedance of the air spaces (which is inversely related to the air-space volume) dominates the middle ear input impedance and opening the middle ear air spaces to the outside produces large increases in middle ear sound transfer in response to low sound frequencies, whereas in others (e.g., human, cat, kangaroo rat [*Dipodomys merriami*]) the impedance of the TM and the attached ossicles and cochlea is generally larger than the impedance of the middle ear air spaces, and opening the middle ear air spaces produces small changes in middle ear sound transfer (Dallos 1970; Rosowski 1994).

The middle ear air spaces of different species are often broken into several compartments by partial bony partitions (Kampen 1905; Werner 1960; Møller 1965). These partitions include: the incomplete bony septum that separates the middle ear air spaces of cat-like carnivores, feloids, into two compartments connected by a narrow foramen (e.g., Møller 1965; Huang et al. 1997), the bony wall around the facial nerve that nearly separates the human tympanic cavity from the generally larger air volume within the mastoid antrum and air cells (Zwislocki 1962; Molvær et al. 1978; Stepp and Voss 2005), and the many interconnected air spaces of the chinchilla (Browning and Granich 1978) and gerbil (*Meriones unguiculatus*: Lay 1972) middle ear. There is also a wide variation among species in the total volume of air enclosed in the middle ear air spaces (Rosowski 1994); not surprisingly this volume correlates with body size (Huang et al. 1997). However, individual species of similar body size can have significantly different middle ear air volumes: Compare the 0.25-cc air spaces of the guinea pig with the 2.0-cc middle ear air volume of the chinchilla (Rosowski 1994). There is also significant variation in the middle ear air volumes within species; this has been best demonstrated in humans (e.g., Molvær et al. 1978), in whom the total middle ear volume in an adult human ear varies by a factor of 10 among individuals.

The bones of the skull that enclose the middle ear air spaces differ from species to species (Kampen 1905; Werner 1960; Novacek 1977) and include the tympanic, endotympanic, petrous, and the squamous among others. These differences have been used in the past as an indicator of the phylogenetic relationship between different mammalian species (e.g., Hunt 1974; Novacek 1977). This role has been superseded by direct genetic analyses (e.g., Johnson et al. 2006).

In many mammalian species the bones covering the middle ear air spaces form smooth nearly egg-shaped bony protrusions on the posteroventral surface of the skull (van der Klaauw 1931; Keen and Grobbelaar 1941). This protrusion is often called the bulla (e.g., Legouix and Wisner 1955), though in some instances of common use "bulla" is used to refer to the entire middle ear apparatus within the bulla. Not all mammalian species have a smooth egg-shaped bulla surrounding their middle ear; e.g., in many primates (including humans) the middle ear air spaces are more deeply embedded in the bones of the posterodorsal skull and these species are sometimes said to have no bulla (e.g., Keen and Grobbelaar 1941).

Any discussion of the mammalian middle ear air spaces needs to point out that the spaces remain aerated due to the periodic opening of the Eustachian tube (e.g., Ingelstedt 1976; Elner 1976). In mammals the bilateral Eustachian tubes are supported by cartilage walls and open via muscular contraction and the generation of static pressure differences between the middle ear and the pharynx (Swarts and Rood 1990; Doyle 2000). A similar mechanism must be in place for opening the Eustachian tube in crocodylids and birds, where this tube can simultaneously aerate both of the connected middle ears (Wever and Vernon 1957; Rosowski and Saunders 1980). In lizards and anurans the middle ears are aerated through large open connections to the pharynx. Blockage of the mammalian Eustachian tube via experimental manipulation or pathology leads to replacement of the air with body fluids (e.g., Wiederhold et al. 1978; Doyle et al. 1980; Alper et al. 1997).

3.3.5.6 Correlations Between Mammalian Middle Ear Structure and Hearing Capabilities

Although all mammalian middle ears have some basic features in common, including three ossicles, a TM and air-filled middle ear spaces, there are large differences in the form and size of these components within the order of mammals (Fleischer 1973; Rosowski 1994; Nummela 1995). Differences in size should not be surprising, as the entire body of the smallest mammals (e.g., the dwarf shrew [*Suncus etruscus*] with a body mass of 0.01 kg) would fit into the middle ear air spaces of the largest terrestrial mammals (the African elephant [*Loxodonta africans*]with a body mass of 6,000 kg; Rosowski 1994). It is worthwhile noting that although ear size varies regularly with body size, the relationship is not isometric: the ratio of the body weights between the elephant and shrew above is about 6×10^5:1, while the ratio of the areas of their TMs is about 100:1 (Hunt and Korth 1980). If the TM areas were to scale with body weight, we might expect the two TMs to have a ratio of $(6 \times 10^5)^{2/3}$:1, or 7,100:1. Therefore, in smaller animals the ear is larger relative to body size than it is in larger animals (Nummela 1995). (The use of the 2/3 power in the preceding relationship is a dimensional factor to compensate for comparing the change in body mass—a measure of body volume—with TM area).

It is also known that there are large variations in the hearing capabilities of different mammalian species (Fay 1988; Heffner and Heffner 1992, 2010), and significant correlations have been observed between the frequency range of hearing, body size, and form and size of the middle ear structures across mammals of different species (Heffner and Heffner 1992; Hemilä et al. 1995; Coleman and Colbert 2010). These correlations suggest that species of large body size have larger middle ear structures than species of small body size, are more sensitive to sounds at frequencies below 1 kHz than smaller mammals, and are less sensitive to sounds of frequencies greater than 10 kHz.

There also appears to be significant correlation between some basic structural types of the middle ear and the frequency range of hearing (Rosowski 1992). Fleischer (1973, 1978) separated the middle ears of a large collection of mammals into three groups based on ossicular structure, including (see Fig. 3.2): (1) a "free-standing" ossicular group in which the malleus and incuse are only supported by mobile ligaments, as exemplified by humans, chinchillas, and guinea pigs; (2) a "micro-type" middle ear in which the middle ear structures are small and there is a large bony or firm ligamentous connection between the malleus and the tympanic bone, as exemplified by bats, shrews, marsupials, and many murid rodents, including rats of the genus *Rattus* and mice of the genus *Mus*; (3) an "intermediate" or "transitional" form in which there is a small but firm connection bony between malleus and the tympanic ring, for example, small bony connections between the anterior malleus and the tympanic ring have been described in cats and gerbil (Rosowski et al. 1999). These classes of ossicular structure are associated with differences in hearing capabilities: Ears with free-standing ossicular chains tend be more sensitive to sounds of frequencies below 500 Hz than other ears and less sensitive to sounds above 20 kHz; microtype ears tend be more sensitive to sounds above 10 kHz and less sensitive to sounds below 2 kHz; intermediate ears, however, seem to show high sensitivity to both low and higher sounds (Rosowski 1992). Of course the preceding relationships are complicated by the observations that small animals with small middle ear structures tend to have micro-type ears.

3.3.5.7 Middle Ear Function in Marine Mammals

Marine mammals evolved from mammals that arose on land and then re-entered the sea (Nummela et al. 2007). In doing so they returned to an environment in which the animal was surrounded by a fluid with an acoustic impedance more similar to that of the animal's body and inner ear. In truly aquatic animals, such as cetaceans, the external ear appears vestigial and the pathway by which sound is conducted from the environment to the inner ear is not well defined (Ketten 1994, 2000; Rosowski 1994). Although the ears of cetaceans still maintain the presence of an ossicular chain, their size and structure is much different from that observed in terrestrial mammals (Ketten 1992, 2000; Nummela et al. 1999). Further, the analog of the tympanic membrane is much different in structure from the fairly thin and fragile tympanic membranes of terrestrial mammals (Ketten 1992, 2000). Nonetheless most models of middle-function in cetaceans assume ossicular conduction of sound to the inner ear (Hemilä et al. 1999; Tubelli et al. 2012), and use measurements and estimates of ossicular mass, and stiffness to define sound transmission through these middle ears (Hemilä et al. 2001; Miller et al. 2006; Zosuls et al. 2012). Such models have been used to compute audiograms for different cetaceans (Hemilä et al. 1999, 2001; Tubelli et al. 2012).

3.4 Issues in Comparative Middle Ear Function

3.4.1 The Frequency Dependence of the Middle Ear Mechanism

Comparisons of the frequency dependence of middle ear sound transfer in terrestrial vertebrates with aerated middle ears and ossicular coupling of the TM to the inner ear demonstrate significant differences between columellar and three-ossicle ears (Saunders and Johnstone 1972; Rosowski 2003; Manley 2010): The sound-induced velocity of the columella in frogs, reptiles, and birds is limited in response at sound frequencies greater than 4 or 5 kHz (Manley 1972b; Saunders and Johnstone 1972; Gummer et al. 1989b), whereas the velocity of the stapes in many mammalian species continues at a high level past 10–20 kHz (Saunders and Johnstone 1972; Wilson and Bruns 1983; Ruggero and Temchin 2002; Ravicz et al. 2008). Manley (1972b) suggested that the root of the limitation in high-frequency response in the nonmammalian ossicular system is flexibility in the coupling between the TM and columella that acts as a low-pass filter. As described previously, mammals also have flexible couplings within their ossicular chains that limit the high-frequency response of the middle ear (Zwislocki 1962; Guinan and Peake 1967; Willi et al. 2002), but this limitation occurs above 10 kHz in many mammals (though at lower frequencies in humans: Willi et al. 2002). Whether the rotational lever action of the mammalian ossicular system is better able to transfer high-frequency sound energy from the TM to the inner ear (Manley 2010) needs further study.

3.4.1.1 Are Three Ossicles Better than One?

Although it has been suggested that the three-ossicle mammalian middle ear is inherently superior in the conduction of high-frequency sound to the inner ear, as has been discussed previously, a precise mechanism for this advantage has never been defined, although there are several hypotheses that consider alterations in the mode of motion of the ossicles at higher frequencies (e.g., Fleischer 1978; Puria and Steele 2010). Three-ossicle ears and columellar ears all have nonrigid ossicular components that produce reductions in the high-frequency response of middle ear, and both also contain lever mechanisms to assist in the transformation of sound in air to sound in the inner ear. Indeed, because it is generally thought that the middle ear of mammals developed independently of the columellar ears of birds and reptiles, arguments that one form of middle ear is superior to the other may not be productive. Instead, one might argue that both columellar and three-ossicle middle ears allow the animals they serve to function in the environmental niche to which they have adapted.

As discussed previously, one known difference provided by multiple ossicles is the ability for independent muscular control of the position of the TM-malleus and the stapes. Because of the presence of the ossicular joints, the tensor tympani can produce significant changes in the static position of the tympanic membrane without producing large displacements of the stapes (Marquet 1981; Hüttenbrink 1988). Similarly, the stapedius muscle can significantly stiffen the stapedial--vestibular joint (the annular ligament between the stapes footplate and the oval window of the inner ear) without producing large changes in the position of the malleus and tympanic membrane (Pang and Peake 1986). The joints also help isolate the inner ear from sudden changes in air pressure that produce large inward and outward motions of the TM (Hüttenbrink 1988), where the potential hazard to changes in air pressure were increased when the middle ear became enclosed by bone.

Another hypothesis concerning the benefit of three-ossicles in the mammalian ear is the presence of a moment of inertia about the rotational axis of the ossicular chain (e.g., Puria and Steele 2010; Lavender et al. 2011). Coincidence of the rotational axis and the center of gravity of the ossicle will greatly reduce the contribution of ossicular mass to the middle ear function. On the other hand, ossicular specializations that lead to significant differences between the location of the ossicular center of gravity and the rotational axes could enhance the middle ear's response to sounds conducted through the bone by substrate and other vibrations (Bárány 1938; Mason 2003; Puria and Rosowski 2012).

3.4.1.2 The Middle Ear and Audibility

It has long been noted that the middle ear controls which sounds reach the inner ear with enough stimulus power to produce an auditory sensation (e.g., Waetzman and Keibs 1936). Indeed, the concept of middle ear pathology producing a "conductive" hearing loss is built on the concept that sounds must pass through the middle ear before they can be sensed by the inner ear. An academic question that is often asked is, How does the action of the normal middle ear shape the frequency-dependence of hearing? More specifically, does the function of the middle ear define the frequency range to which the entire ear is most sensitive?

The best answer to the question above is "Not entirely." After all, the sensitive range of the inner ear must play some role (e.g., Ruggero and Temchin 2002), and in mammals, the frequency dependence of sound transfer through the external ear also contributes to the transfer of sound from the environment to the cochlea (e.g., Wiener et al. 1966; Shaw 1974). Further, the acousto-mechanical load of the inner ear on the stapes plays a significant role in determining the frequency dependence of the middle ear (e.g., Møller 1965; Allen 1986; Rosowski et al. 2006). Comparisons of middle ear transfer characteristics that are corrected for the action of the external ear do accurately predict the range of best hearing in a number of mammalian species (e.g., Dallos 1973; Rosowski 1991; Puria et al. 1997), but usually do a poorer job of defining the extreme limits of low and high-frequency sensitivity. (One exception is Ravicz et al. (2008), who found a good match between the high-frequency roll off of the audiogram and middle ear transfer function in the gerbil.)

3.4.2 The Role of the TM in Sound Conduction

One of the biggest current questions in the study of the middle ear concerns the precise mechanisms involved in the coupling of sound to the ossicular system by the TM. The complex motion patterns on the surface of the mammalian TM that are induced by sounds of frequencies above a few kHz (Khanna and Tonndorf 1972; Tonndorf and Khanna 1972; Rosowski et al. 2009) have been interpreted in a number of ways. Tonndorf and Khanna (1970) suggested that these complex patterns indicated the breakup of the motion of the TM into multiple out-of-phase modal maxima, which would lead to a reduction of the average motion of the TM and a decrease in the efficacy of middle ear function at higher frequencies (Rosowski et al. 2009). Others (Puria and Allen 1998; Parent and Allen 2007, 2010) have suggested that the conical mammalian TM acts like an acoustic horn that is better matched to the impedance of air at the external edges of the cone, and that sound energy travels from the edges toward the ossicular attachment at the center of the TM. A third hypothesis is that the presence of multiple closely spaced natural frequencies of the TM ensures that the membrane can be set into motion by nearly any sound frequency (Fay et al. 2006). A related but fundamentally different hypothesis is based on observations of a combination of traveling waves on the mammalian TM surface and larger more uniform simple motions of the TM (Cheng et al. 2010; de La Rochefoucauld and Olson 2010; Rosowski et al. 2011). These studies suggest that simple low-order natural response of the TM to sound dominate the motion of the TM, even at high sound frequencies. Which of these hypotheses is most correct has yet to be established. Also, although there are significant differences in the shape and ultrastructure of the TM between mammals and nonmammalian terrestrial vertebrates, there have been no studies of the detailed motion of the TMs of nonmammalian vertebrates. The latter might be useful in evaluating the effect of the significant structural differences.

3.4.3 The Two-Window Hypothesis: Evidence for and Against a Normal "Third Window"

One of the most fundamental hypotheses of middle ear function in terrestrial vertebrates is that the inner ear is sensitive to waves of fluid displacement produced by a difference in sound pressure across the two windows between the inner ear and the middle ear. It is this near instantaneous fluid wave that deforms the hair-cell support or accessory structures tied to the ciliary bundles in amphibians; displaces the basilar membrane in reptiles, birds, and mammals; and is responsible for the launching of the traveling wave in the mammalian cochlea. The two-window hypothesis has been used to explain not only the function of the normal middle ear, but also the effects of middle ear pathology on hearing function and some forms of reconstructive middle ear surgery (e.g., Peake et al. 1992; Rosowski et al. 1995;

Merchant and Rosowski 2010). The evidence for this hypothesis includes (1) experiments in which the window-pressure difference was externally controlled (e.g., Wever and Lawrence 1950; Voss et al. 1996) that demonstrate the largest responses when the stimuli at the two windows are of equal level and opposite phase; (2) observations of near equality between the sound-induced volume displacements of the stapes and round window during stimulation with air-conducted sound (Kringlebotn 1995; Stenfelt et al. 2004); and (3) the positive results of middle ear reconstructive procedures designed to maximize the difference in the sound pressures at two functional cochlear windows (e.g., Shambaugh 1954; Wüllstein 1959; Merchant et al. 1995).

Although the two-window hypothesis can explain much of the inner ear's response to sound conducted via the normal middle ear, and in cases of specific middle ear pathologies—e.g., interruption of the ossicular chain (Peake et al. 1992)—there are inner-ear pathologies that it does not readily explain. One such case is pathological or developmental immobilization of the round window. The two-window hypothesis would predict that round-window immobilization would prevent stimulation of the inner ear by the direct middle ear route; however, while humans with such a disorder demonstrate significant hearing loss, the hearing loss is smaller than expected (e.g., Harrison et al. 1964; Linder et al. 2003; Mansour et al. 2011). Attempts to block the round window in animal experiments also produced hearing losses that are smaller than expected (e.g., Tonndorf and Tabor 1962).

One possible explanation for the failure of round-window immobilization to produce large hearing losses is the presence of either other paths for sound energy to leave the inner ear, or compressible elements for energy storage within the inner ear (Shera and Zweig 1992). These other paths or compressible elements would act like normal "third windows" into the inner ear that would allow a difference in the magnitude of the motion of the round and oval window and allow stimulation of the inner ear when either the stapes or round-window was immobilized (Ranke 1953; Tonndorf and Tabor 1962;). Obvious candidates for such pathways are direct communications between the fluid spaces of the inner ear and the brain cavity, including the cochlear and vestibular aqueducts, the vascular channels to the brain and the interstitial spaces around the neurons innervating the inner ear (Békésy 1960; Tonndorf and Tabor 1962; Rask-Andersen et al. 1977; Nakashima et al. 2000).

3.5 Summary

This chapter summarizes the structural adaptations that help conduct sound energy within the environment to the inner ears of different vertebrate species. It also describes some of the inner-ear adaptations required to sensitize the inner ear in terrestrial vertebrates to sound conducted by the middle ear. One lesson of this chapter is that adaptations within the different parts of the ear do not occur in isolation. Put another way, the efficacy of the varied middle ears of terrestrial vertebrates is tightly tied to the adaptations to the periotic system of the inner ear in these same species that enabled the different middle ears to produce significant sound-induced fluid motion within the inner ear. This coupling of adaptations in the inner ear and the middle ear also leads to parallelisms in the sensitivity of the inner ear and middle ear to sounds of different frequencies. For example, in common laboratory mice whose inner ear contains neurons tuned to frequencies between 2 and 60 kHz (Taberner and Liberman 2005) the middle ear is very stiff and the velocity of the stapes is small (<0.1 mm-s^{-1}-Pa^{-1}) at frequencies less than 1 kHz but is maintained at high levels out to 30–40 kHz (Saunders and Summers 1982). Another example is the human inner ear, which is thought to contain hair cells tuned to frequencies from near 50 Hz to near 20 kHz (Merchant 2010) and has a highly compliant middle ear that produces stapes velocities greater than 0.1 mm-s^{-1}-Pa^{-1} in response to sound stimulus frequencies as low as 100 Hz but not at frequencies above 10 kHz (Aibara et al. 2001). The similarity between the frequency range of inner-ear sensitivity and the frequency range of effective middle ear sound transfer is a common feature in terrestrial vertebrates and supports the notion that the inner and middle ears of vertebrates adapted in parallel to meet the demands required for species survival.

Acknowledgments Work on this chapter was funded by an R01 from the NIDCD. Many students, co-workers, and colleagues through the years have made indirect contributions to this work. Significant contributions to my understanding of comparative middle ear mechanism were made by my former research supervisors and old friends Jim Saunders of the University of Pennsylvania and Bill Peake of MIT and the Eaton-Peabody Laboratory at the Massachusetts Eye and Ear Infirmary. If given the chance, both of them would have suggested many corrections and additions to this chapter.

References

Aertsen, A. M. H. J., Vlaming, M. S. M. G, Eggermont, J. J., & Johannesma, P. I. M. (1986). Directional hearing in the grassfrog (*Rana temporaria* L.). II. Acoustics and modeling of the auditory periphery. *Hearing Research*, 21, 17–40.

Aibara, R., Welsh, J. T., Puria, S., & Goode, R. L. (2001). Human middle-ear sound transfer function and cochlear input impedance. *Hearing Research*, 152, 100–109.

Allen, J. B. (1986). Measurements of eardrum acoustic impedance. In J. B. Allen, J. H. Hall, A. Hubbard, S. T. Neely & A. Tubis (Eds.), *Peripheral auditory mechanisms* (pp. 44–51). New York: Springer-Verlag.

Allin, E. F., & Hopson, J. A. (1992). Evolution of the auditory system in synapsida ("mammal-like" reptiles and primitive mammals) as seen in the fossil record. In D. B. Webster, R. R. Fay, & A. N. Popper (Ed.), *The evolutionary biology of hearing* (pp. 587–614). New York: Springer-Verlag.

Alper, C. M., Tabari, R., Seroky, J. T., & Doyle, W. J. (1997). Magnetic resonance imaging of the development of otitis media with effusion caused by functional obstruction of the eustachian tube. *Annals of Otology, Rhinology & Laryngology*, 106, 422–431.

Bárány, E. (1938). A contribution to the physiology of bone conduction. *Acta Oto-Laryngologica, Supplementum*, 26, 1–223.

Bennett, M. (1984). Impedance concepts relating to the acoustic reflex. In S. Silman (Ed.), *The acoustic reflex: Basic principles and clinical applications* (pp. 35–61). New York: Academic Press.

Beranek, L. L. (1993). *Acoustics*. New York: Acoustical Society of America.

Bolt, J. R., & Lombard, R. E. (1992). Nature and quality of the fossil evidence for otic evolution in early tetrapods. In D. B. Webster, R. R. Fay & A. N. Popper (Eds.), *The evolutionary biology of hearing* (pp. 377–403). New York: Springer-Verlag.

Bolz, A., & Lim, D. J. (1972). Morphology of the stapediovestibular joint. *Acta Oto-Laryngologica*, 73, 10–17.

Borg, E., & Zakrisson, J. E. (1974). Stapedius reflex and monaural masking. *Acta Oto-Laryngologica*, 78, 155–161.

Borg, E., & Zakrisson, J. E. (1975). The activity of the stapedius muscle in man during vocalization. *Acta Oto-Laryngologica*, 79, 325–333.

Borg, E., & Nilsson, R. (1984). Acoustic reflex in industrial noise. In S. Silman (Ed.), *The acoustic reflex: Basic principles and clinical applications* (pp. 413–440). New York: Academic Press.

Borg, E., Counter, S. A., & Rösler, G. (1984). Theories of middle-ear muscle function. In S. Silman (Ed.), *The acoustic reflex: Basic principles and clinical applications* (pp. 63–99). New York: Academic Press.

Browning, G. G., & Granich, M. S. (1978). Surgical anatomy of the temporal bone in the chinchilla. *Annals of Otology, Rhinology & Laryngology*, 87, 875–882.

Carmel, P. W., & Starr, A. (1963). Acoustic and nonacoustic factors modifying middle ear muscle activity in waking cats. *Journal of Neurophysiology*, 26, 598–616.

Cheng, J. T., Aarnisalo, A. A., Harrington, E., Hernandez-Montes, M. dS., Furlong, C., Merchant, S. N., & Rosowski, J. J. (2010). Motion of the surface of the human tympanic membrane measured with stroboscopic holography. *Hearing Research*, 263, 66–77.

Christensen-Dalsgaard, J., & Narins, P. M. (1993). Sound and vibration sensitivity of VIIIth nerve fibers in the frogs *Leptodactylus albilabris* and *Rana pipiens pipiens*. *Journal of Comparative Physiology A: Neuroethology, Sensory, Neural, and Behavioral Physiology*, 172, 653–662.

Christensen-Dalsgaard, J., & Manley, G. A. (2005). Directionality of the lizard ear. *Journal of Experimental Biology*, 208, 1209–1217.

Coffin, A. C., Kelley, M., Manley, G. A., & Popper, A. N. (2004). Evolution of sensory hair cells. In G. A. Manley, A. N. Popper & R. R. Fay (Eds.), *Evolution of the vertebrate auditory system* (pp. 55–94). New York: Springer-Verlag.

Coleman, M. N., & Ross, C. F. (2004). Primate auditory diversity and its influence on hearing performance. *Anatomical Record*, 281, 1123–1137.

Coleman, M. N., & Colbert, M. W. (2010). Correlations between auditory structures and hearing sensitivity in non-human primates. *Journal of Morphology*, 271, 511–532.

Coles, R. B., & Guppy, A. (1988). Directional hearing in the barn owl (*Tyto alba*). *Journal of Comparative Physiology, A: Neuroethology, Sensory, Neural, and Behavioral Physiology*, 163, 117–133.

Counter, S. A, & Borg, E. (1979). Physiological activation of the stapedius muscle in *Gallus gallus*. *Acta Oto-Laryngologica*, 403, 13–19.

Counter, S. A., & Borg, E. (1982). The avian stapedius muscle. *Acta Oto-Laryngologica*, 94, 267–274.

Dahmann, H. (1929). Zur Physiologie des Hörens; experimentelle Untersuchungen über die Mechanik der Gehörknöchelchenkette, sowie über deren Verhalten auf Ton und Luftdruck. *Zeitschrift fur Hals- Nasen- Ohrenheilkunde*, 24, 462–497.

Dallos, P. (1970). Low frequency auditory characteristics: Species dependence. *Journal of the Acoustical Society of America*, 48, 489–499.

Dallos, P. (1973). *The auditory periphery*. New York: Academic Press.

Decraemer, W. F., & Dirckx, J. J. J. (1998). Pressure regulation due to displacement of the pars flaccida and pars tensa of the tympanic membrane. *Otorhinolaryngology Nova*, 8, 277–281.

Decraemer, W. F., & Funnell, W. R. J. (2008). Anatomical and mechanical properties of the tympanic membrane. In B. Ars (Ed.), *Chronic otitis media: Pathogenesis-oriented therapeutic management* (pp. 51–84). The Hague: Kugler.

de La Rochefoucauld, O., & Olson, E. S. (2010). A sum of simple and complex motions on the eardrum and manubrium in gerbil. *Hearing Research*, 263, 9–15.

Dirckx, J. J. J., Decraemer, W. F., Unge, M. von, & Larsson, C. (1998). Volume displacement of the gerbil eardrum pars flaccida as a function of middle ear pressure. *Hearing Research*, 118, 35–46.

Doran, A. H. G. (1879). Morphology of the mammalian ossicula auditus. *Journal of the Transactions of the Linnean Society*, 1, 371–497.

Doyle, W. J. (2000). Middle ear pressure regulation. In J. J. Rosowski & S. N. Merchant (Eds.), *The function and mechanics of normal, diseased and reconstructed middle ears* (pp. 3–21). The Hague: Kugler.

Doyle, W. J, Cantekin, E. I., & Bluestone, C. D. (1980). Eustachian tube function in cleft palate children. *Annals of Otology, Rhinology & Laryngology*, 89 (*Supplement* 68), 34–40.

Ehret, G., Tautz, J., Schmitz, B., & Narins, P. M. (1990). Hearing through the lungs: Lung-eardrum transmission of sound in the frog *Eleutherodactylus coqui*. *Naturwissenschaften*, 77, 192–195.

Elner, A. (1976). Normal gas exchange in the human middle ear. *Annals of Otology, Rhinology & Laryngology*, 85 (*Supplement* 25, *part* 2), 161–164.

Erlandsson, B., Håkanson, H., Ivarsson, A., & Nilsson, P. (1980). The effect of static middle ear pressures on the hearing threshold. *Acta Oto-Laryngologica*, 90, 324–331.

Fay, J. P., Puria, S., & Steele, C. R (2006). The discordant eardrum. *Proceedings of the National Academy of Sciences of the USA*, 103, 19743–19748.

Fay, R. R. (1984). The goldfish ear codes the axes of particle motion in three dimensions. *Science*, 225, 951–953.

Fay, R. R. (1988). *Hearing in vertebrates: A psychophysics databook*. Winnetka, IL: Hill-Fay Associates.

Fay, R. R., & Edds-Walton, P. L. (1997). Directional response properties of saccular afferents of the toadfish. *Hearing Research*, 113, 235–246.

Feng, A. S., & Shofner, W. P. (1981). Peripheral basis of sound localization in anurans. Acoustic properties of the frog's ear. *Hearing Research*, 5, 201–216.

Feng, A. S., Narins, P. M., & Capranica, R. R. (1975). Three populations of primary auditory fibers in the bullfrog (*Rana catesbeiana*): Their peripheral origins and frequency selectivity. *Journal of Comparative Physiology*, 100, 221–229.

Fleischer, G. (1973). Studien am Skelett des Gehörorgans der Säugetiere, einschliesslich des Menschen. *Säugetierkundl. Mitteilungen (München)*, 21, 131–239.

Fleischer, G. (1978). Evolutionary principles of the mammalian middle ear. *Advances in Anatomy, Embryology and Cell Biology*, 55, 3–69.

Fritzsch, B. (1992). The water-to-land transition: Evolution of the tetrapod basilar papilla, middle ear, and auditory nuclei. In D. B. Webster, R. R. Fay, & A. N. Popper (Eds.), *The evolutionary biology of hearing* (pp. 351–375). New York: Springer-Verlag.

Fumagalli, Z. (1949). Morphological research on the sound transmission apparatus. *Archivio Italiano di Otologia, Rinologia e Laringologia*, 60(*S*1), 1–323.

Funnell, W. R., & Laszlo, C. A. (1982). A critical review of experimental observations on ear-drum structure and function. *ORL*, 44, 181–205.

Gaudin, E. P. (1968). On the middle ear of birds. *Acta Oto-Laryngologica*, 65, 316–326.

Gerhardt, H. C., & Rheinlaender, J. (1982). Localization of an elevated sound source by the green tree frog. *Science*, 217, 663–664.

Gleich, O., & Manley, G. A. (2000). The hearing organ of birds and crocodilia. In R. J. Dooling, R. R. Fay & A. N. Popper (Eds.), *Comparative hearing: Birds and reptiles* (pp. 70–138). New York: Springer-Verlag.

Goode, R. L., Ball, G., Nishihara, S., & Nakamura, K. (1996). Laser Doppler vibrometer (LDV): A new clinical tool for the otologist. *American Journal of Otology*, 17, 813–822.

Gridi-Papp, M., Feng, A. S., Shen, J.-X., Yu, Z.-L., Rosowski, J. J., & Narins, P. M. (2008). Active control of ultrasonic hearing in frogs. *Proceedings of the National Academy of Sciences of the USA*, 105, 11013–11018.

Guinan, J. J., Jr, & Peake, W. T. (1967). Middle-ear characteristics of anesthetized cats. *Journal of the Acoustical Society of America*, 41, 1237–1261.

Gummer, A. W., Smolders, J. W. T., & Klinke, R. (1989a). Mechanics of a single-ossicle ear: I. The extra-stapedius of the pigeon. *Hearing Research*, 39, 1–14.

Gummer, A. W., Smolders, J. W. T., & Klinke, R. (1989b). Mechanics of a single-ossicle ear: II. The columella footplate of the pigeon. *Hearing Research*, 39, 15–26.

Harrison, W. H., Shambaugh, G. E., & Derlacki, E. L. (1964). Congenital absence of the round window: Case report with surgical reconstruction by cochlear fenestration. *Laryngoscope*, 74, 967–978.

Heffner, R. S., & Heffner, H. E. (1992). Evolution of sound localization in mammals. In D.B. Webster, R. R. Fay, & A. N. Popper (Eds.), *The evolutionary biology of hearing* (pp. 691–716). New York: Springer-Verlag.

Heffner, R. S., & Heffner, H. E. (2010). Explaining high frequency hearing. *The Anatomical Record*, 293, 2080–2082.

Hellstrom, S., & Stenfors, L.-E. (1983). The pressure equilibrating function of the pars flaccida in middle ear mechanics. *Acta Physiologica Scandinavia*, 118, 337–341.

Helmholtz, H. L. von. (1868). Die Mechanik der Gehörknöchelchen und des Trommelfells. *Pflügers Archive*, 1, 1–60.

Hemilä, S., Nummela, S., & Reuter, T. (1995). What middle ear parameters tell about impedance matching and high frequency hearing. *Hearing Research*, 85, 31–44.

Hemilä, S., Nummela, S., & Reuter, T. (1999). A model of the odontocete middle ear. *Hearing Research*, 133, 82–97.

Hemilä, S., Nummela, S., & Reuter, T. (2001). Modeling whale audiograms: Effects of bone mass on high-frequency hearing. *Hearing Research*, 151, 221–226.

Henson, O. W., Jr. (1965). The activity and function of the middle-ear muscles in echo-locating bats. *Journal of Physiology*, 180, 871–887.

Henson, O. W., Jr. (1970). The ear and audition. In W. A. Wimsatt (Ed.), *The biology of bats* (pp. 181–262). New York: Academic Press.

Henson, O. W., Jr. (1974). Comparative anatomy of the middle ear. In W. D. Kiedel & W. D. Neff (Eds.), *Handbook of sensory physiology: The auditory system* (Vol. V/1, pp. 39–110). New York: Springer-Verlag.

Hetherington, T. E. (1985). Role of the opercularis muscle in seismic sensitivity in the bullfrog *Rana catesbeiana*. *Journal of Experimental Zoology*, 235, 27–34.

Hetherington, T. E. (1987). Physiological features of the opercularis muscle and their effects on vibration sensitivity in the bullfrog *Rana catesbeiana*. *Journal of Experimental Biology*, 131, 189–204.

Hetherington, T. E. (1988). Biomechanics of vibration reception in the bullfrog, Rana catesbeiana. *Journal of Comparative Physiology*, 163, 43–52.

Hetherington, T. E. (1992). The effect of body size on functional properties of middle ear systems of anuran amphibians. *Brain Behavior and Evolution*, 39, 133–142.

Hetherington, T. E. (1994). The middle ear muscle of frogs does not modulate tympanic responses to sound. *Journal of the Acoustical Society of America*, 95, 2122–2125.

Hill, K. G., Lewis, D. B., Hutchings, M. E., & Coles, R. B. (1980). Directional hearing in the Japanese quail (*Coturnix coturnix japonica*). I. Acoustic properties of the auditory system. *Journal of Experimental Biology*, 86, 135–151.

Huang, G. T., Rosowski, J. J., Flandermeyer, D. T., Lynch, T. J. III, & Peake, W. T. (1997). The middle ear of a lion: Comparison of structure and function to domestic cat. *Journal of the Acoustical Society of America*, 101, 1532–1549.

Hunt, R. M., Jr. (1974). The auditory bulla in carnivora: An anatomical basis for reappraisal of carnivore evolution. *Journal of Morphology*, 143, 21–76.

Hunt, R. M., & Korth, W. W. (1980). The auditory region of Dermoptera: Morphology and function relative to other living mammals. *Journal of Morphology*, 164, 167–211.
Hüttenbrink, K. B. (1988). The mechanics of the middle-ear at static air pressures. *Acta Oto-Laryngologica Supplementum*, 451, 1–35.
Ingelstedt, S. (1976). Physiology of the Eustachian tube. *Annals of Otology, Rhinology & Laryngology*, 85 (*Supplement* 25, part 2), 156–160.
Ingelstedt, S., & Jonson, B. (1966). Mechanisms of gas exchange in the normal human middle ear. *Acta Oto-Laryngologica Supplementum*, 224, 452–461.
Jerger, J. (1970). Clinical experience with impedance audiometry. *Archives of Otolaryngology*, 92, 311–324.
Johnson, W. E., Eizirik, E., Pecon-Slattery, J., Murphy, W. J., Antunes, A., Teeling, E., & O'Brien, S. J. (2006). The Late Miocene radiation of modern felidae: A genetic assessment. *Science*, 311, 73–77.
Kalmijn, A. J. (1988). Hydrodynamic and acoustic field detection. In J. Atema, R. R. Fay, A. N. Popper & W. N. Tavolga (Eds.), *Sensory biology of aquatic animals* (pp. 83–130). New York: Springer-Verlag.
Kalmijn, A. J. (1989). Functional evolution of lateral line and inner ear systems. In S. Coombs, P. Görner & P. Münz (Eds.), *The mechanosensory lateral line-neurobiology and evolution* (pp. 187–216). New York: Springer-Verlag.
Keen, J. A., & Grobbelaar, C. S. (1941). The comparative anatomy of the tympanic bulla and auditory ossicles, with a note suggesting their function. *Transactions of the Royal Society of South Africa*, 28, 307–329.
Ketten, D. R. (1992). The marine mammal ear: Specializations for aquatic audition and echolocation. In D. B. Webster, R. R. Fay, & A. N. Popper (Eds.), *The evolutionary biology of hearing* (pp. 717–754). New York: Springer-Verlag.
Ketten, D. R. (1994). Functional analysis of whale ears: Adaptations for underwater hearing. *IEEE Proceedings on Underwater Acoustics*. 1, 264–270.
Ketten, D. R. (2000). Cetacean ears. In W. W. L. Au, A. N. Popper, & R. R. Fay (Eds.), *Hearing by whales and dolphins* (pp. 43–108). New York: Springer-Verlag.
Khanna, S. M. (1970). A holographic study of tympanic membrane vibrations in cats. Ph. D. dissertation, City University, New York.
Khanna, S. M., & Tonndorf, J. (1972). Tympanic membrane vibrations in cats studied by time-averaged holography. *Journal of the Acoustical Society of America*, 51, 1904–1920.
Kinsler, L. E., Frey, A. R., Coppens, A. B., & Sanders, J. V. (1982). *Fundamentals of acoustics*. New York: John Wiley & Sons.
Kirikae, I. (1960). *The structure and function of the middle ear*. Tokyo: University of Tokyo Press.
Kohllöffel, L. U. E. (1984). Notes on the comparative mechanics of hearing. III. On Shrapnell's membrane. *Hearing Research*, 13, 83–88.
Kringlebotn, M. (1995). The equality of volume displacements in the inner ear windows. *Journal of the Acoustical Society of America*, 98, 192–196.
Lavender, D., Taraskin, S. N., & Mason, M. J. (2011). Mass distribution and rotational inertia of "microtype" and "freely mobile" middle ear ossicles in rodents. *Hearing Research*, 218, 97–107.
Lay, D. M. (1972). The anatomy, physiology, functional significance and evolution of specialized hearing organs of gerbilline rodents. *Journal of Morphology*, 138, 41–120.
Legouix, J. P., & Wisner, A. (1955). Role fonctionnel des bulles tympaniques géantes de certains rongeurs (Meriones). *Acoustica*, 5, 208–216.
Lewis, E. R., & Narins, P. M. (1999). The acoustic periphery of amphibians: Anatomy and Physiology. In R. R. Fay & A. N. Popper (Eds.), *Comparative hearing: Fish and amphibians* (pp. 101–154). New York: Springer.
Lidén, G., Peterson, J. L., & Björkman, G. (1970). Tympanometry. *Archives of Otolaryngology*, 92, 248–257.
Lim, D. J. (1968a). Tympanic membrane. Part I. Pars tensa. *Acta Oto-Laryngologica*, 66, 181–198.

Lim, D. J. (1968b). Tympanic membrane. Part II. Pars flaccida. *Acta Oto-Laryngologica*, 66, 515–532.
Linder, T. E., Ma, F., & Huber, A. (2003). Round window atresia and its effect on sound transmission. *Otology & Neurotology*, 24, 259–263.
Lombard, R. E., & Straughan, I. R. (1974). Functional aspects of anuran middle ear structure. *Journal of Experimental Biology*, 61, 71–93.
Manley, G. A. (1972a). The middle ear of the Tokay gecko. *Journal of Comparative Physiology*, 81, 239–250.
Manley, G. A. (1972b). Frequency response of the middle ear of geckos. *Journal of Comparative Physiology*, 81, 251–258.
Manley, G. A. (1990). *Peripheral hearing mechanisms in reptiles and birds*. Berlin: Springer-Verlag.
Manley, G. A. (2000). The hearing organs of lizards. In R. J. Dooling, R. R. Fay, & A. N. Popper (Eds.), *Comparative hearing: Birds and reptiles* (pp. 139–196). New York: Springer-Verlag.
Manley, G. A. (2010). An evolutionary perspective on middle ears. *Hearing Research*, 263, 3–8.
Manley, G. A., Yates, G. K., & Köppl, C. (1988). Auditory peripheral tuning: Evidence for a simple resonance phenomenon in the lizard Tiliqua. *Hearing Research*, 33, 181–190.
Mansour, S., Nicolas, K., & Ahmad, H. H. (2011). Round window otosclerosis: Radiological classification and clinical correlations. *Otology & Neurotology*, 32, 384–392.
Margolis, R. H., & Shanks, J. E. (1985). Tympanometry. In J. Katz (Ed.), *Handbook of clinical audiology* (pp. 438–475). Baltimore: Williams & Wilkins.
Marquet, J. (1981). The incudo-malleal joint. *Journal of Laryngology and Otology*, 95, 542–565.
Mason, M. J. (2003). Bone conduction and seismic sensitivity in golden moles (Chrysochloridae). *Journal of Zoology, London*, 260, 405–413.
Mason, M. J. (2007). Pathways for sound transmission to the inner ear in amphibians. In P. M. Narins, A. S. Feng, R. R. Fay & A. N. Popper (Eds.), *Hearing and sound communication in amphibians* (pp. 147–183). New York: Springer Science+Business Media LLC.
Masterton, B., Heffner, H. E., & Ravizza, R. (1969). The evolution of human hearing. *Journal of the Acoustical Society of America*, 45, 966–985.
McPherson, D. L., Miller, J. M., & Axelsson, A. (1976). Middle ear pressure: Effects on the auditory periphery. *Journal of the Acoustical Society of America*, 59, 135–142.
Merchant, S. N. (2010). Methods of removal, preparation and study. In S. N. Merchant & J. B. Nadol, Jr. (Eds.), *Schuknecht's pathology of the ear* (3rd ed., pp. 3–53). Shelton, CT: People's Medical Publishing House-USA.
Merchant, S. N., & Rosowski, J. J. (2008). Conductive hearing loss caused by third-window lesions of the inner ear. *Otology & Neurotology*, 29, 282–289.
Merchant, S. N., & Rosowski, J. J. (2010). Acoustics and mechanics of the middle ear. In A. J. Gulya, L. B. Minor, & D. S. Poe (Eds.), *Surgery of the ear* (pp. 49–72). Shelton, CT: People's Medical Publishing House-USA.
Merchant, S. N., Rosowski, J. J., & Ravicz, M. E. (1995). Middle-ear mechanics of type IV and type V tympanoplasty. II. Clinical analysis and surgical implications. *American Journal of Otology*, 16, 565–575.
Miller, B. S., Zosuls, A. L., Ketten, D. R. & Mountain, D. C. (2006). Middle-ear stiffness of the bottlenose dolphin. *IEEE Journal of Oceanic Engineering*, 3, 87–94.
Miller, M. R. (1980). The reptilian cochlear duct. In A. N. Popper & R. R. Fay (Eds.), *Comparative studies of hearing in vertebrates* (pp. 169–204). Berlin: Springer-Verlag.
Moiseff, A., & Konishi, M. (1981). The owl's interaural pathway is not involved in sound localization. *Journal of Comparative Physiology*, 144, 299–304.
Møller, A. R. (1964). Effect of tympanic muscle activity on movement of the eardrum, acoustic impedance and cochlear microphones. *Acta Oto-Laryngologica*, 58, 1–10.
Møller, A. R. (1965). Experimental study of the acoustic impedance of the middle ear and its transmission properties. *Acta Oto-Laryngologica*, 60, 129–149.

Møller, A. R. (1974). The acoustic middle-ear muscle reflex. In W. D. Keidel & W. D. Neff (Eds.), *Handbook of sensory physiology: Auditory system* (Vol. V/1, pp. 519–548). New York: Springer-Verlag.
Molvær, O., Vallersnes, F. M., & Kringlebotn, M. (1978). The size of the middle ear and the mastoid air cells. *Acta Oto-Laryngologica*, 85, 24–32.
Nakashima, T., Ueda, H., Furuhashi, A., Sato, E., Asahi, K., Naganawa, S., & Beppu, R. (2000). Air-bone gap and resonant frequency in large vestibular aqueduct syndrome. *American Journal of Otology*, 21, 671–674.
Narins, P. M., & Lewis, E. R. (1984). The vertebrate ear as an exquisite seismic sensor. *Journal of the Acoustical Society of America*, 76, 1384–1387.
Narins, P. M., Ehret, G., & Tautz, J. (1988). Accessory pathway for sound transfer in a neotropical frog. *Proceedings of the National Academy of Sciences of the USA*, 85, 1508–1512.
Novacek, M. J. (1977). Aspects of the problem of variation, origin and evolution of the eutherian auditory bulla. *Mammal Review*, 7, 131–150.
Nummela, S. (1995). Scaling of the mammalian middle ear. *Hearing Research*, 85, 18–30.
Nummela, S., Wagar, T., Hemilä, S., Holmberg, P., & Paukka, P. (1999). Scaling of the cetacean middle ear. *Hearing Research*, 133, 71–81.
Nummela, S., Thewissen, J. G. M., Bajpai, S., Hussain, S. T., & Kumar, K. (2007). Sound transmission in archaic and modern whales: Anatomical adaptations for underwater hearing. *Anatomical Record*, 290, 716–733.
Nuttall, A. L. (1974). Tympanic muscle effects on middle-ear transfer characteristics. *Journal of the Acoustical Society of America*, 56, 1239–1247.
Pang, X. D., & Guinan, J. J., Jr. (1997). Effects of stapedius-muscle contractions on the masking of auditory nerve responses. *Journal of the Acoustical Society of America*, 102(6), 3576–3586.
Pang, X. D., & Peake, W. T. (1986). How do contractions of the stapedius muscle alter the acoustic properties of the middle ear? In J. B. Allen, J. L. Hall, A. Hubbard, S. T. Neely, & A. Tubis (Eds.), *Peripheral auditory mechanisms* (pp. 36–43). New York: Springer-Verlag.
Paparella, M. M., Schachern, P. A., & Choo, Y. B. (1983). The round window membrane: Otological observations. *Annals of Otology, Rhinology & Laryngology*, 92, 629–634.
Parent, P., & Allen, J. B. (2007). Wave model of the cat tympanic membrane. *Journal of the Acoustical Society of America*, 122, 918–931.
Parent, P., & Allen, J. B. (2010). Time-domain "wave" model of the human tympanic membrane. *Hearing Research*, 263, 152–167.
Peake, W. T., Rosowski, J. J., & Lynch, T. J III. (1992). Middle-ear transmission: Acoustic vs ossicular coupling in cat and human. *Hearing Research*, 57, 245–268.
Popper, A. N., & Fay, R. R. (1999). The auditory periphery in fishes. In R. R. Fay & A. N. Popper (Eds.), *Comparative hearing: Fish and amphibians*. New York: Springer.
Popper, A. N., & Schilt, C. R. (2008). Hearing and acoustic behavior (basic and applied). In J. F. Webb, R. R. Fay, & A. N. Popper (Eds.), *Fish bioacoustics* (pp. 17–48). New York: Springer Science+Business Media, LLC.
Popper, A. N., & Fay, R. R. (2011). Rethinking sound detection by fishes. *Hearing Research*, 273, 25–36.
Popper, A. N., Fay, R.R., Platt, C., & Sand, O. (2003). Sound detection mechanisms and capabilities of teleost fishes. In S. P. Collin & N. J. Marshall (Eds.), *Sensory processing in aquatic environments* (pp. 3–38). New York: Springer.
Purgue, A. P., & Narins, P. M. (2000a). Mechanics of the inner ear of the bullfrog (*Rana catesbeiana*): The contact membranes and the periotic canal. *Journal of Comparative Physiology A: Neuroethology, Sensory, Neural, and Behavioral Physiology*, 186, 481–488.
Purgue, A. P., & Narins, P. M. (2000b). Mechanics of the inner ear of the bullfrog (*Rana catesbeiana*): The contact membranes and the periotic canal. *Journal of Comparative Physiology A: Neuroethology, Sensory, Neural, and Behavioral Physiology*, 186, 481–488.

Puria, S., & Allen, J. B. (1998). Measurements and model of the cat middle ear: Evidence of tympanic membrane acoustic delay. *Journal of the Acoustical Society of America*, 104, 3463–3481.

Puria, S., & Steele, C. (2010). Tympanic-membrane and malleus-incus-complex co-adaptations for high-frequency hearing in mammals. *Hearing Research*, 263, 183–190.

Puria, S., & Rosowski, J. J. (2012). Békésy's contributions to our present understanding of sound conduction to the inner ear. *Hearing Research*, 293, 21–30.

Puria, S., Peake, W. T., & Rosowski, J. J. (1997). Sound-pressure measurements in the cochlear vestibule of human cadavers. *Journal of the Acoustical Society of America*, 101, 2745–2770.

Ranke, O. F. (1953). Physiologie des Gehörs. In O. F. Ranke & H. Lullies (Eds.), *Gehör-Stimme-Sprache* (pp. 3–110). Berlin: Springer.

Rask-Andersen, H., Stahle, J., & Wilbrand, H. (1977). Human cochlear aqueduct and its accessory canals. *Annals of Otology, Rhinology & Laryngology*, 86, 1–16.

Ravicz, M. E., & Rosowski, J. J. (1997). Sound power collection by the auditory periphery of the Mongolian gerbil Meriones unguiculatus: III. Effect of variations in middle-ear volume. *Journal of the Acoustical Society of America*, 101, 2135–2147.

Ravicz, M. E., Cooper, N. P., & Rosowski, J. J. (2008). Gerbil middle-ear sound transmission from 100 Hz to 60 kHz. *Journal of the Acoustical Society of America*, 124, 363–380.

Reijnen, C. J., & Kuijpers, W. (1971). The healing pattern of the drum membrane. *Acta Oto-Laryngologica Supplementum*, 287, 1–74.

Rosowski, J. J. (1991). The effects of external- and middle-ear filtering on auditory threshold and noise-induced hearing loss. *Journal of the Acoustical Society of America*, 90, 124–135.

Rosowski, J. J. (1992). Hearing in transitional mammals: Predictions from the middle-ear anatomy and hearing capabilities of extant mammals. In D. B. Webster, A. N. Popper, & R. R. Fay (Eds.), *The evolutionary biology of hearing* (pp. 615–631). New York: Springer-Verlag.

Rosowski, J. J. (1994). Outer and middle ear. In A. N. Popper & R. R. Fay (Eds.), *Comparative hearing: Mammals* (pp. 172–247). New York: Springer-Verlag.

Rosowski, J. J. (2003). The middle and external ears of terrestrial vertebrates as mechanical and acoustic transducers. In F. G. Barth, J. A. C. Humphrey, & T. W. Secomb (Eds.), *Sensors and sensing in biology and engineering* (pp. 59–69). New York: Springer-Verlag.

Rosowski, J. J. (2010). External and middle ear function. In P. A. Fuchs (Ed.), *The Oxford handbook of auditory science: The ear* (pp. 49–91). Oxford: Oxford University Press.

Rosowski, J. J., & Saunders, J. C. (1980). Sound transmission through the avian interaural pathways. *Journal of Comparative Physiology*, 136, 183–190.

Rosowski, J. J., & Graybeal, A. (1991). What did Morganucodon hear? *Zoological Journal of the Linnean Society*, 101, 131–168.

Rosowski, J. J., Peake, W. T., Lynch, T. J. III, Leong, R., & Weiss, T. F. (1985). A model for signal transmission in an ear having hair cells with free-standing stereocilia, II. Macromechanical stage. *Hearing Research*, 20, 139–155.

Rosowski, J. J., Davis, P. J., Merchant, S. N., Donahue, K. M., & Coltrera, M. D. (1990). Cadaver middle ears as models for living ears: Comparisons of middle-ear input immittance. *Annals Otology Rhinology and Laryngology*, 99, 403–412.

Rosowski, J. J., Merchant, S. N., & Ravicz, M. E. (1995). Middle ear mechanics of type IV and V tympanoplasty. I. Model analysis and predictions. *American Journal of Otology*, 16, 555–564.

Rosowski, J. J., Teoh, S. W., & Flandermeyer, D. T. (1997). The effect of the pars flaccida of the tympanic membrane on the ear's sensitivity to sound. In E. R. Lewis, G. R. Long, R. F. Lyon, P. M. Narins, C. R. Steele, & E. Hect-Poiner (Eds.), *Diversity in auditory mechanics* (pp. 129–135). Hackensack, NJ: World Scientific.

Rosowski, J. J., Ravicz, M. E., Teoh, S. W., & Flandermeyer, D. T. (1999). Measurements of middle-ear function in the Mongolian gerbil, a specialized mammalian ear. *Audiology and Neurotology*, 4, 129–136.

Rosowski, J. J., Ravicz, M. E., & Songer, J. E. (2006). Structures that contribute to middle-ear admittance in chinchilla. *Journal of Comparative Physiology A: Neuroethology, Sensory, Neural, and Behavioral Physiology*, 192, 1287–1311.

Rosowski, J. J., Cheng, J. T., Ravicz, M. E., Hulli, N., Harrington, E. J., Hernandez-Montes, M. S., & Furlong, C. (2009). Computer-assisted time-averaged holography of the motion of the surface of the tympanic membrane with sound stimuli of 0.4 to 25 kHz. *Hearing Research*, 253, 83–96.

Rosowski, J. J., Cheng, J. T., Merchant, S. N., Harrington, E., & Furlong, C. (2011). New data on the motion of the normal and reconstructed tympanic membrane. *Otology & Neurotology*, 32, 1559–1567.

Ruggero, M. A., & Temchin, A. N. (2002). The roles of the external, middle and inner ears in determining the bandwidth of hearing. *PNAS*, 99, 13206–13210.

Sadé, J., Fuchs, C., & Luntz, M. (1996). The pars flaccida middle ear pressure and mastoid pneumatization index. *Acta Oto-Laryngologica*, 116, 284–287.

Salih, W. H., Buytaert, J. A., Aerts, J. R., Vanderniepen, P., Dierick, M. & Dirckx, J. J. (2012). Open access high-resolution 3D morphology models of cat, gerbil, rabbit, rat and human ossicular chains. *Hearing Research*, 284, 1–5.

Saunders, J. C., & Johnstone, B. M. (1972). A comparative analysis of middle-ear function in non-mammalian vertebrates. *Acta Oto-Laryngologica*, 73, 353–361.

Saunders, J. C., & Summers, R. M. (1982). Auditory structure and function in the mouse middle ear: An evaluation by SEM and capacitive probe. *Journal of Comparative Physiology A: Neuroethology, Neural, Sensory, and Behavioral Physiology*, 146, 517–525.

Saunders, J. C., Duncan, R. K., Doan, D. E, & Werner, Y. L. (2000). The middle ear of reptiles and birds. In R. J. Dooling, R. R. Fay, & A. N. Popper (Eds.), *Comparative hearing: Birds and reptiles* (pp. 13–69). New York: Springer-Verlag.

Schwartzkopf, J. (1973). Mechanoreception: IV Labyrinth (Auditory organ). In D. S. Farner & J. R. King (Eds.), *Avian biology* (pp. 440–477). New York: Academic Press.

Shambaugh, G. E., Jr. (1954). Technique for increasing sound pressure differential between fenestra and round window to enhance hearing improvement following successful fenestration. *Transactions of the American Academy of Ophthalmology and Otolaryngology*, 58, 454–459.

Shaw, E. A. G. (1974). The external ear. In W. D. Keidel & W. D. Neff (Eds.), *Handbook of sensory physiology*, Vol. V/1: *Auditory system* (pp. 455–490). New York: Springer-Verlag.

Shera, C. A., & Zweig, G. (1992). An empirical bound on the compressibility of the cochlea. *Journal of the Acoustical Society of America*, 92, 1382–1388.

Shrapnell, H. J. (1832). On the form and structure of the membrana tympani. *London Medical Gazette*, 10, 120–124.

Silman, S. (1984). *The acoustic reflex: Basic principles and clinical applications*. New York: Academic Press.

Slepecky, N. B. (1996). Structure of the mammalian cochlea. In P. Dallos, A. N. Popper, & R. R. Fay (Eds.), *The cochlea* (pp. 44–129). New York: Springer-Verlag.

Smith, C. A. (1985). Inner ear. *Form and Function in Birds*, 3, 273–310.

Songer, J. E., & Rosowski, J. J. (2007). A mechano-acoustic model of the effect of superior canal dehiscence on hearing in chinchilla. *Journal of the Acoustical Society of America*, 122, 943–950.

Stenfelt, S., Hato, N., & Goode, R. L. (2004). Fluid volume displacement at the oval and round windows with air and bone conduction stimulation. *Journal of the Acoustical Society of America*, 115, 797–812.

Stepp, C. E., & Voss, S. E. (2005). Acoustics of the human middle-ear air space. *Journal of the Acoustical Society of America*, 118, 861–871.

Swarts, J. D., & Rood, S. W. (1990). The morphometry and three-dimensional structure of the adult eustachian tube: Implications for function. *The Cleft Palate Journal*, 27(4), 374–381.

Taberner, A. M, & Liberman, M. C. (2005). Response properties of single auditory nerve fibers in the mouse. *Journal of Neurophysiology*, 93, 557–569.
Teoh, S. W., Flandermeyer, D. T., & Rosowski, J. J. (1997). Effects of pars flaccida on sound conduction in ears of Mongolian gerbil: Acoustic and anatomical measurements. *Hearing Research*, 106, 39–65.
Tonndorf, J., & Tabor, J. R. (1962). Closure of the cochlear windows: Its effects upon air and bone-conduction. *Annals of Otology Rhinology and Laryngology*, 71, 5–29.
Tonndorf, J., & Khanna, S. M. (1970). The role of the tympanic membrane in middle ear transmission. *Annals of Otology Rhinology and Laryngology*, 79, 743–753.
Tonndorf, J., & Khanna, S. M. (1972). Tympanic-membrane vibrations in human cadaver ears studied by time-averaged holography. *Journal of the Acoustical Society of America*, 52, 1221–1233.
Tubelli, A., Zosuls, A., Ketten, D., & Mountain, D.C. (2012). Prediction of a mysticete audiogram via finite element analysis of the middle ear. *Advances in Experimental Medicine and Biology*. 730, 57–59.
van der Klaauw, C. J (1931). The auditory bulla in some fossil mammals: With a general introduction to this region of the skull. *Bulletin of the American Museum of Natural History*, 62, 352.
van Kampen, P. N. (1905). Die Tympanalgegend des Säugetierschädels. *Gegenbaurs Jb*, 34, 321–722.
Vlaming, M. S. M. G., Aertsen, A. M. H. J., & Epping, W. J. (1984). Directional hearing in the grassfrog (*Rana temporaria* L.). I. Mechanical vibration of the tympanic membrane. *Hearing Research*, 14, 191–201.
von Békésy, G. (1960). *Experiments in hearing*. New York: McGraw-Hill.
Voss, S. E., Rosowski, J. J., & Peake, W. T. (1996). Is the pressure difference between the oval and round windows the effective acoustic stimulus for the cochlea? *Journal of the Acoustical Society of America*, 100, 1602–1616.
Vrettakos, P. A., Dear, S. P., & Saunders, J. C. (1988). Middle-ear structure in the chinchilla: A quantitative study. *American Journal of Otology*, 9, 58–67.
Waetzmann, E. von, & Keibs, L. (1936). Theoretischer und experimenteller Vergleigh von Hörschwellenmessungen. *Akustische Zeitschrift*, 1, 1–12.
Weiss, T. F., & Leong, R. (1985). A model for signal transmission in an ear having hair cells with free-standing stereocilia. III. Micromechanical stage. *Hearing Research*, 20, 157–174.
Werner, C. F. (1960). *Das Gehörorgan der Wirlbeltiere und des Menschen*. Leipzig: V.G. Thieme.
Wever, E. G. (1971). The mechanics of hair-cell stimulation. *Annals Otology Rhinology and Laryngology*, 80, 786–805.
Wever, E. G. (1974). The evolution of vertebrate hearing. In W. D. Keidel & W. D. Neff (Eds.), *Auditory system: Anatomy and physiology (ear)* (pp. 423–454). New York: Springer Verlag.
Wever, E. G. (1978). *The Reptile Ear*. Princeton, NJ: Princeton University Press.
Wever, E. G. (1979). Middle ear muscles of the frog. *Proceedings of the National Academy of Sciences of the USA*, 76(6), 3031–3033.
Wever, E. G. (1985). *The amphibian ear*. Princeton NJ: Princeton University Press.
Wever, E. G., & Lawrence, M. (1950). The acoustic pathway to the cochlea. *Journal of the Acoustical Society of America*, 22, 460–467.
Wever, E. G., & Lawrence, M. (1954). *Physiological acoustics*. Princeton, NJ: Princeton University Press.
Wever, E. G., & Vernon, J. A. (1957). Auditory responses in the spectacled caiman. *Journal of Cellular and Comparative Physiology*, 50, 333–339.
Wever, E. G., & Vernon, J. A. (1961). The protective mechanisms of the bat's ear. *Annals of Otology, Rhinology & Laryngology*, 70, 5–17.
Wever, E. G., Lawrence, M., & Smith, K. R. (1948). The effects of negative air pressure in the middle ear. *Annals of Otology, Rhinology & Laryngology*, 57, 418–428.

Wiederhold, M. L., Martinez, S. A., Scott, R. E. C., & deFries, H. O. (1978). Effects of Eustachian tube ligation on auditory nerve responses to clicks. *Annals of Otology, Rhinology & Laryngology*, 87, 12–20.
Wiener, F. M., Pfeiffer, R. R., & Backus, A. S. N. (1966). On the sound pressure transformation by the head and auditory meatus of the cat. *Acta Oto-Laryngologica*, 61, 255–269.
Wiggers, H. C. (1937). The functions of the intra-aural muscles. *American Journal of Physiology*, 120, 771–780.
Willi, U. B., Ferrazzini, M. A., & Huber, A. M. (2002). The incudo-malleolar joint and sound transmission loss. *Hearing Research*, 174, 32–44.
Wilson, J. P., & Bruns, V. (1983). Middle-ear mechanics in the CF-bat *Rhinolopus ferrumequinum*. *Hearing Research*, 10, 1–13.
Wüllstein, H. (1959). Tympanoplasty: Audiology, indications, technique and results. In H. G. Kobrak (Ed.), *The middle ear* (pp. 174–190). Chicago: University of Chicago Press.
Zakrisson, J. E., Borg, E., Lidén, G., & Nilsson, R. (1980). Stapedius reflex in industrial noise: Fatigability and role for temporary threshold shift (TTS). *Scandinavian Audiology, Supplement*, 12, 326–334.
Zosuls, A., Newburg, S. O., Ketten, D. R., & Mountain, D. C. (2012). Reverse engineering the cetacean ear to extract audiograms. *Advances in Experimental Medicine and Biology*, 730, 61–63.
Zwislocki, J. (1962). Analysis of the middle-ear function. Part I: Input impedance. *Journal of the Acoustical Society of America*, 34, 1514–1523.

Chapter 4
Function and Acoustics of the Normal and Diseased Middle Ear

Susan E. Voss, Hideko Heidi Nakajima, Alexander M. Huber, and Christopher A. Shera

Keywords Aerated middle ear • Incus motion • Malleus motion • Middle ear cavity • Middle ear disease • Middle ear fluid • Middle ear impedance • Middle ear model • Middle ear pathologies • Middle ear reflectance • Middle ear transfer functions • Ossicular disarticulation • Ossicular fixation • Ossicular motion • Otitis media • Stapes motion • Tympanic-membrane atelectasis • Tympanic-membrane function • Tympanic-membrane perforations • Tympanic-membrane structure • Tympanosclerosis • Tympanostomy tubes

S.E. Voss (✉)
Picker Engineering Program, Smith College, 100 Green St. Ford Hall, Northampton, MA 01063, USA
e-mail: svoss@smith.edu

H.H. Nakajima
Department of Otology and Laryngology, Harvard Medical School, Massachusetts Eye and Ear Infirmary, 243 Charles Street, Boston, MA 02114, USA

Eaton-Peabody Laboratories, Massachusetts Eye and Ear Infirmary, 243 Charles Street, Boston, MA 02114, USA
e-mail: Heidi_Nakajima@meei.harvard.edu

A.M. Huber
Department of Otorhinolaryngology, University Hospital of Zurich, Frauenklinikstrasse 24, 8091 Zurich, Switzerland
e-mail: alex.huber@usz.ch

C.A. Shera
Eaton-Peabody Laboratories, Massachusetts Eye and Ear Infirmary, 243 Charles Street, Boston, MA 02114, USA

S. Puria et al. (eds.), *The Middle Ear: Science, Otosurgery, and Technology*, Springer Handbook of Auditory Research 46, DOI 10.1007/978-1-4614-6591-1_4,

4.1 Introduction

Reviews such as this one usually begin by stating that the primary function of the middle ear is to transfer sound from the air in the ear canal to the fluid in the cochlea. Although the middle ear and its cochlear load do act as an acoustic transformer to provide pressure gain, the system is far better thought of as a wave *transducer*, a device that converts one type of wave at the input into a completely different type on the output. In the ear canal, sound energy propagates as longitudinal (or compressional) waves; in the cochlea, the functionally important motions—those responsible for stimulating the hair cells—are not sound waves in the fluid but transverse, fluid-membrane waves visible in the vibrations of the cochlear partition. These two different types of waves—compressional sound waves in the ear canal and fluid-membrane (or "surface") waves in the cochlea—have very different properties (e.g., amplitudes, wavelengths, wave speeds, and modes of excitation). When driven by sound in the ear canal, the middle ear and cochlea convert the sound into basilar-membrane traveling waves, and vice versa: When driven in reverse by cochlear traveling waves, the middle ear converts these waves into sound in the ear canal (e.g., otoacoustic emissions). This chapter outlines a framework for how this conversion occurs in normal ears and then discusses how a range of middle ear pathologies affect middle ear function. For simplicity, a lumped-element model for the middle ear is employed, modified as necessary to describe various pathologies, to understand measurements of middle ear function in both the normal and the diseased states. The overall goal is to use measurements and models to determine how structural changes in the middle ear are related to changes in its transmission via air conduction pathways. For information on the effect of middle ear pathologies on bone-conduction transmission, see Stenfelt (Chap. 6).

4.2 The Normal Middle Ear

Figure 4.1 shows a conceptual model of the normal middle ear (Peake et al. 1992; Shera and Zweig 1992). The arrival of a sound pressure wave at the eardrum (P_{TM}) triggers a series of events in the middle ear, which consists of those structures within and facing onto the tympanic cavity. In brief, the eardrum oscillates, driven by the pressure difference between the ear canal and tympanic cavity ($P_{TM} - P_{CAV}$). Motion of the eardrum both changes the pressure in the cavity and moves the ossicular chain. Suspended from ligaments and muscles attached to the walls, the three bones of the ossicular chain span the cavity like an arch and transmit the motion of the eardrum to the oval window, where the vibration of the stapes sets the cochlear fluids into motion and generates a pressure difference across the basilar membrane ($P_{OW} - P_{RW}$). The pressure difference drives a fluid-membrane wave that travels along the cochlear partition. Volume displacements of the stapes footplate are relieved by displacement of the round window. Although coupling

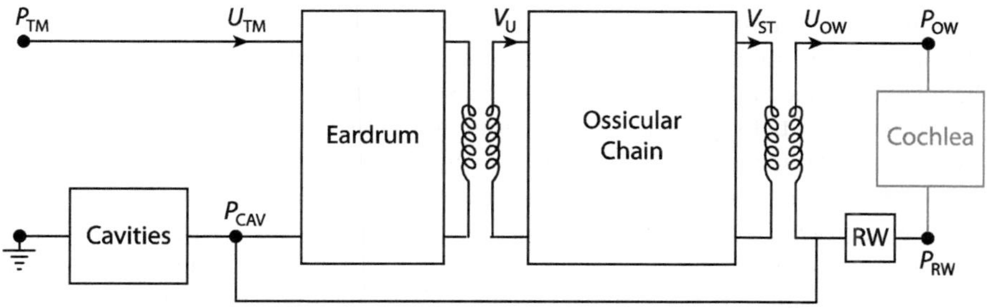

Fig. 4.1 A conceptual model representing the structure and function of the normal middle ear. Components of the middle ear are shown in *black* and the cochlear load is shown in *gray*. When driven in the forward direction, the middle ear converts sound pressure at the tympanic membrane (P_{TM}) into a fluid-membrane wave in the cochlea driven by the intracochlea pressure difference between the oval and round windows ($P_{OW} - P_{RW}$). The eardrum is driven by the pressure difference across its surface, $P_{TM} - P_{CAV}$, where P_{CAV} is the pressure in the tympanic cavity. The cavities, round-window membrane, and cochlear input impedance are represented as one-port impedances (e.g., lumped elements); the eardrum and ossicular chain as more general two-port networks. The transformer inserted between the eardrum and ossicular chain represents the conversion of acoustical to mechanical variables performed by the eardrum (e.g., pressure to force); the transformer between the ossicular chain and cochlea represents the conversion back to acoustic variables performed by the stapes footplate (e.g., velocity to volume velocity). U_{TM} represents the volume velocity of the tympanic membrane and V_U represents the velocity of the umbo, and V_{ST} represents the velocity of the stapes and U_{OW} represents the volume velocity of the oval window (Adapted from Shera and Zweig 1992, and Peake et al. 1992)

to the cochlea through the ossciular chain is stronger, pressure variations in the tympanic cavity also affect the motion of both the stapes and the round window. The middle ear can also be driven "in reverse" by the arrival at the stapes and round window of waves generated or reflected within the cochlea.

4.2.1 *Measures of Normal Middle Ear Function*

Because the focus of this chapter is the human middle ear, measurements from live and cadaveric human preparations are primarily discussed here. Although live humans provide physiologically ideal preparations, invasive measurements of pressure and motions are limited and often require substantial interpretation. For example, it is not always possible to control parameters such as middle ear pressure or to access measurement locations that would require invasive entry into the middle ear. Measurements on human cadaveric ears have been made to describe middle ear function in normal, diseased, and reconstructed states for more than 100 years (e.g., Helmholtz 1868; von Békésy 1960; Voss et al. 2000). Both Rosowski et al. (1990) and Goode et al. (1993) compared mechanical measurements at the input of the middle ear (acoustic impedance and umbo velocity) made on human cadaveric ears and live ears and showed no statistical differences between the two groups. More recently, there has been contradictory work regarding whether or not the output of the middle ear (e.g., stapes motion or cochlear pressure) is comparable between live and cadaveric ears, raising question about the possibility of mechanical differences between the two preparations (Huber et al. 2001; Ruggero and Temchin 2003; Chien et al. 2006). However, Chien et al. (2009) showed that the output of the middle ear is comparable for live and cadaveric ears as long as the measurements of the stapes motion are made at the same angle. Thus, when measurements on live human ears are not feasible, it appears that measurements on cadaveric preparations can provide estimates for human middle ear transmission.

The literature includes many examples of middle ear input measures (umbo velocity, impedance, reflectance), middle ear output measures (stapes velocity, intracochlear pressures, audiometry), and combinations of input and output measures that define transfer functions of middle ear output with respect to middle ear input. It is widely recognized that substantial variation exists within a population of normal-hearing ears for any of these measures. Figures 4.2 through 4.4 show representative measurements to demonstrate the general appearance of some of these measures.

Figure 4.2 plots impedance measurements and the corresponding power reflectance from 12 cadaveric ears made within about 3 mm of the tympanic membrane. As numerous reports demonstrate (e.g., Onchi 1961; Zwislocki 1962; Voss et al. 2000), the impedance is compliance dominated at frequencies below about 1,000 Hz, with a magnitude that decreases with increasing frequency at about 20 dB/decade and an angle that is approximately flat with frequency and approaches −0.25 cycles. Above about 1 kHz, the behavior is more complicated, with contributions from both damping and mass-dominated features, and multiple local minima and maxima, with details dependent on the individual ear and likely resulting from both the sound-transmission system of the ear (i.e., tympanic membrane and ossicles) as well as the structure of the middle ear cavities (Stepp and Voss 2005).

The power reflectance is a measure of the amount of sound power reflected from the tympanic membrane. It can be calculated from the impedance and an estimate of the ear-canal cross-sectional area (e.g., Allen 1986; Keefe et al. 1993; Voss et al. 2008). A power reflectance of 1 means that all sound is reflected, and a power reflectance of zero means that all sound is absorbed by the middle ear and cochlea. The example plots here are typical of additional measurements in the literature that show a power reflectance near 1 at the lower, compliance-dominated frequencies; a lower power reflectance near 1–4 kHz where the middle ear absorbs more power; and more variability in power reflectance at the higher frequencies. These middle frequencies (1–4 kHz) where power is most absorbed also correspond to the frequency region where human hearing is known to be most sensitive. Limited measurements of reflectance have been reported at higher frequencies up to 15 kHz (Farmer-Fedor and Rabbitt 2002; Rasetshwane and Neely 2011); the reflectance generally approaches 1 above about 10 kHz, suggesting that the middle ear limits the transmission of pressure waves at these higher frequencies. These higher

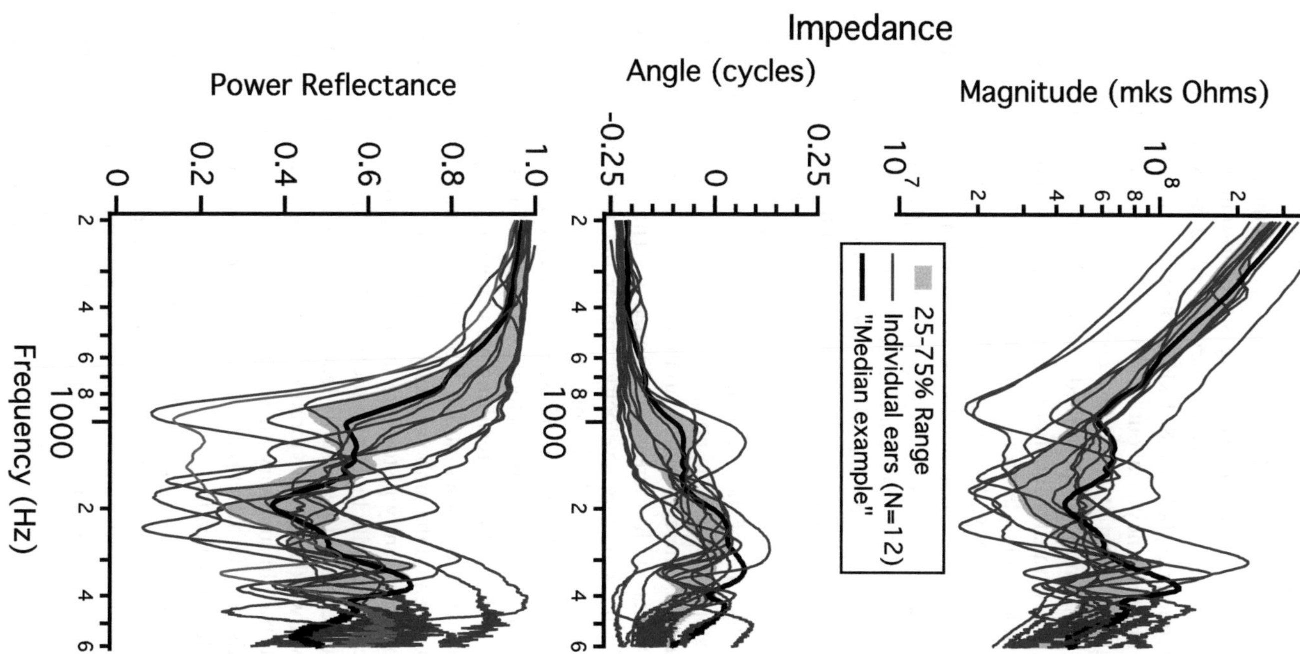

Fig. 4.2 Measurements of impedance magnitudes (*upper row*) and angles (*middle row*) and corresponding calculations of power reflectance (*lower row*) from 12 cadaveric ears described in detail by Voss et al. (2000). Individual ears are plotted with gray lines, the shaded region represents the 25–75% range of all data at each frequency, and the individual ear plotted in black is a representative measurement that approximates the median at most frequencies. The impedance is the ratio between the sound pressure and the volume velocity at the tympanic membrane, and the mks Ohms have fundamental units of N-s/m^5. The power reflectance is calculated from the impedance as $\mathcal{R}(f) = \left|\frac{Z(f)-1}{Z(f)+1}\right|^2$, where $Z(f)$ is the impedance normalized by the ear-canal's characteristic impedance $\rho c/A$, where A is the cross-sectional area of the ear canal, ρ is the density of air, and c is the speed of sound in air

frequency measurements also show substantial variability across subjects; this is an area where more measurements and interpretation are needed.

Figure 4.3 plots the transfer functions of the umbo velocity with respect to the ear-canal sound pressure (left column), the stapes velocity and the ear-canal sound pressure (middle column), and the corresponding stapes-to-umbo-velocity ratio (right column) from a population of cadaveric ears described by Nakajima et al. (2005a). At the lower frequencies, below 0.8–1 kHz, both velocity transfer-function magnitudes increase with frequency at about 20 dB per decade, and both transfer functions have an angle of about 0.25 cycles, consistent with the compliance-dominated impedance measurements described earlier. As frequency increases, the behavior of both velocity transfer functions becomes more complicated; generally the magnitudes exhibit multiple local minima and maxima and the angles decrease with increasing frequency. The stapes-velocity transfer function's angle decreases at a faster rate than the umbo-velocity transfer function's angle, thus leading to an increasing difference in these angles with frequency above about 0.8–1 kHz. The divergence of the stapes and umbo velocity angles here could potentially result from complex three-dimensional motion of the stapes at higher frequencies, as the stapes angle has more variation than that of the umbo. The right column of Fig. 4.3 plots the ratio between the stapes and umbo transfer functions and provides a measure for the velocity gain of the middle ear. At 0.5 kHz, the magnitude gain for the 25–75% of the data range is from 0.17 to 0.36, indicating that the magnitude of stapes velocity is about 9–15 dB smaller than that of the umbo. At higher frequencies there is more variability in this ratio, but the ratio generally decreases further with increasing frequency. The angle difference between the umbo velocity and stapes velocity transfer functions indicates that the stapes and umbo move "in phase" with one another at the lowest frequencies (below about 0.5 kHz), where the angle difference is nearly zero. However, as frequency increases, the angle of the stapes velocity decreases with frequency faster than the angle of the umbo velocity, resulting in a negative angle difference between the two. Thus, at frequencies above 0.5 kHz, the angle of the stapes velocity increasingly lags behind the umbo velocity. In addition, while the vibration mode of the stapes is predominantly piston-like in the low frequencies, its motion becomes increasingly complex with higher frequencies (Sim et al. 2010). Although tilting motions of the stapes do not necessarily lead to bulk movements of cochlear fluids away from the oval window, such vibration modes may also lead to cochlear activity (Huber et al. 2008a). In summary, these measurements of umbo and stapes velocities suggest that the transfer function from the ear canal to the cochlea is both frequency dependent and ear dependent.

Figure 4.4 shows measurements from Puria et al. (1997), Aibara et al. (2001), and Nakajima et al. (2009) of the transfer function between the intracochlear pressure within the scala vestibuli and the ear-canal pressure at the tympanic membrane. This transfer function is largest in the middle frequencies (1–4 kHz), where it approaches a gain of about 20–25 dB. This is the same frequency ranges where the energy reflectance is smallest; thus both measurements are consistent with the largest stimuli reaching the cochlea for 1–4 kHz stimuli.

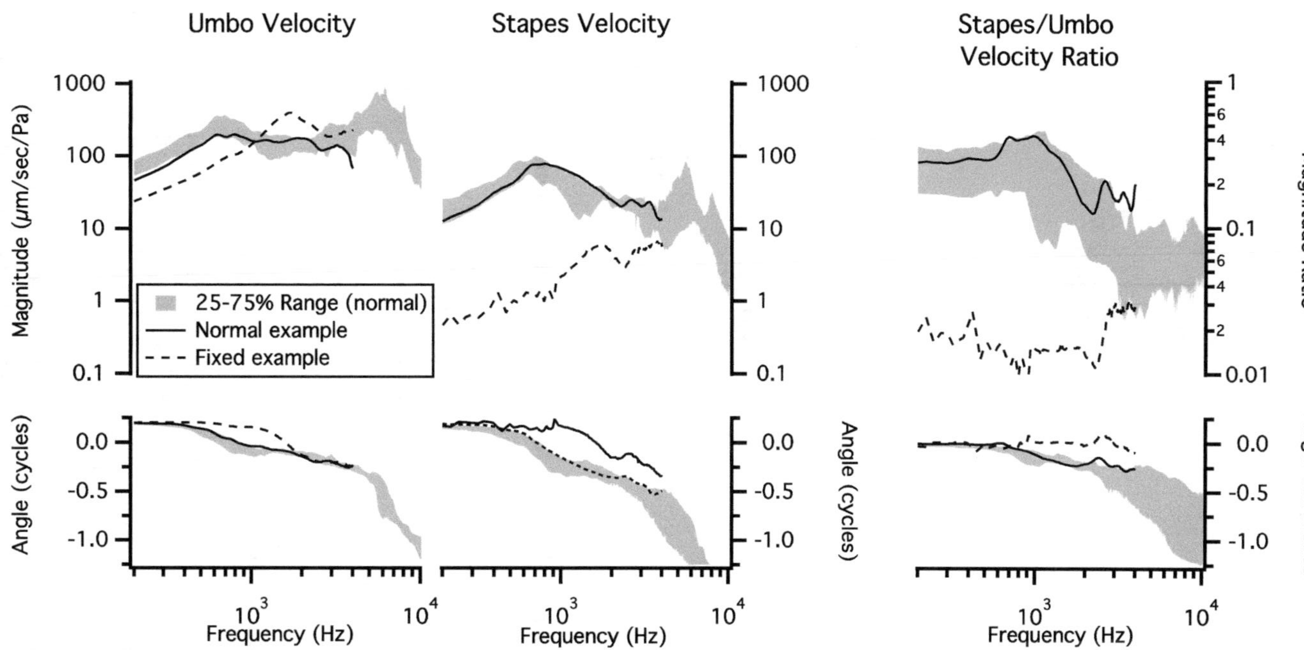

Fig. 4.3 Measurements of umbo velocities (*left column*), stapes velocities (*middle column*), and the ratio stapes-to-umbo-velocity (*right column*). The shaded regions are the 25–75% ranges from 14 normal cadaveric ears described in detail by Nakajima et al. (2005a). One representative individual measurement is plotted with the *solid line* in the normal state and the *dashed line* with the stapes fixed (Discussed later in this chapter)

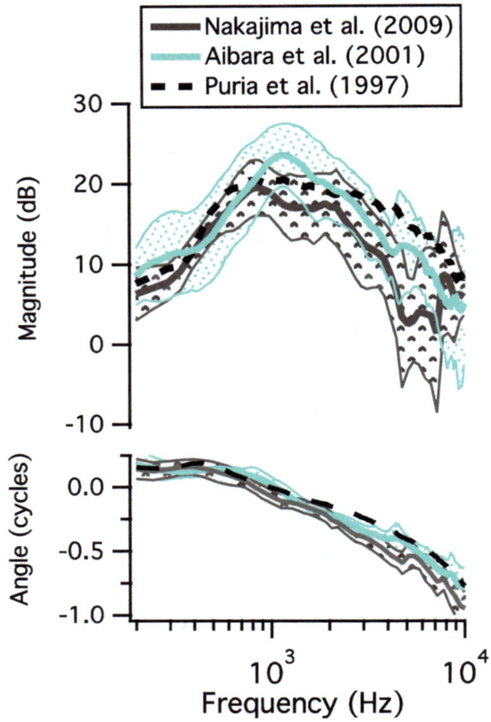

Fig. 4.4 Middle ear pressure gain of scala vestibule pressure to ear-canal pressure from Puria et al. (1997), Aibara et al. (2001), and Nakajima et al. (2009). *Shaded regions* indicate means plus and minus standard deviations

4.2.2 *Function of Individual Parts of the Normal Middle Ear*

The measurements presented earlier provide a system-level picture for how sound is modified and processed by the middle ear. This section describes the motion of the individual pieces that come together to produce the relatively smooth transfer functions pictured earlier.

4.2.2.1 Tympanic Membrane

It is at the tympanic membrane (TM) where the initial transformation of sound from the ear canal to the cochlea takes place. The unique anatomical shape and material properties of the TM contribute to its motion. Thus the function of the TM relies on the specific motion of the TM and how that motion transduces sound to the middle ear ossicles.

To understand human TM motion, various measurements and studies have been performed. For example, time averaged holograms have been used by Tonndorf and Khanna (1972) and Løkbergand et al. (1979) and more recently by Rosowski et al. (2009).

Speckle holography was used by Wada et al. (2002); scanning laser Doppler vibrometry (LDV) by Konrádsson et al. (1987), Ball et al. (1997), and Decraemer et al. (1999); and stroboscopic holography by Cheng et al. (2010).

Observations of the spatial patterns of TM motion in response to sound suggest an increase in the modes of motion of the surface of the TM with frequency, where the spatial patterns of motion become more complex with higher frequencies, as would be expected of a circular membrane such as a microphone diaphragm (Tonndorf and Khanna 1970; Khanna and Tonndorf 1972). Below 1 kHz, the entire surface of the human and cat TM moves in phase with largest motion in the posterior aspect (Decraemer et al. 1999; Cheng et al. 2010). As frequency increases (2–6 kHz), the motion patterns appear more complicated (Tonndorf and Khanna 1972; Rosowski et al. 2009). At even higher frequencies (8 kHz and above), Rosowski et al. (2009) found that the pattern appears more "ordered," with noticeable circular and radial patterns of motion. Before 2010, many reports of the spatial variations in TM motion depended on time-averaged holography, which is insensitive to the phase of motion, and the complex and ordered motions were hypothesized to represent different modal patterns where in individual modal maxima would move out of phase with surrounding maxima. However, more recent data from high-spatial-density laserdoppler measurements (de La Rochefoucauld and Olson 2010) and stroboscopic holography (Cheng et al. 2010) demonstrate that the phase variations seen along the surface of the TM are more consistent with a traveling wave. de La Rochefoucauld and Olson (2010) noted motion that appeared to be a combination of "wavy" and "piston-like" motion. The stroboscopic holographic data have been interpreted as a combination of both modal motion (standing waves) and traveling waves on the TM surface (Cheng et al. 2010).

Although various studies have measured TM motion, how this motion contributes to the transduction of sound to the cochlea, especially at higher frequencies, is not fully understood. Various theories of how TM motion results in the transduction of sound have been proposed. Helmholtz (1868) proposed that the curved shaped of the TM works as a catenary lever, where large displacements near the annular ring (the outer edge) produce small displacements of the malleus. Later, von Békésy (1941) proposed that the transformation of TM motion to the middle ear was dependent on the area ratio between the TM and stapes footplate and that the curvature of the TM was unimportant. The question of the curvature is still a point of study. The idea that the magnitude of the motion of the manubrium of the malleus is much less than the motion of other areas of the TM has been repeatedly demonstrated with holography (Khanna and Tonndorf 1972; Cheng et al. 2010) and with laser-Doppler measurements (Goode et al. 1996; Decraemer et al. 1999).

Although the modal interpretations of TM motion suggest that those TM regions that move with the manubrium are coupled to it, the presence of traveling waves on the TM surface (Cheng et al. 2010; de La Rochefoucauld and Olson 2010) have led to an alternative theory. If surface waves on the TM actually carry sound energy from the periphery to the center of the TM, it would contribute to a delay in the middle ear (Olson 1998; Puria and Allen 1998; Parent and Allen 2010). There does seem to be a delay between the sound pressure within the ear canal near the TM and the motion of the stapes or cochlear sound pressure that is characterized by a group delay of 0.04–0.09 ms (O'Connor and Puria 2008; Nakajima et al. 2009). This delay, however, is an order of magnitude faster than the estimated 0.3–1.3 ms for the traveling wave along the human TM measured by the stroboscopic holographic technique (Chen et al. 2012). Similarly, a long traveling-wave delay of 0.18 ms was found on the TM of the gerbil versus a middle ear transmission delay of only 0.025–0.03 ms (de La Rochefoucauld and Olson 2010).

The best described theory of TM traveling waves is that of Puria and Allen (1998), who have proposed that the power of sound in the ear canal is matched to the outer rim of the TM, and that power collected at the rim is conducted to the umbo at the center of the TM via waves that travel on the TM surface. O'Connor and Puria (2008) and Parent and Allen (2007, 2010) further investigated this idea with a transmission line model that matches the impedance between the ear-canal air and TM and between the TM and the ossicles. These idealized models lacked reflections and mode-like standing waves. Measurements show TM surface motions consistent with a combination of two types of motion: de La Rochefoucauld and Olson (2010) described both a wavy and piston-like motion with LDV in the gerbil, and Cheng et al. (2010) described a traveling wave and modal motion with stroboscopic holography in the human. Estimates of the ratio of the magnitude of the two components generally find that the modal component is larger. Because the umbo impedance cannot possibly be matched to the TM impedance perfectly, it is not surprising that both modes are present. There would have to be some reflections at the umbo due to the traveling wave. Whether both, or which one of these components contributes to TM-ossicular sound transduction is still unknown, but is being addressed using computational model techniques.

There is a long history of the use of finite element models to investigate the transduction of sound by the TM (Gan et al. 2002; Koike et al. 2002). Recently, Fay et al. (2006) used a finite-element model to investigate the consequences of the shape of the TM and known material properties and spatial variations in the thickness of the TM; their work suggests that the shape and mechanical variations enable broad impedance matching from the low impedance of the air in the ear canal to the high mechanical impedance at the ossicles. If the eardrum is shaped too deep, high-frequency transmission is lost, while if the eardrum is too shallow, low-frequency transmission is lost. They also predicted that at low frequencies, the eardrum moves in unison, while at higher frequencies, there are many modes of motion. Furthermore, they propose that the close frequency spacing of the natural frequencies of these many modes sum to enable efficient transfer of power with a smooth frequency response. This suggestion is related but different from the observations that TM motion patterns appear dominated by low-order modes of motion. Funnell et al. (1987) also suggested that a spatial integration is likely taking place over the eardrum, allowing for a relatively smooth frequency response in transferring sound.

Additional experiments have been performed to determine how the eardrum's ultrastructure of its radial fibers influences sound transmission. Experiments have demonstrated that for frequencies below 4 kHz, slits in the TM fibers have little effect on the sound transmission, as patching the TM allows the response to return nearly to normal (Voss et al. 2001a; O'Connor et al. 2008). However, O'Connor et al. (2008) also show experiments that are consistent with their conclusion that "Radial collagen fibers in the tympanic membrane play an important role in the conduction of sound above 4 kHz." Thus, understanding the structure–function relationship of the eardrum is an ongoing area of work. Continued detailed experimental measurements in conjunction with realizable models will aid in determining how the TM transduces sound to the ossicles, especially at high frequencies, which is generally unique to mammalian hearing.

4.2.2.2 Malleus and Incus Complex Motion

Two classical hypotheses regarding motion of the malleus–incus complex (MIC) are: (1) the MIC moves as a rigid body without relative motion between the malleus and the incus (Wever and Lawrence 1954; von Békésy 1960) and (2) prevalent motion of the MIC is a hinged rotation about the anterior–posterior axis of the MIC that passes through the center of gravity of the ossicles (Manley and Johnstone 1974). In measurements on cat and human, motions of the MIC are well adapted to the classical hypotheses at low frequencies, and motions of the MIC showed more complicated patterns with change of the rotational axis in all three-dimensional motion components at high frequencies (Decraemer et al. 1991; Decraemer and Khanna 1994; Sim et al. 2004). Relative motions between the malleus and the incus have also been observed (Dahmann 1930; Hüttenbrink 1988; Willi et al. 2002). Puria and Steele (2010) hypothesize that slippage between the malleus and incus evolved as part of several mechanisms that allow for more efficient middle ear function at higher frequencies. A complete description of the actual motion of the MIC and its corresponding importance for sound transmission to the cochlea is an active area of research.

4.2.2.3 Stapes Motion

Though it is often assumed that the stapes moves in a piston-like manner, spatial modes of the stapes vibration have been measured. von Békésy (1960) described rotational motion around an axis near the posterior edge of the footplate, and Kirikae (1960) reported hinged rotation around a posterior axis and rotation around the long axis of the footplate in measurements with a drained cochlea. Recent developments in measurement techniques and methods have shown that motion of the stapes is almost piston-like at low frequencies and contains complex spatial modes at high frequencies (Gyo et al. 1987; Decraemer et al. 2007). In recent studies with human temporal bones, assuming that anatomical features of the stapes annular ring restrict motions of the stapes footplate in the plane of the footplate, piston-like motion and two rocking-like motions of the stapes (i.e., the two rotational motions along the long and short axes of the footplate) were considered as primary motion components of the human stapes (Hato et al. 2003; Sim et al. 2010). However, it has not been proven that the other components of the stapes motion are insignificant in human middle ear transmission, and Decraemer et al. (2007) reported non-negligible motions through the plane of the footplate in the gerbil stapes.

4.3 The Diseased Middle Ear

4.3.1 Overview

The measurements presented earlier focus on transmission through the normal middle ear. Almost all middle ear models assume that the only mechanism for transmission to the cochlea is from the ossicular chain being driven by the pressure difference across the tympanic membrane and the stapes moving in and out of the oval window; this mode of transmission was termed "ossicular coupling" by Peake et al. (1992). Peake et al. (1992) also emphasize that when the ear is not normal, a mode they term "acoustic coupling" can become important. Acoustic coupling refers to the response of the cochlea to the pressure difference between the pressures adjacent to the oval and round windows, and its contribution is about 60 dB below ossicular coupling when the ear is normal; thus acoustic coupling is important in some specific disease states but has a negligible contribution in the normal ear.

In this section, several middle ear disorders and their effects on middle ear transmission are described. In some cases, the model of Fig. 4.5 can be adapted to fit the disorder, allowing model predictions to be compared to available data. In other cases, no model exists for the specific situation. The specific disorders discussed are grouped into three categories: (1) primarily affecting the middle ear cavity, (2) primarily affecting the tympanic membrane, and (3) primarily affecting the ossicles.

4.3.2 The Middle Ear Cavity

When the ear is normal, the impedance of the middle ear cavity plays a relatively small role in determining transmission through the middle ear; the volume of the cavity is large enough to translate into an impedance that is small compared to the other relevant impedances in the system. In terms of Fig. 4.5, for the normal ear, the combined impedance of the tympanic membrane, ossicular system, and cochlea (often termed Z_{TOC}) has a much larger magnitude than that of the cavity

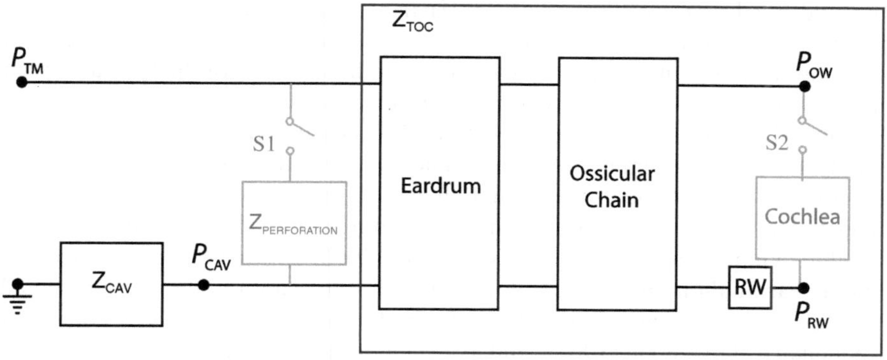

Fig. 4.5 A lumped-element analog model of the middle ear, adapted from Kringlebotn (1988). The two switches, S1 and S2, are included to model the effect of middle ear pathologies. For the normal ear, S1 is open and S2 is closed. With the exception of the middle ear cavities, parameter values are the same as those published by Kringlebotn. For the cavities, two of the values are derived from more recent measurements of Voss et al. (2000) and Stepp and Voss (2005) and are $R_{ad} = 5 \times 10^4 \sqrt{f}$ and $M_{ad} = 1{,}000$. When the middle ear cavity is open to the environment during measurements, the impedance that represents the antrum of the middle ear cavity would be set to 0. When switch S1 is closed to model a perforation, the model values differ from those used by Kringlebotn (1988); see Voss et al. (2001c) for details

Z_{CAV} (e.g., Fig. 2 of Voss et al. 2001c). However, there are several middle ear disorders for which the middle ear cavity becomes important, resulting from the relative magnitudes of the abnormal Z_{TOC} and Z_{CAV}. At its limit, if $|Z_{CAV}| >> |Z_{TOC}|$, then the pressure driving Z_{TOC} approaches zero because $P_{TM} - P_{CAV}$ approaches zero, leading to no movement of the tympanic membrane or ossicular chain. Changes in the relative sizes of $|Z_{CAV}|$ and $|Z_{TOC}|$ can come about by either increases in $|Z_{CAV}|$ or reductions in $|Z_{TOC}|$.

$|Z_{CAV}|$ is increased when the volume of compressible air in the middle ear cavity is reduced. Merchant et al. (1997) refer to the condition of the loss of compressible air within the middle ear cavity as "nonaeration of the middle ear," and point out that this is a common condition within both diseased ears and some postsurgical ears. Diseases such as Eustachian-tube dysfunction can lead to fluid within the middle ear cavity, which reduces the volume of air by exchanging compressible air for incompressible fluid. Rosowski and Merchant (1995) calculated that the middle ear cavity volume should be at least 0.5 cm^3 in order for the ossicular system to be within 10 dB of normal transmission.

$|Z_{TOC}|$ can be reduced in several ways, including TM perforations, TM atelectasis, and interruption of the ossicular chain. These disorders are discussed in subsequent sections, and the middle ear cavity can play an important role in describing transmission within their presence.

The description of middle ear transmission in the presence of middle ear fluid is complicated because the fluid has at least two fundamental effects: (1) reduction of the middle ear cavity volume and (2) mass loading on the TM, ossicles, and windows of the cochlea. Perhaps the most thorough study of transmission through the fluid-filled ear is by Ravicz et al. (2004); they used measurements on a cadaveric preparation of the ear to draw several important conclusions related to middle ear function with fluid, including: (1) the effects of the viscosity of the fluid are either nonexistent or so small that they were not measurable; (2) for low frequencies (less than 0.8 kHz), changes in umbo velocity result from a decrease in the volume of the middle ear air space and not from mechanical loading of the tympanic membrane; and (3) for higher frequencies (at and above 2 kHz), the primary mechanism for reduction in umbo velocity with fluid is the loading of the tympanic membrane and not the volume of air in the cavity. Gan et al. (2006) also measured the effects of middle ear fluid on umbo velocity in a cadaveric preparation; they demonstrated that for cavities filled at about 50% the umbo motion is reduced primarily at frequencies above about 1 kHz, and as the cavity becomes nearly fully filled the umbo displacement is reduced across all frequencies. These results are consistent with the interpretation that the reduction in air volume will affect the lower frequencies once the air volume is reduced enough. Voss et al. (2012) measured the effects of middle ear fluid on the power reflectance measured in the ear canals of cadavers; as in the umbo velocity measurements, large variations occurred that depended on both the fluid level and the volume of the middle ear cavities. Collectively, this work shows that middle ear transmission can be substantially reduced by middle ear fluid, but there is not a simple description or model for how the change in volume and the loading of fluid on the ossicles affects middle ear function. At its limit of the cavity being completely filled with fluid, the model of Fig. 4.5 would predict no transmission to the cochlea because it would be impossible to move the tympanic membrane and ossicular system within the incompressible middle ear space (i.e., $P_{TM} = P_{CAV}$ because $|Z_{CAV}| >> |Z_{TOC}|$). Indeed, moving forward, finite-element models such as those proposed by Gan and Wang (2007) will likely be helpful in further understanding middle ear function with fluid.

4.3.3 Disorders that Involve the Tympanic Membrane

The tympanic membrane (TM) can be affected by a number of factors, including perforations, the insertion of tympanostomy tubes, scarring, and TM atelectasis. Most of these disorders result from chronic middle ear disease, but perforations can also result from trauma.

4.3.3.1 TM Perforations and Tympanostomy Tubes

Extensive measurements and corresponding models of transmission with TM perforations have been reported (Voss et al. 2001b, c; Mehta et al. 2006). The Voss et al. (2001c) model for the ear with a tympanic-membrane perforation is equivalent to that shown in Fig. 4.5 with switches S1 and S2 closed. This model has a topology based on the Kringlebotn (1988) model but with an additional component to represent a TM perforation; a similar model for the effect of a tympanostomy tube would share the same topology. Briefly, the impedance of the perforation $Z_{PERFORATION}$, which depends on the thickness of the tympanic membrane and the diameter of the perforation, acts as a shunt for volume velocity to flow directly from the ear canal to the middle ear cavity. Similarly, the impedance for a tympanostomy tube would depend on the tube's length and diameter. Voss et al. (2001b, c) demonstrated that the major mechanism for changes in transmission with most perforations (except very large ones) is the loss of pressure difference across the tympanic membrane that occurs as volume velocity travels through the perforation and not through the ossicular chain. The physical reduction in the tympanic membrane area or other mechanical changes to the membrane have little effect on the transmission changes, thus leaving the eardrum portion of this model intact.

Figure 4.6 shows several types of measurements made on an example cadaveric ear with perforations introduced. All measurements behave in a systematic manner as the perforation size increases. The impedance remains compliant dominated at lower frequencies but with a reduced magnitude. At the lower frequencies the hole in the TM introduces a shunt path for volume velocity to flow directly into the middle ear cavity, and it is the compliant middle ear cavity that dominates the impedance. As frequency increases above about 0.5–1 kHz, the perforation forms a resonance between the mass of air within the perforation and the air volume of the middle ear cavities (analogous to a Helmholz resonator), and at frequencies above the resonant frequency, the ear's response approaches its normal value. The low-frequency power reflectance is substantially reduced from normal with perforations; this reduction does not mean the cochlea is absorbing more energy but instead that the middle ear cavity is absorbing the energy (Voss et al. 2012). Measurements of the stapes velocity show a systematic low-frequency reduction in magnitude with perforations for the lower frequencies, a slight increase in magnitude at the resonant frequency, and they hover around their normal value at higher frequencies. In summary, the measurements and model are consistent in predicting the following general behavior with TM perforations:

1. Loss is largest at the lowest frequency and decreases with increasing frequency.
2. Loss increases as perforation size increases.
3. Loss does not depend on perforation location [an assumption of this model proven experimentally by Voss et al. (2001a)], disproving what had traditionally been assumed within the clinical literature (e.g., Glasscock and Shambaugh 1990; Schuknecht 1993).
4. The dominant loss mechanism is a reduction in the pressure difference across the tympanic membrane.

Voss et al. (2001c) derived an equation to estimate hearing loss under certain conditions as

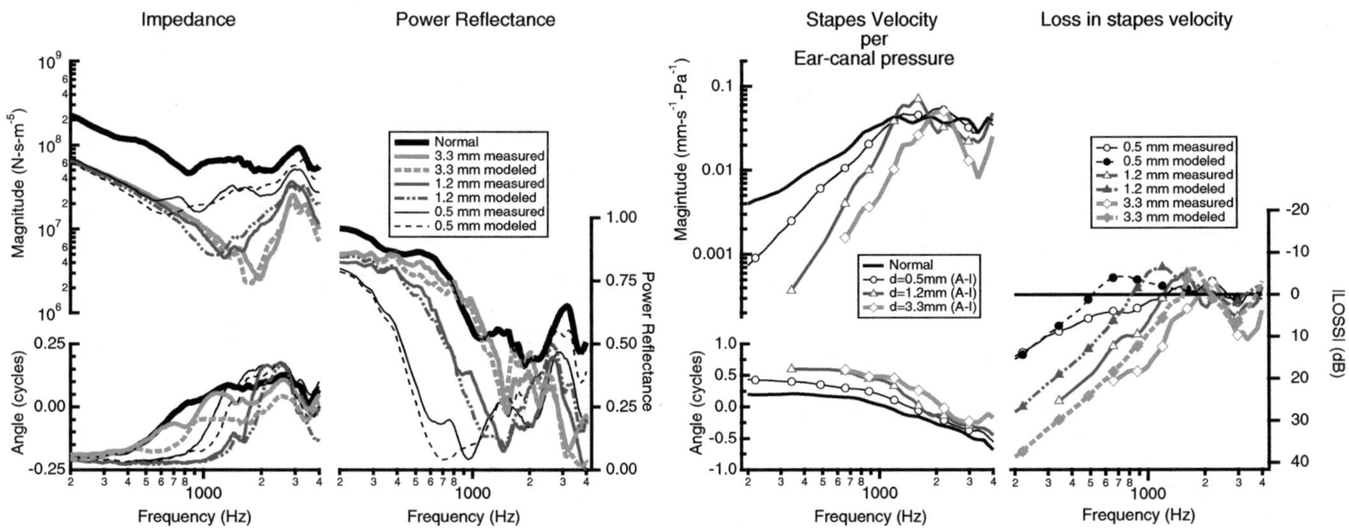

Fig. 4.6 Measurements and model results for an ear with three different sized perforations introduced. (*Left*) Impedance and power reflectance measurements and their corresponding model calculations. (*Right*) Measurements of the stapes velocity per ear-canal sound pressure and measurements and models of the corresponding losses in stapes velocity for each perforation

$$\text{LOSS} = 20\log\left(\left|1 - \frac{\kappa d}{f^2 V}\right|\right), \tag{4.1}$$

where perforation diameter (*d*) is in millimeters and $d > 1$, frequency (*f*) is in Hertz and $f < 500$ Hz, middle ear cavity volume *V* is in cubic centimeters, LOSS is in decibels, and the constant κ equals 2.9×10^6 cm^3 mm^{-1} s^{-2}.

4.3.3.2 TM Atelectasis

TM atelectasis refers to a condition in which the tympanic membrane is displaced medially (retracted). The atelectasis may be of varying severity. It is often the sequella of Eustachian-tube dysfunction producing negative pressure in the middle ear cavity. TM atelactasis can include physical changes to the TM that can affect its coupling to the middle ear system, resulting in a wide range of hearing loss. Merchant et al. (1997) report that hearing loss with atelectasis can range from 0 to 50 dB. Severe TM atelectasis can result in retraction pockets that can be associated with chronic otitis media, cholesteatoma formation, and erosion of the ossicles.

4.3.3.3 Tympanosclerosis

Tympanosclerosis is the formation of white plaques due to hyaline deposits. Such deposits may occur within the TM or in other parts of the middle ear. Tympanosclerosis limited to the TM (also called myringosclerosis) often occurs after chronic inflammation or traumatic events such as TM perforation due to tympanostomy tube placement. This plaque formation can result in a thicker, stiffer eardrum resulting in an abnormal tympanogram. However, tympanosclerosis of the TM may not necessarily affect hearing. Rosowski et al. (2012) showed several examples of abnormal reflectance and umbo-velocity measurements from ears with tympanosclerosis of the TM, although these ears have normal audiograms. Similarly, experimental application of cartilage on the TM produced significant changes in TM motion measured by holography (especially > 4 kHz), but surprisingly did not produce significant changes in transduction of sound measured by stapes motion for the measured frequencies between 0.5 and 8 kHz (Aarnisalo et al. 2009, 2010).

4.3.4 Disorders that Involve the Ossicles

A range of disease processes can affect the ossicles. Disarticulation of the ossicular chain (partial or complete) can be caused by various entities: congenital deformity, as sequelae of chronic otitis media (with or without cholesteatoma), and traumatic injuries. Ossicular discontinuity commonly occurs at the level of the distal incus near the incudo–stapedial joint, but may also affect other parts of the ossicular chain. Fixation of one or more of the ossicles can result from disease processes such as otosclerosis, as sequelae of chronic otitis media, and from congenital abnormalities. Otosclerosis is a disorder affecting remodeling of the human otic capsule of the temporal bone. Etiology yet remains to be fully explained; however genetic, viral, inflammatory, autoimmune, environmental, and hormonal factors have been implicated (Karosi and Sziklai 2010). Most commonly, the stapes footplate becomes immobilized by otosclerotic bone growth, subsequently reducing sound transmission.

4.3.4.1 Ossicular Disarticulation

Complete ossicular disarticulation, in the presence of an intact tympanic membrane, leads to a reduction of middle ear transmission on the order of 40–60 dB depending on frequency (Merchant et al. 1997; Nakajima et al. 2012). Peake et al. (1992) showed that such a loss in ossicular coupling is consistent with the cochlea responding only to the pressure difference at its oval and round windows (i.e., acoustic coupling). In cadaveric preparations, complete ossicular discontinuity produced reduction in the differential pressure across the partition at the cochlear base (the input signal to the cochlea) similar to clinical audiologic findings, as shown in Fig. 8 of Nakajima et al. (2009). Partial ossicular disarticulation, where there is an insecure connection, often consisting of fibrous tissue, generally results in less conductive hearing loss at low frequencies as compared to high frequencies (Nakajima et al. 2012).

The model in Fig. 4.5 represents the case of incudo–stapedial joint disarticulation when switches S1 and S2 are both open (Voss et al. 2012). Specifically, incus–stapes disarticulation is modeled by connecting the malleus and incus directly to the middle ear air space and bypassing the connection to the cochlea and windows. [In this case, the stapes superstructure remains attached to the cochlea and so the box labeled ossicular chain represents only the malleus and incus; it is assumed that the lack of the relatively small stapes mass here has a negligible effect on the two-port representation of the ossicular chain.] Figure 4.7 shows model predictions and measurements for this disarticulated case. The model predictions for the impedance have a lower magnitude than the normal ear and a lower power reflectance (left column); this behavior results because the malleus and incus are no longer connected to the cochlea but are instead hanging in the middle ear cavity. The low-frequency reduction in power reflectance might naively suggest that sound is absorbed and not reflected, but the sound is actually dissipated within the middle ear cavity and not transferred to the cochlea. Measurements of power reflectance are consistent with the model prediction (right column of Fig. 4.7).

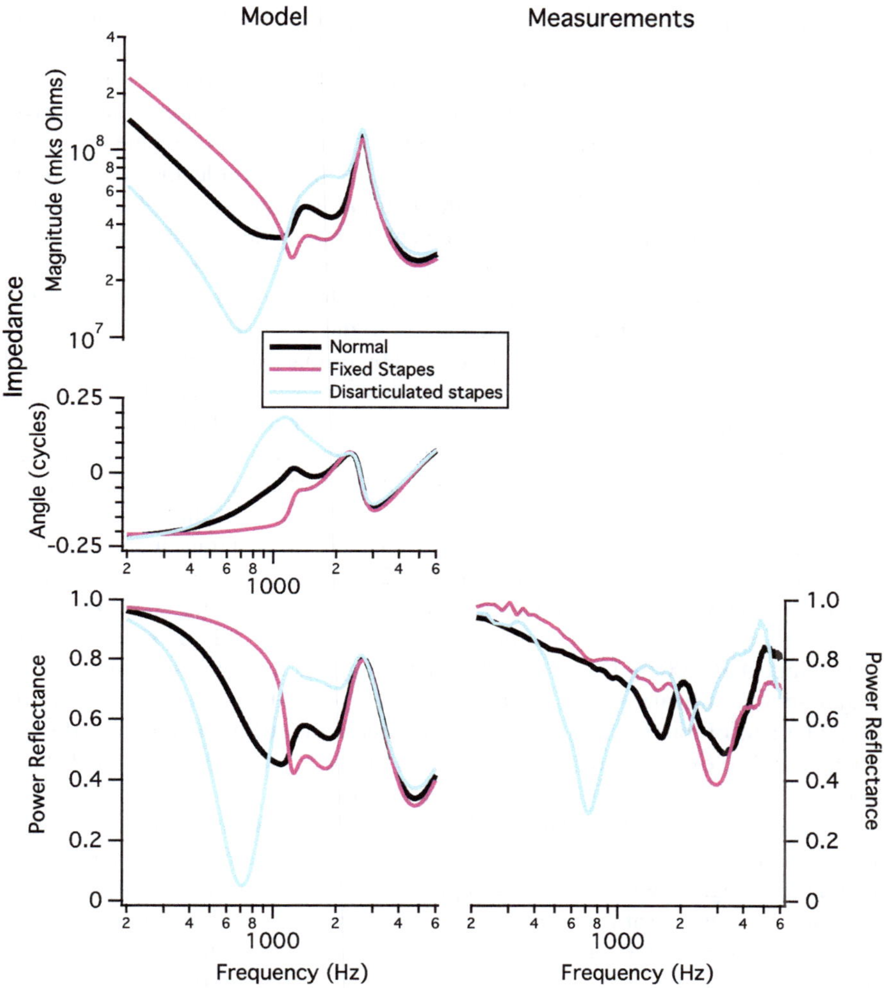

Fig. 4.7 Impedance magnitudes and angles and power reflectance calculations from the models of the ears with the disarticulated incudo–stapedial joint and fixated stapes (*left column*) and corresponding measurements of power reflectance from a cadaveric ear (*right column*) (Voss et al. 2012). Note, measurements are not included for impedances because the measurement was not taken at the TM

4.3.4.2 Fixation of the Ossicles

Stapes fixation most commonly occurs at the level of the footplate and can be caused by otosclerosis or as a result of deposition of fibrous tissue or new bone in cases of chronic otitis media. It is often progressive and can result in conductive hearing loss up to 60 dB, depending on the extent of fixation (Cherukupally et al. 1998). It has been suggested that Ludwig von Beethoven suffered from otosclerosis (Shearer 1990), but unfortunately for him it was not until 1956 that Shea developed modern stapedectomy surgery with replacement prostheses. Stapedectomy results in surgical reconstruction of the ossicular chain in otosclerosis that offers excellent results with correction of the transmission loss in the vast majority (Huber et al. 2012). In stapedectomy (more accurately stapedotomy), a piston prosthesis drives the oval window through an opening made in the stapes footplate. One end of the prosthesis is attached to the long process of the incus, while the other end of the prosthesis acts as a piston driving through the hole in the footplate, allowing for the piston to produce volume displacement of the cochlear fluid. The properties of this prosthesis play a major role in the functional outcome, with the most important factors being the prosthesis diameter (Rosowski and Merchant 1995; Laske et al. 2011) and fixation characteristics to the incus (Huber et al. 2008b).

Fixation of the malleus head can be caused by formation of fibrous tissue or new bone due to chronic otitis media or as a result of a congenital anomaly. Such malleus head fixation can result in a conductive hearing loss of up to 60 dB (Harris et al. 2002). The malleus can also be "fixed" by increased stiffness of the anterior malleal ligament (which connects the neck of the malleus near the short process anteriorly to the bony wall). It was reported that stiffening of the anterior malleal ligament by hyalinization or calcification results in decreased transduction of sound to the cochlea (Fisch et al. 2001; Huber et al. 2003). However, it was subsequently shown in cadaveric preparations that artificially stiffening the anterior malleal ligament resulted in only insignificant decrease (about 5 dB) of stapes velocity (Nakajima et al. 2005b; Dai et al. 2007). This suggests that such stiffening is not a major source of hearing loss. Because the anterior malleal ligament is along the ossicular axis of rotation for low-frequency stimuli, the stiffening torque at the axis itself would be expected to be small, and this torque is proportional to the distance between the stiffening element and the axis of rotation.

Voss et al. (2012) modified the model of Fig. 4.5 (with S1 open and S2 closed) to represent the case of stapes fixation by increasing the impedance of the annular ligament through a decrease in the compliance of the annular ligament (C_c); as this impedance goes toward infinity, the volume velocity within the circuit that represents stapes motion goes toward zero. Because fixation occurs in varying degrees, Voss et al. (2012) altered this compliance C_c instead of simply making it an open circuit, and here it is reduced by a factor of 0.01. Figure 4.7 shows that model predictions for the fixed stapes case have a higher impedance magnitude and higher power reflectance than a normal ear at the lower frequencies (left column). Voss et al. (2012) also show that as the model's annular-ligament impedance approaches infinity, the changes from normal in reflectance reach a limit; the interpretation is that there are multiple compliances in the middle ear system that affect the impedance and reflectance so that even when the stapes is effectively immobile there is still movement within the middle ear system. Nakajima et al. (2005a) show that fixation of the stapes does not result in large decreases in umbo velocity due to the significant flexibility in the ossicular joints (Fig. 4.3). One shortcoming of current lumped-element models (Kringlebotn 1988; Voss et al. 2012) is that they do a poor job of predicting the umbo velocity for the fixed stapes manipulation, possibly due to the need for better representation of the compliances between the ossicular joints.

4.4 Summary

This chapter outlines a framework for how sound in the ear canal is converted from a compressional sound wave to a fluid-membrane wave in the cochlea. In particular, several types of middle ear measures are discussed (e.g., impedance, ossicular vibration, cochlear fluid pressure) and shown to be relatively smooth with frequency. At the same time, complicated and multidimensional motions of the TM and ossicular subsystems have been measured. It remains an active research area to understand how these complicated subsystem motions come together to form the seemingly simple and smooth transfer functions between the cochlea and ear canal. This chapter also employs a lumped-element model to provide a theoretical framework for understanding how some middle ear disorders influence middle ear function for frequencies below about 6 kHz. (e.g., TM perforations, stapes fixation, and stapes disarticulation); the model is less helpful for the conditions of middle ear fluid or cholesteatoma.

Acknowledgments We thank the late Dr. Saumil N. Merchant for many helpful discussions relating to the clinical aspects of the work discussed here. We also thank Mike Ravicz for his extensive help in putting together older data sets for our use in some of the figures. Several colleagues provided substantial help in clarifying the most recent and important aspects of middle ear function: Dr. Jae Hoon helped us to summarize work related to ossicular motion, and Drs. John Rosowski and Lisa Olson helped us to describe recent work on tympanic-membrane motion. We thank Dr. Sunil Puria for his thorough and thoughtful comments that led to great improvements in the overall presentation. This work was supported by the National Institutes of Health (S. E. Voss, H. H. Nakajima, and C. A. Shera) and the National Science Foundation (S. E. Voss).

References

Aarnisalo, A. A., Cheng, J. T., Ravicz, M. E., Hulli, N., Harrington, E. J., Hernandez-Montes, M. S., Furlong, C., Merchant, S. N., & Rosowski, J. J. (2009). Middle ear mechanics of cartilage tympanoplasty evaluated by laser holography and vibrometry. *Otology & Neurotology*, 30, 1209–1214.

Aarnisalo, A. A., Cheng, J. T., Ravicz, M. E., Furlong, C., Merchant, S. N., & Rosowski, J. J. (2010). Motion of the tympanic membrane after cartilage tympanoplasty determined by stroboscopic holography. *Hearing Research*, 263, 78–84.

Aibara, R., Welsh, J. T., Puria, S., & Goode, R. L. (2001). Human middle-ear sound transfer function and cochlear input impedance. *Hearing Research*, 152, 100–109.

Allen, J. B. (1986). Measurement of eardrum acoustic impedance. In J. B. Allen, J. L. Hall, A. Hubbard, S. T. Neely, & A. Tubis (Eds.), *Peripheral auditory mechanisms* (pp. 44–51). New York: Springer-Verlag.

Ball, G. R., Huber, A., & Goode, R. L. (1997). Computerized laser doppler interferometric scanning of the vibrating tympanic membrane. *Ear, Nose, & Throat Journal*, 76, 213–218.

Chen, J. T., Hamade, M., Harrington, E., Furlong, C., Merchant, S. N., & Roswoski, J. J. (2012). Wave motion on the surface of the human membrane: holographic measurement and modeling analysis. *Journal of the Acoustical Society of America*, in press.

Cheng, J. T., Aarnisalo, A. A., Harrington, E., Hernandez-Montes, M. S., Furlong, C., Merchant, S. N., & J. Rosowski, J. (2010). Motion of the surface of the human tympanic membrane measured with stroboscopic holography. *Hearing Research*, 263, 66–77.
Cherukupally, S., Merchant, S., & Rosowski, J. J. (1998). Correlations between pathologic changes in the stapes and conductive hearing loss in otosclerosis. *The Annals of Otology, Rhinology, and Laryngology*, 107, 319–326.
Chien, W., Ravicz, M., Merchant, S., & Rosowski, J. (2006). The effect of methodological differences in the measurement of stapes motion in live and cadaver ears. *Audiology & Neuro-otology*, 11, 183–197.
Chien, W., Rosowski, J. J., Ravicz, M. E., Rauch, S. D., Smullen, J., & Merchant, S. N. (2009). Measurements of stapes velocity in live human ears. *Hearing Research*, 249, 54–61.
Dahmann, H. (1930). Zur physiologie des hörens; experimentelle untersuchungen über die mechanik der gehörknöchelchenkette sowie über deren verhalten auf ton und luftdruck. *Hals–Nasen–Ohrenheilkunde*, 27, 329–368.
Dai, C., Cheng, T., Wood, M. W., & Gan, R. Z. (2007). Fixation and detachment of superior and anterior malleolar ligaments in human middle ear: Experiment and modeling. *Hearing Research*, 230, 24–33.
Decraemer, W. F., Khanna, S. M., & Funnell, W. R. J. (1991). Malleus vibration model changes with frequency. *Hearing Research*, 54, 305–318.
Decraemer, W. F., & Khanna, S. M. (1994). Modelling the malleus vibration as a rigid body motion with one rotational and one translational degree of freedom. *Hearing Research*, 72, 1–18.
Decraemer, W., Khanna, S., & Funnell, W. (1999). Measurement and modeling of the three-dimensional vibration of the stapes in cat. In *Abstracts of the Symposium on Recent Developments in Auditory Mechanics*.
Decraemer, W. F., de La Rochefoucauld, O., Dong, W., Khanna, S. M., & Olson, E. S. (2007). Scala vestibuli pressure and three-dimensional stapes velocity measured in direct succession in gerbil. *Journal of the Acoustical Society of America*, 121, 2774–2791.
de La Rochefoucauld, O., & Olson, E. S. (2010). A sum of simple and complex motions on the eardrum and manubrium in gerbil. *Hearing Research*, 263, 9–15.
Farmer-Fedor, B. L., & Rabbitt, R. D. (2002). Acoustic intensity, impedance and reflection coefficient in the human ear canal. *Journal of the Acoustical Society of America*, 112, 600–620.
Fay, J. P., Puria, S., & Steele, C. R. (2006). The discordant eardrum. *Proceedings of the National Academy of Sciences of the USA*, 103, 19743–19748.
Fisch, U., Acar, G. O., & Huber, A. M. (2001). Malleostapedotomy in revision surgery for otosclerosis. *Otology & Neurotology*, 22, 776–785.
Funnell, W. R., Decraemer, W. F., & Khanna, S. M. (1987). On the damped frequency response of a finite-element model of the cat eardrum. *Journal of the Acoustical Society of America*, 81, 1851–1859.
Gan, R. Z., & Wang, X. (2007). Multifield coupled finite element analysis for sound transmission in otitis media with effusion. *Journal of the Acoustical Society of America*, 122, 3527–3538.
Gan, R. Z., Sun, Q., Dyer, R. K., Jr., Chang, K. H., & Dormer, K. J. (2002). Three-dimensional modeling of middle-ear biomechanics and its applications. *Otology & Neurotology*, 23, 271–280.
Gan, R. Z., Sun, Q., Feng, B., & Wood, M. W. (2006). Acoustic-structural coupled finite element analysis for sound transmission in human ear—Pressure distributions. *Medical Engineering & Physics*, 28, 395–404.
Glasscock, M. E., & Shambaugh, G. E. (1990). *Surgery of the ear*, 4th edition. Philadelphia: W. B. Saunders.
Goode, R. L., Ball, G., & Nishihara, S. (1993). Measurement of umbo vibration in human subjects—method and possible clinical applications. *American Journal of Otology*, 14(3), 247–251.

Goode, R. L., Ball, G., Nishihara, S., & Nakamura, K. (1996). Laser doppler vibrometer (ldv)—a new clinical tool for the otologist. *American Journal of Otology*, 17, 813–822.
Gyo, K., Aritomo, H., & Goode, R. L. (1987). Measurement of the ossicular vibration ratio in human temporal bones by use of a video measuring system. *Acta Oto-Laryngologica*, 103, 87–95.
Harris, J. P., Mehta, R. P., & Nadol, J. B. (2002). Malleus fixation: Clinical and histopathologic findings. *The Annals of Otology, Rhinology, and Laryngology*, 111, 246–254.
Hato, N., Stenfelt, S., & Goode, R. L. (2003). Three-dimensional stapes footplate motion in human temporal bones. *Audiology & Neuro-otology*, 8, 40–152.
Helmholtz, H. L. (1868). Die mechanik der gehörknöchelchen und des trommelfells. *Pflüg Archives European Journal of Physiology*, 1–60.
Huber, A., Linder, T., Ferrazzini, M., Schmid, S., Dillier, N., Stoeckli, S., & Fisch, U. (2001). Intraoperative assessment of stapes movement. *The Annals of Otology, Rhinology, and Laryngology*, 110, 31–35.
Huber, A., Sequeira, D., Breuninger, C., & Eiber, A. (2008a). The effects of complex stapes motion on the response of the cochlea. *Otology & Neurotology*, 29, 1187–1192.
Huber, A., Veraguth, D., Schmid, S., Roth, T., & Eiber, A. (2008b). Tight stapes prosthesis fixation leads to better functional results in otosclerosis surgery. *Otology & Neurotology*, 3292, 893–899.
Huber, A., Schrepfer, T., & Eiber, A. (2012). Clinical evaluation of the NiTi-BOND stapes prosthesis, an optimized shape memory alloy design. *Otology & Neurotology*, 33, 132–136.
Huber, A. M., Koike, T., Wada, H., Nandapalan, V., & Fisch, U. (2003). Fixation of the anterior mallear ligament: Diagnosis and consequences for hearing results in stapes surgery. *The Annals of Otology, Rhinology, and Laryngology*, 112, 348–355.
Hüttenbrink, K. (1988). The mechanics of the middle ear at static air pressures. *Acta Oto-Laryngologica Supplement*—, 451, 1–35.
Karosi, T. & Sziklai, I. (2010). Etiopathogenesis of otosclerosis. *European Archives of Oto-Rhino-Laryngology*, 267, 1337–1349.
Keefe, D. H., Bulen, J. C., Arehart, K. H., & Burns, E. M. (1993). Ear-canal impedance and reflection coefficient in human infants and adults. *Journal of the Acoustical Society of America*, 94(5), 2617–2638.
Khanna, S. M., & Tonndorf, J. (1972). Tympanic membrane vibrations in cats studied by time-averaged holography. *Journal of the Acoustical Society of America*, 51(6), 1904–1920.
Kirikae, I. (1960). *The structure and function of the middle ear*. Tokyo: University of Tokyo Press.
Koike, T., Wada, H., & Kobayashi, T. (2002). Modeling of the human middle ear using the finite-element method. *Journal of the Acoustical Society of America*, 111, 1306–1317.
Konrádsson, K. S., Ivarsson, A., & Bank, G. (1987). Scanning laser doppler vibrometry of the middle ear ossicles. *Scandinavian Audiology*, 16, 159–166.
Kringlebotn, M. (1988). Network model for the human middle ear. *Scandinavian Audiology*, 17, 75–85.
Laske, R., Röösli, C., Chatzimichalis, M., Sim, J., & Huber, A. M. (2011). The influence of prosthesis diameter in stapes surgery: A meta-analysis and systematic review of the literature. *Otology & Neurotology*, 32, 520–528.
Løkbergand, O. J., Høgmoen, K., & Gundersen, T. (1979). On holographic-interferometric investigations of the membrana tympani (living man). In G. v. Bally (Ed.), *Holography in medicine and biology* (11. 212–217). Berlin: Springer-Verlag.
Manley, G. A., & Johnstone, B. M. (1974). Middle-ear function in the guinea pig. *Journal of the Acoustical Society of America*, 56, 571–576.
Mehta, R. P., Rosowski, J. J., Voss, S. E., O'Neil, E., & Merchant, S. N. (2006). Determinants of hearing loss in perforations of the tympanic membrane. *Otology & Neurotology*, 27, 136–143.
Merchant, S. N., Ravicz, M. E., Puria, S., Voss, S. E., Whittemore, K. R., Jr., Peake, W. T., & Rosowski, J. J. (1997). Analysis of middle-ear mechanics and application to diseased and reconstructed ears. *American Journal of Otology*, 18, 139–154.

Nakajima, H. H., Ravicz, M. E., Merchant, S. N., Peake, W. T., & Rosowski, J. J. (2005a). Experimental ossicular fixations and the middle ear's response to sound: Evidence for a flexible ossicular chain. *Hearing Research*, 204, 60–77.

Nakajima, H. H., Ravicz, M. E., Rosowski, J. J., Peake, W. T., & Merchant, S. N. (2005b). Experimental and clinical studies of malleus fixation. *The Laryngoscope*, 115, 147–154.

Nakajima, H. H., Dong, W., Olson, E. S., Merchant, S. N., Ravicz, M. E., & Rosowski, J. J. (2009). Differential intracochlear sound pressure measurements in normal human temporal bones. *Journal of the Association for Research in Otolaryngology*, 10, 23–36.

Nakajima, H. H., Pisano, D. V., Roosli, C., Hamade, M. A., Merchant, G. R., Mafoud, L., Halpin, C. F., Rosowski, J. J., & Merchant, S. N. (2012). Comparison of ear-canal reflectance and umbo velocity in patients with conductive hearing loss: A preliminary study. *Ear and Hearing*, 33, 35–43.

O'Connor, K. N., & Puria, S. (2008). Middle-ear circuit model parameters based on a population of human ears. *Journal of the Acoustical Society of America*, 123(1), 197–211.

O'Connor, K. N., Tam, M., Blevins, N. H., & Puria, S. (2008). Tympanic membrane collagen fibers: A key to high-frequency sound conduction. *The Laryngoscope*, 118, 483490.

Olson, E. S. (1998). Observing middle and inner ear mechanics with novel intracochlear pressure sensors. *Journal of the Acoustical Society of America*, 103, 3445–3463.

Onchi, Y. (1961). Mechanism of the middle ear. *Journal of the Acoustical Society of America*, 33, 794–805.

Parent, P., & Allen, J. B. (2007). Wave model of the cat tympanic membrane. *Journal of the Acoustical Society of America*, 122, 918–931.

Parent, P., & Allen, J. B. (2010). Time-domain wave model of the human tympanic membrane. *Hearing Research*, 263, 152–167.

Peake, W., Rosowski, J. J., & Lynch, T. J. (1992). Middle-ear transmission: Acoustic versus ossicular coupling in cat and human. *Hearing Research*, 57, 245–268.

Puria, S., & Allen, J. B. (1998). Measurements and model of the cat middle ear: Evidence of tympanic membrane acoustic delay. *Journal of the Acoustical Society of America*, 104, 3463–3481.

Puria, S., & Steele, C. (2010). Tympanic membrane and malleus-incus-complex co-adaptations for high-frequency hearing in mammals. *Hearing Research*, 263, 183–190.

Puria, S., Peake, W. T., & Rosowski, J. J. (1997). Sound-pressure measurements in the cochlear vestibule of human-cadaver ears. *Journal of the Acoustical Society of America*, 101, 2754–2770.

Rasetshwane, D. M., & Neely, S. T. (2011). Inverse solution of ear-canal area function from reflectance. *Journal of the Acoustical Society of America*, 130, 3873–3881.

Ravicz, M., Rosowski, J., & Merchant, S. (2004). Mechanisms of hearing loss resulting from middle-ear fluid. *Hearing Research*, 195, 103–130.

Rosowski, J. J., & Merchant, S. N. (1995). Mechanical and acoustic analysis of middle ear reconstruction. *American Journal of Otology*, 16, 486–497.

Rosowski, J. J., Davis, P. J., Merchant, S. N., Donahue, K. M., & Coltrera, M. D. (1990). Cadaver middle ears as models for living ears: Comparisons of middle ear input impedance. *The Annals of Otology, Rhinology, and Laryngology*, 99(5), 403–412.

Rosowski, J. J., Cheng, J. T., Ravicz, M. E., Hulli, N., Harrington, E. J., Mdel, S. H.-M., & Furlong, C. (2009). Computer-assisted time-averaged holography of the motion of the surface of the tympanic membrane with sound stimuli of 0.4 to 25 khz. *Hearing Research*, 253, 83–96.

Rosowski, J. J., Nakajima, H. H., Hamade, M. A., Mafoud, L., Merchant, G. R., Halpin, C. F., & Merchant, S. N. (2012). Ear-canal reflectance, umbo velocity and tympanometry in normal hearing adults. *Ear and Hearing*, 33, 19–34.

Ruggero, M. A., & Temchin, A. N. (2003). Middle-ear transmission in humans: Wide-band, not frequency-tuned? *Acoustic Research Letters Online*, 4, 53–58.

Schuknecht, H. F. (1993). *Pathology of the ear*, 2nd edition. Malvern, PA: Lea & Febiger.

Shearer, P. (1990). The deafness of Beethoven: An audiologic and medical overview. *American Journal of Otology*, 11, 370–374.
Shera, C. A., & Zweig, G. (1992). Middle-ear phenomenology: The view from the three windows. *Journal of the Acoustical Society of America*, 92(3), 1356–1370.
Sim, J., Puria, S., & Steele, C. R. (2004). Three-dimensional measurement and analysis of the isolated malleus-incus complex. In K. Gyo & H. Wada (Eds.), *The 3rd International Symposium on Middle Ear Mechanics in Research and Otology*, (pp. 61–67). Singapore: World Scientific.
Sim, J., Chatzimichalis, M., Lauxmann, M., Rsli, C., Eiber, A., & Huber, A. (2010). Complex stapes motions in human ears. *Journal of the Association for Research in Otolaryngology*, 11, 329–341.
Stepp, C. E., & Voss, S. E. (2005). Acoustics of the human middle-ear air space. *Journal of the Acoustical Society of America*, 118, 861–871.
Tonndorf, J., & Khanna, S. M. (1970). The role of the tympanic membrane in middle ear transmission. *Journal for Oto-Rhino-Laryngology and Its Related Specialties*, 79, 743–753.
Tonndorf, J., & Khanna, S. M. (1972). Tympanic membrane vibrations in human cadaver ears studied by time-averaged holography. *Journal of the Acoustical Society of America*, 52, 1221–1233.
von Békésy, G. (1941). On the measurement of the amplitude of vibration of the ossicles with a capacitive probe [in German]. *Akustische Zeitschrift*, 6.
von Békésy, G. (1960). *Experiments in hearing*. Edited by E. G. Wever. New York: McGraw-Hill.
Voss, S., Horton, N., Woodbury, R., & Sheffield, K. (2008). Sources of variability in reflectance measurements on normal cadaver ears. *Ear and Hearing*, 29, 651–655.
Voss, S., Merchant, G. R., & Horton, N. J. (2012). Effects of middle-ear disorders on power reflectance measured in cadaveric ear canals. *Ear and Hearing*, 33, 195–208.
Voss, S. E., Rosowski, J. J., Merchant, S. N., & Peake, W. T. (2000). Acoustic responses of the human middle ear. *Hearing Research*, 150, 43–69.
Voss, S. E., Rosowski, J. J., Merchant, S. N., & Peake, W. T. (2001a). How do tympanic-membrane perforations affect human middle-ear sound transmission? *Acta Oto-Laryngologica*, 121, 169–173.
Voss, S. E., Rosowski, J. J., Merchant, S. N., & Peake, W. T. (2001b). Middle-ear function with tympanic-membrane perforations. I. Measurements and mechanisms. *Journal of the Acoustical Society of America*, 110, 1432–1444.
Voss, S. E., Rosowski, J. J., Merchant, S. N., & Peake, W. T. (2001c). Middle-ear function with tympanic-membrane perforations. II. A simple model. *Journal of the Acoustical Society of America*, 110, 1445–1452.
Wada, H., Ando, M., Takeuchi, M., Sugawara, H., & Koike, T. (2002). Vibration measurement of the tympanic membrane of guinea pig temporal bones using time averaged speckle pattern interferometry. *Journal of the Acoustical Society of America*, 111, 2189 2199.
Wever, E. G., & Lawrence, M. (1954). *Physiological acoustics*. Princeton, NJ: Princeton University Press.
Willi, U. B., Ferranzzini, M. A., & Huber, A. M. (2002). The incudo-malleolar joint and sound transmission losses. *Hearing Research*, 174, 32–44.
Zwislocki, J. (1962). Analysis of the middle-ear function. Part 1: Input impedance. *Journal of the Acoustical Society of America*, 34, 1514–1523.

Chapter 5
Quasi-static Pressures in the Middle Ear Cleft

Joris J.J. Dirckx, Yael Marcusohn, and Michael L. Gaihede

Keywords Baroreceptors • Cholesteatoma • Eustachian tube • Gas exchange • Obliteration • Otitis • Pars flaccida • Perforation • Retraction pockets • Ventilation tubes

5.1 Introduction

The primary function of the middle ear (ME) is to allow efficient transfer of sound waves from the air-filled external ear canal to the inner ear cochlear fluid. The closed ME cavity prevents acoustic shortcut over the tympanic membrane and thus contributes to sound sensitivity, especially at low frequencies. The volume of the cavity needs to be large enough so that the eardrum can vibrate freely. Modern ME research has been aimed predominantly at investigating its acoustic function. The ME is also a semirigid biological gas pocket that is closed most of the time and therefore is subject to slower quasi-static variations between the pressure in the external ear canal and the ME cavity. These quasi-static pressure changes can be several orders of magnitude larger than the loudest tolerable dynamic sound pressures. Many researchers agree that there is a close relationship between ME

J.J.J. Dirckx (✉)
Laboratory of Biomedical Physics, University of Antwerp, Groenenborgerlaan 171, B-2020 Antwerp, Belgium
e-mail: Joris.dirckx@ua.ac.be

Y. Marcusohn
P. O. Box 4575, 30900 Zichron Ya'acov, Israel
e-mail: yael.mcs@gmail.com

M.L. Gaihede
Department of Otolaryngology, Head and Neck Surgery, Aalborg Hospital, Aalborg University Hospital, Hobrovej 18-22, DK-9000 Aalborg, Denmark
e-mail: mlg@rn.dk

S. Puria et al. (eds.), *The Middle Ear: Science, Otosurgery, and Technology*, Springer Handbook of Auditory Research 46, DOI 10.1007/978-1-4614-6591-1_5,

pressure (de)regulation and pathology of the tympanic membrane (TM) as well as the ME, but many questions still remain regarding the underlying mechanisms.

This chapter deals with several aspects of quasi-static pressures in the ME. It explains how ME pressure can be measured, discusses the different factors involved in regulating ME pressure, and comments on the connections between ME pressure deregulation and ME pathology. Lastly, some recent results are discussed that explore the role of neural mechanisms in regulating quasi-static pressures in the ME.

5.1.1 Units and Definitions

In young normal hearing humans, hearing sensitivity is limited to frequencies ranging between approximately 20 Hz and 20 kHz. Apart from audible sound, which is a periodically alternating pressure signal, a "DC" pressure component may also be present. Such static pressure differences between the outside world and the ME cavity never stays exactly constant. Hence, because the pressure continuously changes, "static pressure" in the ME does not exist in practice. ME pressure is therefore "quasi-static," albeit over a broad and varying range of time scales. In the literature, the terms static and quasi-static pressures are used interchangeably to describe these slower pressure changes. Whether such pressure changes are referred to as very low frequency sound or as slow pressure variation is rather arbitrary. In this chapter, the term "quasi-static" pressures is used, meaning pressure variations on time scales that are slower than the lowest frequency of audible sounds, say 20 Hz.

In acoustics, sound pressure level, or pressure amplitude, is given in decibels sound pressure level (dB SPL) relative to a reference sound pressure level of 20 μPa (in air). The threshold of hearing at 1,000 Hz is on the order of 0 dB SPL for a normal hearing young adult, while the loudest tolerable sound pressure is on the order of 120 dB SPL. The thresholds of both hearing and pain strongly depend on the frequency. For quasi-static pressure variations it is less customary to use the dB SPL scale; the amplitude of the pressure level is instead usually given on a linear scale. In the past, several different units have been used to indicate the amplitude of pressure in the ME. Traditionally, in tympanometry, many authors used "millimeters of water pressure" (mm H_2O), where 1 mm of water column corresponds to approximately 9.8 Pa (slightly depending on temperature). As clinicians grew accustomed to reading tympanograms on the "millimeter water scale," the change to SI units was made by using "decaPascals" (daPa), which was very close to the previously used unit. In the scientific literature, pressures are expressed in Newtons per square meter, or the appropriate SI unit: Pascal (Pa).

5.1.2 The Presence of ME Pressures

Quasi-static pressure variations, which lead to pressure differences between the ME cavity and the surrounding environment, are an essential part of everyday life. During the course of a day, gradual meteorological changes in ambient pressures can easily be on the order of several kiloPasacals. Under artificial circumstances, such as taking a fast elevator, airplane travel, or in a ski descent, ambient pressure can also change by several kiloPascals over a period of a few seconds. In suddenly submerging the head in water, a dive of just 1 m deep causes a pressure increase of 10 kPa in a fraction of a second.

The preceding examples are extrinsic sources of pressure difference between the ME cavity and the outside world. Physiologic processes within the ME itself can also cause a buildup of pressure differences. First, the exchange of gases between the ME cleft (the combined structure of ME cavity and mastoid) and the gases dissolved in the blood perfusing the mucosa can gradually deplete the gas content of the ME cleft (Loring and Butler 1987). Second, changes in mucosa volume can alter the ME cleft volume and consequently alter the pressure (Magnuson 2003).

From the preceding examples it is clear that the ME has to constantly deal with pressure changes on the order of tens to thousands of Pascals. The pressure changes can occur on very different time scales, from fractions of seconds to hours. These pressure changes, which occur under normal circumstances of everyday life, are far larger than the highest pressure amplitudes of the acoustic pressure range. At 120 dB SPL, the pressure amplitude is just a mere 10 Pa, whereas submerging the head 10 cm into a bathtub already causes a pressure load of 1 kPa.

5.1.3 Mechanisms of ME Pressure Changes

Figure 5.1a is an artist's impression of some of the mechanisms involved in ME pressure regulation. This section provides a short overview of the most important processes involved in ME pressure regulation.

The ME cavity is connected to a porous part of the temporal bone called the mastoid that consists of many air-filled cells. The connected volume of the ME cavity and the mastoid is referred to as the ME cleft. Both the interior wall of the ME and the air cells are covered with mucosa that is perfused by blood vessels. The Eustachian tube (ET) forms a connection between the ME cavity and the nasopharynx, and under normal circumstances it is closed. The TM forms a nonrigid wall of the cavity. Gases can leave or enter the cavity either through the ET or by gas exchange with the blood through the mucosa. The volume of the ME cleft can change due to deformations of the TM as well as to changes of the mucosa thickness. The process of gas exchange may be part of a centrally controlled feedback mechanism, with the TM as a possible pressure detector. Figure 5.1b

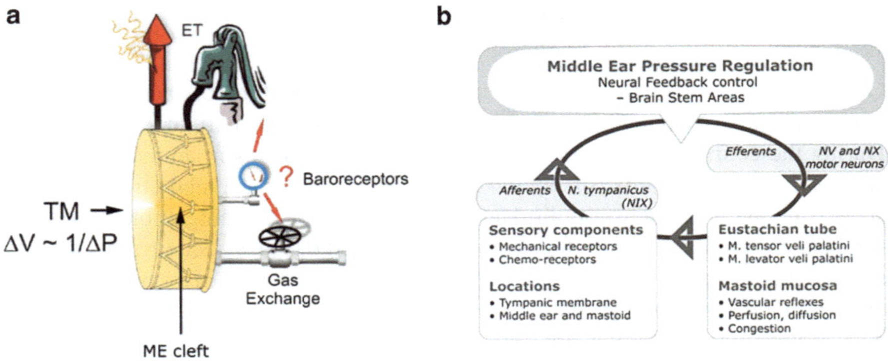

Fig. 5.1 (**a**) An artist's impression of the pressure regulation mechanisms in the middle ear (*ME*). The ME ear cavity together with the mastoid (a porous part of the skull bone) is called the ME cleft. This cleft forms a gas-filled volume (represented by the *yellow drum*). The drum has a fixed volume content, except that the drum membrane, the tympanic membrane (*TM*), can be displaced. For a fixed amount of gas, the product of volume and pressure is constant, so volume change ΔV by deformation of the TM leads to pressure change ΔP. The Eustachian tube (*ET*) opens at high pressure differences between the ME and ambient pressure, acting as a vent or safety valve. It may also act as a pump, injecting small boluses of air into the ME. Gas exchange between the ME cleft and the blood can change ME pressure significantly over long time scales. Indications exist that baroreceptors are present in the ME, which may be involved in an active feedback mechanism controlling ET opening and gas exchange. (**b**) Diagram illustrating the factors involved in neural feedback mechanism for the overall control of ME pressure. Afferent stimuli containing information about the ME pressure project to brain stem areas (nucleus of the solitary tract) that communicate with the nucleus ambiguus and the trigeminal motor nucleus. Efferents projects via NV and NX motor neurons to the tensor veli palatini and levator veli palatini muscles. In addition, mastoid perfusion may be influenced by vascular reflexes by afferents of the n. tympanicus deriving from the motor nucleus of the dorsal respiratory group (*NX*)

gives an overview of the neural components involved in the process. The details are explained in Sect. 5.8.

If the ME was an all-rigid and closed cavity, any pressure change in the outside world would result in a pressure difference between the cavity and the outside of exactly the same magnitude. Because the TM exhibits viscoelastic properties, it can deform when it is loaded with pressure. Part of this load is compensated by a volume change in the ME cavity. Simultaneously, strain is developed in the TM and in ME structures. The TM consists of two major parts: the pars tensa, which is connected to the manubrium, and the much more elastic pars flaccid, which is situated superiorly to the malleus. The size of the pars flaccida differs strongly between different species; in some, it is nearly absent, whereas in others it is even as large as the pars tensa (Decraemer and Funnell 2008). The pars flaccida can deform very easily, but in most species, including humans, it is far smaller than the pars tensa. Owing to its conical and curved shape, the pars tensa may also easily bend. These membrane deformations will change the volume enclosed in the ME cavity. At larger pressure differences, strain and stress in the pars flaccida and the pars tensa increase, and beyond $\pm$ 2 kPa the TM becomes increasingly stiff. At these pressures the role of the TM as a regulator disappears, and the membrane becomes a victim of the pressure load, which may have clinical consequences, as discussed later.

In addition to its role in the clearance of ME fluid, the ET plays an important role in pressure regulation. Although nowadays it is known that the ET is not an open tube at all, but rather a tissue fold that opens only momentarily, its name suggests the old concept of a constant vent. Under normal circumstances it is closed and during swallowing the tube opens actively, allowing small amounts of air into the ME. For larger positive pressure differences between the ME and the nasopharynx, the ET opens passively, allowing gas under overpressure to leave the ME. In this way one may divide the ET action in two functions. On the one hand it operates as a slow regulator of minor pressure differences by gradual (quantized) admittance of small amounts of gas into the ME, while on the other hand it acts as an emergency valve protecting the ear from excessive pressure loads. Direct measurements of ME pressure have shown both slow continuous changes of pressure and more sudden stepwise changes related to ET openings. In general, large pressure differences will result in ET openings and pressure equilibration whereas smaller differences will not (Gaihede et al. 2010).

Finally, there is the important role of gas exchange and the ME cleft mucosa. The gases in the ME cleft are in equilibrium with the gases dissolved in the venous blood perfusing the mucosa. This equilibrium is disturbed whenever the ET opens and small amounts of gas are introduced from the nasopharynx. This leads to a continuous net absorption of gases before a new equilibrium is reached, which ultimately leads to the formation of underpressure in the ME cleft. This process is discussed in further detail in Sect. 5.4.

The rate at which this process occurs depends on several parameters such as the solubility of the gases, their diffusion coefficient, the amount of blood that can absorb gas, the effective surface area of the blood vessels, the effective tissue thickness, and other physical parameters of the gasses and the vessels. In this way, the process can be limited either by the diffusion rate or by the perfusion rate of the mucosal blood supply. Which of the two is the limiting factor is still a point of ongoing discussion. If diffusion is the limiting factor, then gas exchange is a constant process that cannot be changed on a short time scale by the body itself. If perfusion is the limiting factor then, in principle, an active control mechanism could exist that may change the perfusion rate of the mucosa by contraction or dilatation of blood vessels. Changes in perfusion may therefore actively influence the gas exchange: if blood flow increases, larger amounts of gas can enter or leave the ME cleft per unit of time.

Apart from gas exchange, swelling and de-swelling of the mucosa may also influence ME pressure by changing the effective ME gas volume (Magnuson 2003). This process is clearly perfusion bound. It may act quickly and therefore also has the potential to be part of a regulatory mechanism.

ET function is triggered by either voluntary or involuntary actions. Pressure regulation by gas exchange and mucosal congestion might also be an actively controlled mechanism. For active feedback mechanisms to exist, the ME needs to have some way of detecting the presence of pressure differences between it and the environment. Stenfors et al. (1979) have argued that the pars flaccida may act as a pressure receptor, a view supported by the discovery of nerve endings in the pars flaccida (Lim 1970). If such mechanoreceptors are present, they may constitute the afferent components in an active feedback system that controls ME pressure. Recent work has shown distinct brain stem activation, in response to static pressure changes, that demonstrates that distinct afferent pathways exist that can be related to such actively controlled feedback regulation of ME pressure (Sami et al. 2009). If such an overall control mechanism can be demonstrated, it may have an important impact on future clinical management of those ME diseases that are connected to ME pressure deregulation.

5.2 Clinical Importance of ME Pressures

ME pressure is an important pathogenic factor which plays a major role in diseases of the ME, and can cause hearing loss and need for ME surgery. This section reviews a series of ME diseases and related sequelae in which negative ME pressure plays a crucial role.

5.2.1 Otitis Media with Effusion

In childhood, negative ME pressure is frequently encountered in cases of otitis media with effusion, with 80 % of all children experiencing at least one episode before the age of 4 years (Zielhuis et al. 1990). Traditionally, the pathogenic events have been explained by the ex vacuo theory proposed by Politzer (1867), where gas absorption in the ME cavity is insufficiently counterbalanced by a decreased gas supply resulting from an impaired function of the ET. Due to the negative pressure, a fluid effusion is formed that fills the ME cavity and leads to the clinical symptoms of hearing impairment and a sensation of pressure in the ear in some cases. However, the exact details of pathogenic events in otitis media with effusion are still not clear. They may also be related to infection of the ME, so that in some cases the effusion can be interpreted as a resolution phase of acute otitis media (Sadé et al. 2003).

Treatments of otitis media by insertion of ventilation tubes into the TM are considered the most frequent surgical procedure in children in the Western world. In some countries, up to 28 % of children have surgery at least once before the age of 7 years (Gaihede et al. 2007). But although inserting ventilation tubes may restore hearing and prevent systematic negative ME pressure, there are often complications. Permanent perforation of the TM is a common finding after repeated insertions, especially for T-shaped tubes (a special type of ventilation tubes used for long-term treatment in recurrent cases). Up to 24 % of such cases experience persisting permanent perforations, which subsequently need surgical reconstruction of the TM (Strachan et al. 1996).

Other common complications related to otitis media with effusion and treatments with tubes are TM sequelae such as myringosclerosis and atrophy, which are found in 88 % of ears (Gaihede et al. 1997). Myringosclerosis is formed by calcium deposits in the middle layer (lamina propria) of the TM leading to increasing stiffness, while atrophy is related to degeneration and depletion of the lamina propria fibers leading to decreased stiffness (Shanks and Shelton 1991). Whereas myringosclerosis has little clinical importance, because normal hearing is most often preserved, atrophy bears more important implications because it represents weak parts in the TM that are susceptible to pressure loading and formation of retractions pockets.

5.2.2 Tympanic Membrane Atrophy, Atelectasis, and Cholesteatoma

The TM consists of three layers: (1) an outer epidermal layer, (2) an intermediate lamina propria, and (3) an inner mucosal layer. In the pars tensa, the lamina propria contains an elaborate system of collagenous as well as elastin fibers that are arranged in both an outer radially oriented layer and an inner circularly oriented layer as well as intermediate parabolic fibers (Lim 1970). It is believed that this structural arrangement forms the mechanical skeleton of the TM, and that this is important for both its mechanical and acoustic properties (Fay et al. 2006). In the lamina propria of the pars flaccida the fibers are more abundant and have a more irregular and loose organization, whereas the organization of the superficial epidermal and mucosal layers is similar to that of the pars tensa (Lim 1970).

The exact events leading to atrophy and degeneration of the lamina propria are not known. One possibility, suggested by Tos et al. (1984), is that negative ME pressure represents a long-term mechanical loading of the TM resulting in degeneration of fibers. They based their hypothesis on the demonstration of a correlation between the duration of periods with negative ME pressures and the presence of atrophy in children suffering from otitis media with effusion.

Moreover, static experimental pressure loading of the TM has demonstrated that larger deformations are found in areas where retraction pockets are frequently formed (von Unge et al. 1999). Based on these observations, pressure loading leads to a progressive thinning in these areas and to disintegration of the lamina propria resulting in atrophy (Ars et al. 1989). Alternately, intrinsic factors such as increased stress due to inflammatory changes and swelling of the TM outer layers may also contribute to mechanical strain with depletion and degeneration of the fibers (Gaihede 2000).

Once atrophic parts of the TM have been formed, and the original stiffness is lost, the TM becomes less resistant to further pressure loads. This provides the basis

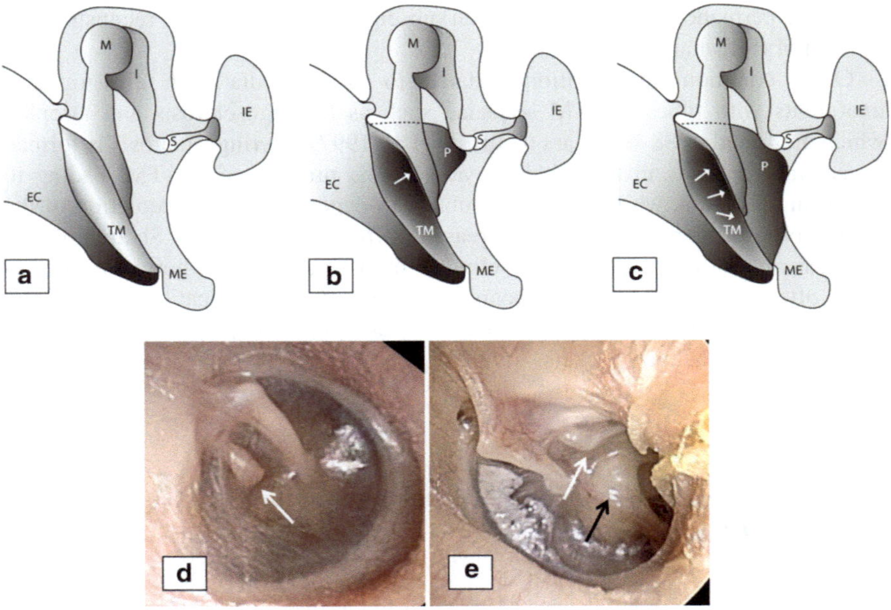

Fig. 5.2 Cross-sectional diagrams of the tympanic membrane (*TM*) and the middle ear (*ME*) cavity. (**a**) Normal TM position with normal aeration of the ME cavity. (**b**) Smaller distinct retraction pocket (*P*) of the posterior part of the TM with contact to the long process of the incus (*I*). (**c**) Pronounced retraction of the TM with contact to the long process of the incus and the medial wall of the ME cavity. *EC*, ear canal; *M*, malleus; *S*, stapes; *IE*, inner ear. The photographs (**d** and **e**) of TMs illustrate the two pathological diagrams (**b** and **c**); (**d**) shows a smaller localized retraction pocket in the TM with contact to the long process of the incus (*white arrow*), and (**e**) shows pronounced TM retraction with contact to the long process of the incus (*white arrow*) and adhesion of the TM to the inner wall of the ME cavity (*black arrow*)

for further progression of the atrophic degeneration and formation of retraction pockets, atelectasis of the TM, and cholesteatoma (Tos et al. 1984; Ars et al. 1989; Sadé 1993). Two examples of these conditions are illustrated by drawings and photographs of the TM shown in Fig. 5.2. Whereas in the first case (Fig. 5.2b, d) the TM shows only a minor retraction with contact to the long process of the incus and incudo-stapedial joint, the latter case (Fig. 5.2c, e) shows a pronounced retraction with contact to the inner wall of the ME cavity. Thus the TM has practically collapsed, illustrating an atelectasis of the posterior part of the cavity. Retraction and atelectasis with TM contact to the ossicles often result in bony erosion, which causes discontinuity and thus conductive hearing loss. In addition, the normal sound transmission of an atelectatic ME is also impaired owing to the mere collapse of the TM. Retractions of the pars flaccida as well as the posterosuperior part of the pars tensa are most commonly encountered.

Cholesteatoma represents a serious complication of retraction pockets in which accumulation of a whitish matrix of cellular debris causes recurrent infections and drainage from the ear and its progressive growth in addition results in bony erosion. The pathogenic events have been described by Sudhoff and Tos (2000). Most frequently, the ossicles are affected, resulting in discontinuity of sound transmission and conductive hearing loss. Moreover, bony erosion may affect other structures in the ME such as the covering of the inner ear. If the inner ear is exposed by invasion of cholesteatoma then severe sensory-neural hearing loss or deafness is almost inevitable. These conditions represent irreversible changes of the ME structures, and surgical resection of the cholesteatoma is needed to prevent its further progression. In addition, the surgical procedures most often include reconstruction of both the TM and the ossicular chain to obtain a dry ear and to restore hearing (the so-called type II and III tympanoplasty), but normal hearing can rarely be obtained at this stage.

The postoperative course, and the long-term success of reconstructive surgery, following these conditions is also dependent on normal ME pressure. If impairment of pressure regulation continues after surgery then new retractions may form and create the basis for new disease. Recurrent cholesteatomas are quite frequent, especially in children, in whom it can amount to 20–25 % of the cases (Edelstein and Parisier 1989). Surgical techniques to prevent problems from sustained negative ME pressure, and thus to improve long-term results, include reconstruction of the TM with cartilage grafts, which have higher stiffness properties than traditional fascia and perichondrium grafts (Zahnert et al. 2000), as well as surgical obliteration of the mastoid, which may prevent sustained negative ME pressure (Takahashi et al. 2007; Vercruysse et al. 2008).

5.2.3 Summary and Future Research

The phenomenon of negative ME pressure is related to a major series of clinical ME conditions that lead to deteriorated hearing and demand surgical intervention and reconstruction of ME structures. Some cases involve simple closures of TM perforations, which generally are able to restore normal hearing, whereas others cause irreversible structural changes. The latter include cholesteatomas and many cases with atelectasis of the ME, in which permanently impaired hearing is inevitable despite optimal TM and ossicular reconstruction.

The behavior of retraction pockets is different. In some cases they remain stable for years, whereas in others they expand deeper into the ME and progress to cholesteatoma. Thus, it would be important to know when progression of the pathology can be expected. Clinical testing of both the topographical mechanical TM properties, as well as capabilities of ME pressure regulation, may reveal future correlations useful to identify such cases. Earlier surgical interventions can then be planned to prevent development of irreversible pathological changes and preserve normal hearing. Experimental studies have been reported in which focal changes in TM elastic properties could be measured quantitatively, but their clinical application requires further study (Dirckx and Decraemer 1992).

5.3 Measurement of ME Pressure

Measurements of ME pressure can be performed directly using various methods, and indirectly using tympanometry. Tympanometry has advantages in that it is a simple and fast procedure, and can be performed without any discomfort for the patient. These advantages make it especially good for measurements on children, in whom it has been employed in large clinical screening studies. Hence, tympanometry is very practical and not limited by ethical restrictions related to direct methods for determination of ME pressure. However, the indirect principle of tympanometry includes a number of methodological limitations and possible measurement errors, especially in diseased ears in which determination of the ME pressure is particularly relevant. This section describes measurements of ME pressure using tympanometry and explains some of these limitations. In addition, methods for direct measurements are shortly reviewed.

5.3.1 Basic Principles and Limitations of Tympanometry

Tympanometry is based on impedance measurements introduced by Metz (1946). Later, Thomsen (1960) described the first tympanogram that recorded the relationship between continuous changes in the ear canal pressure and the impedance. Thomsen demonstrated that the impedance was minimal when the ear canal pressure was equal to the ME pressure. Thus, the ME pressure could be determined from the impedance dip of the tympanogram.

In modern tympanometry the admittance is measured rather than impedance, and a low-frequency probe tone is used so that the compliance component of the admittance can be determined. Shanks and Shelton (1991) provided a review of the basic principles of modern tympanometry. In Fig. 5.3, a normal tympanometric recording is illustrated, in which compliance is depicted as a function of the ear canal pressure. This tympanogram also demonstrates that different curves are obtained for different directions of pressure change, that is, from positive toward negative pressure (negative pressure sweep) and vice versa (positive pressure sweep). For a negative pressure sweep the determination of ME pressure yields a value that is more negative compared with a positive pressure sweep. The corresponding peak pressure difference is illustrated in Fig. 5.3. This peak pressure difference reflects an error of measurements, in which the actual ME pressure probably corresponds to the mean of the two pressure peaks (Decraemer et al. 1984; Hergils et al. 1990).

The peak pressure difference can be explained by phase delay and hysteresis. Phase delay is an instrumental factor related to each instrument, but it can probably be considered negligible in most modern tympanometers (Therkildsen and Gaihede 2005). Hysteresis reflects the viscoelastic properties of the TM and the ME system. In normal ears hysteresis is negligible because it corresponds to only 10–15 daPa

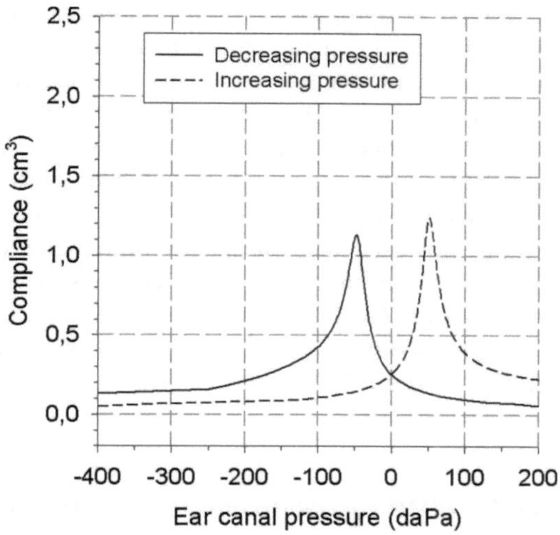

Fig. 5.3 Tympanometric measurements recorded in both negative (+ to −) and positive (− to +) directions. For the negative pressure sweep peak compliance (admittance) appears around −50 daPa, thus indicating a ME pressure of −50 daPa, while for the positive pressure sweep at + 50 daPa. The peak pressure difference (*PPD*) illustrated here amounts to 100 daPa

(Decraemer et al. 1984; Therkildsen and Gaihede 2005). In contrast, in diseased ears with ME effusion the peak pressure difference may increase, so that it introduces an error of greater than 200 daPa (Gaihede et al. 2005).

Further sources of measurement errors include the volume displacement of the TM related to the procedural ear canal pressure changes during tympanometry. Because the ear canal pressure variation displaces the TM, the actual ME pressure is affected. This effect is most prominent in ears with a small ME cleft volume and a flaccid TM (Flisberg et al. 1963; Ingelstedt et al. 1967). Several clinical studies have shown a good agreement between tympanometric estimates and direct measurements of ME pressure; however, these studies are limited in that they were performed only in normal ears, with normal sized ME clefts and normal TM mobility (Thomsen 1960; Takahashi et al. 1987a; Hergils et al. 1990).

Mechanical model experiments have been used to analyze the effect of ME cleft volume on the tympanometric estimate of ME pressure. These experiments suggest that the tympanometric ME pressure approaches $-\infty$ daPa when ME volume approaches 0 cm^3 (Gaihede 2000). Thus, high negative ME pressures can be obtained as an artifact of tympanometry merely due to the depletion of the air volume, which corresponds not only to the situation of a small ME cleft volume per se, but also to cases with ME effusion.

Monitoring changes in ME pressure over time is essential to the understanding of the overall regulation of pressure. Usage of tympanometry may be justified in such studies because the differences between pressure values are important, and any errors of measurements can be assumed to be the same in each recording. However, long-term monitoring of ME pressure using tympanometry is not very practical, and 24-h measurements have been reported in only one study (Bylander et al. 1985). Moreover, the temporal resolution is very poor, as measurements have been reported only at intervals of 3–15 min (Bylander et al. 1985; Grøntved et al. 1989).

Altogether, tympanometric measurements of ME pressure in normal ears have shown good agreement with direct measurements. However, results should be interpreted with caution, especially in diseased ears in which the combination of increased hysteresis and depletion of ME cleft air volume is likely to result in a highly inaccurate estimate of ME pressure. According to the ex vacuo theory described by Politzer (1867), ME effusion has traditionally been interpreted as a transudate resulting from negative ME pressure and not an exudate due to inflammation. However, ME transudate formed due to a hydrostatic pressure difference between the air and blood phase depends on a negative pressure of 50–90 daPa, whereas an exudate can be formed due to inflammation only. Thus, knowing the exact ME pressure is important for a correct interpretation of pathologic events (Sadé and Ar 1997), and this can be obtained only by direct measurements.

5.3.2 Direct Measurements of ME Pressures

Direct methods for clinical measurements of ME pressure include puncturing of the mastoid (Flisberg et al. 1963; Hergils et al. 1990), puncturing of the TM (Buckingham and Ferrer 1973; Sadé et al. 1976), and insertion of a pressure transducer through the ET (Takahashi et al. 1987a). These methods may be more accurate than tympanometry, but they are also limited by various practical problems and obvious ethical restrictions. Moreover, these methods are not suitable for monitoring changes in ME pressure on a day-to-day basis, which is necessary for analysis of the overall and long-term regulation of the ME pressure.

Other researchers have described direct clinical methods suitable for long-term ME pressure measurements. Tideholm et al. (1996) employed a method in which a pressure transducer was incorporated into an ear mold and the pressure could be measured through the ear canal in subjects with either a TM perforation or a ventilation tube; full 24-h monitoring was achieved illustrating normal pressure variations (Tideholm et al. 1998). However, the use of TM perforation, or a ventilation tube, prevents any pressure loading of the TM. Thus, activation of TM mechanoreceptors is avoided, so that possible afferent neural input to regulatory brain stem centers is impaired. Consequently, analysis of overall pressure regulation seems limited.

In more recent clinical experiments, a small pressure transducer was connected to a catheter that was inserted into the mastoid. A recording unit carried by the subject sampled pressure data at a high frequency (0.1 Hz) with the capacity for 48-h recording. Thus, long-term monitoring was possible during normal daily activities (Jacobsen et al. 2007). Because the catheter was inserted into the mastoid, the TM remained intact. Hence, important TM-related neural stimuli remained intact. Figure 5.4 illustrates an example in which a subject was exposed to changes in ambient pressure due to an elevator ride. The results clearly show how changes in ME pressure correlate with changes in altitude. Thus, relatively small variations in ambient pressure, which people are normally unaware of,

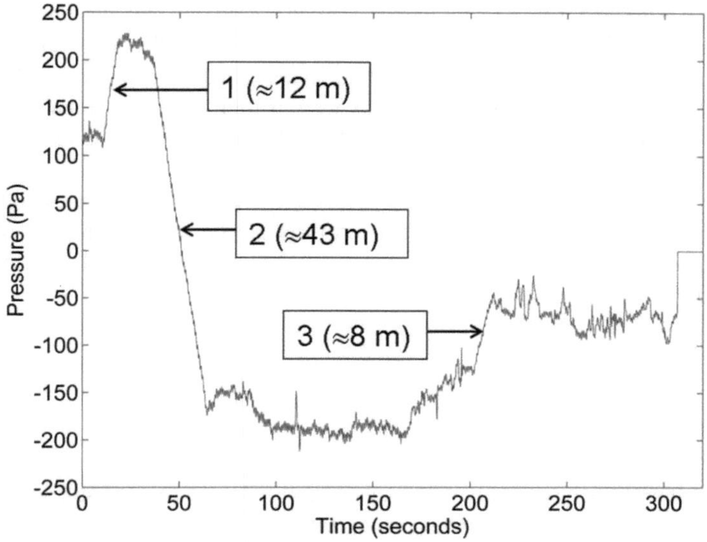

Fig. 5.4 ME pressure changes in a normal human subject in response to changes in altitude. The experiment included three elevator rides illustrated by the *straight lines* pointed out by *arrows*: (1) from the 7th floor to the 10th floor ($\Delta P \approx +\ 100$ Pa); (2) from the 10th floor to the basement ($\Delta P \approx -\ 375$ Pa), and (3) from the basement to the 1st floor ($\Delta P \approx +\ 80$ Pa). Estimated differences in altitudes are indicated. ET openings were not observed during this experiment

can actually be measured. However, the analysis of such detailed variation of ME pressure for long periods of time is very complicated and still awaits future studies (Gaihede et al. 2010).

5.3.3 Summary and Future Research

In summary, tympanometry is a widely used clinical method for determination of ME pressure but it contains methodological limitations decreasing its accuracy in diseased ears. It may be reliable in cases in which temporal changes in pressure are studied and compared, but the time resolution is not satisfactory. Ultimately, direct pressure measurements with an intact TM should be performed over longer periods with methods employing high accuracy and sampling rate in order to analyze the overall ME pressure regulation.

5.4 ME Gas Composition and ME Gas Exchange

As shown in Fig. 5.1a, the ME cleft is a semirigid gas pocket that is closed most of the time. As such, gas diffusion between this (nearly) constant volume cavity and the blood leads to changes in pressure (Loring and Butler 1987). The passage of each gas occurs according to its partial pressure gradient between these different compartments (Piiper 1965). The gas mixture contained in the ME cleft consists of O_2, CO_2, H_2O, N_2, and Ar. These are the same gases that are found in air, blood, and tissue. However, their partial pressures as well as their total pressures are different between the different compartments mentioned previously, as reviewed by Sadé and Ar (1997).

Today it is still not clear whether the exchange of the different gases in the ME is limited by blood perfusion or by diffusion through the mucosa (Marcusohn et al. 2010). A debate exists regarding the role of the TM as well, as explained in the text that follows.

5.4.1 ME Gas Composition

Several researchers have studied ME gas composition in several species (cat, dog, guinea pig, chinchilla) as well as in humans (Ostfeld et al. 1980; Hergils and Magnuson 1990; Sadé and Luntz 1993 and references therein). Riu et al. (1966) used gas chromatography to obtain data on the gas composition of the human ME and found that it contained 9.5 % volume O_2, and 5.5 % volume CO_2.

Results obtained by Ostfeld et al. (1980) in dogs (using gas chromatography) were within the range of values previously obtained in humans (Ostfeld et al. 1980). Felding (1998), using gas electrodes for sampling, reported similar results for normal ears who found partial pressures of $O_2 = 5.7$ % and $CO_2 = 6.6$ %, and concluded they are close to equilibrium with mixed or local ME venous blood ($O_2 = 5.3$ % volume and $CO_2 = 6.1$ % volume). In other studies demonstrating lower CO_2 and higher O_2 values, samples were probably contaminated with atmospheric air. Hergils and Magnuson (1990) emphasized the major difficulty in studies of ME gas composition, namely the ability to obtain samples without contaminating them with ambient air.

Sadé et al. (1995) measured the gas composition in the MEs of guinea pigs continuously using mass spectrometry. They observed increases in P_{CO_2} and decreases in P_{O_2} until the system reached a steady state, and reported that the steady-state values measured in their experiments were similar to results obtained in previous studies by a single measurement. As described previously, these continuous gas composition measurements also provided information on the kinetic pattern of ME gas exchange—a topic explained further in the next section.

5.4.2 Overview of Experiments on Gas Exchange

ME gas exchange has been studied in different animal models as well as in humans. Pressure changes and/or volume changes were measured using various methods.

Ingelstedt and Jonson (1966) connected a pressure transducer directly to the mastoid and a flowmeter to the ear canal of normal subjects. Using their setup they were able to observe and quantify the behavior of the ME for a few hours. They described periods in which underpressure was slowly built and saw how these intervals were interrupted during pressure equilibration events where the ET opened and the TM returned to its normal position. Using these results, they estimated that the ME gas absorption rate in normal human ears is 1–2 mL/day.

In the study by Elner (1972) the subjects did not swallow for 5–10 min. When the ET finally opened and the TM returned from retracted to normal (neutral) position, the volume displacement was determined. Elner's calculations (1972) gave the rate of gas absorption from normal MEs as 0.7–1.1 mL/day.

In various studies performed in animal models (e.g., Doyle et al. 1995, 1999; Kania et al. 2004; Marcusohn et al. 2006), and in humans (Aoki et al. 1998; Uchimizu et al. 2005), pressure or volume changes were monitored after the normal gas composition of the ME was changed. A typical curve included two phases (Marcusohn et al. 2006): In the first phase, "a," an exponential increase (in gas volume) was observed, whereas in the second phase, "b," a gradual decrease was seen. It was assumed that entrance of the very soluble CO_2 (as compared to O_2 and N_2) into the ME from the mucosal blood caused the fast (pressure or volume) increase in the first phase, whereas diffusion of N_2 from the ME to the blood (Kania et al. 2006) was assumed to be the governing factor in the second phase.

5.4.3 Perfusion/Diffusion Limitations of Gas Exchange

The question of whether perfusion or diffusion is the main limiting factor in ME gas exchange is still open. Van Liew (1962) studied gas exchange in subcutaneous gas pockets. His results indicated that the exchange of CO_2 and O_2 was limited by diffusion, while the exchange of N_2 was at least partially limited by perfusion, in accordance with conclusions of previous studies (Van Liew 1962).

Piiper et al. (1962) also addressed the role of perfusion in N_2 exchange. According to their study, in which they used subcutaneous gas pockets, the main factor that governs N_2 exchange is diffusion. Doyle et al. (Doyle and Seroky 1994; Doyle et al. 1995, 1999) studied the exchange of gases in the ME of monkeys. They concluded that the exchange of CO_2 and O_2 is limited by diffusion but the exchange of N_2 is limited by perfusion. Clearly there is not yet agreement among all experimental findings.

Marcusohn et al. (2010) tried to determine the limiting factor of CO_2 exchange using phase "a" data obtained from rabbits and rats and calculations of mass specific cardiac outputs in these animals. They found that the ratio [mass specific initial flow rate in rabbits]/[mass specific initial flow rate in rats] was similar to the ratio [mass specific cardiac output in rabbits]/[mass specific cardiac output in rats].

Under reasonable assumptions this result indicates that the exchange of CO_2 in the ME of mammals is limited mainly by perfusion.

In conclusion, significant evidence supports the notion that CO_2 exchange is mainly perfusion limited, yet other studies still support the idea of a diffusion-limited process. The discussion regarding the different gases is ongoing, and more experiments are needed to draw final conclusions. This debate is highly important because blood perfusion can change quickly (due to constriction or dilatation of vessels), whereas diffusion can change only on much longer time scales (due to thickness changes of the mucosa). A perfusion-limited system can therefore be part of a fast-acting pressure regulation mechanism.

5.4.4 Gas Exchange Across the Tympanic Membrane: Fact or Artifact?

Previous studies reported conflicting results regarding another basic question in ME physiology, namely, whether or not the TM plays an actual role in gas exchange. Elner (1970) studied human TM preparations (taken from cadavers with normal ears) in a diffusion chamber. According to their report, CO_2 diffused through the TM at a very slow rate. An in vivo study by Riu et al. (1966) indicated that there was no exchange (or very low level of exchange) of xenon through the TM. In a more recent study, Yuksel et al. (2009) analyzed samples of gas from a sealed part of the ear canal adjacent to the TM using a mass spectrometer. Their results support the idea that the TM is somewhat permeable to CO_2 and O_2. They also refer to other studies in which trans-TM gas exchange was reported. Still, if there is any gas exchange through the TM, the effect is far smaller than the exchange through the mucosa, so trans-TM gas exchange is not a main factor in ME pressure regulation.

5.5 Tympanic Membrane Deformations as a Pressure Regulator

When the eardrum is deformed due to pressure loading, it changes the volume of the ME cleft. Such volume change in turn has an influence on pressure. In this way, eardrum deformation reduces pressure changes, giving the eardrum a possible regulative role. Such a regulative role has been attributed to the pars flaccida because it can move very easily. In this section quantitative results of eardrum deformation are shown and their contribution to pressure regulation is discussed.

5.5.1 Pressure Regulation by Tympanic Membrane Volume Displacement

When the ET is closed, the ME cleft can be regarded as a closed volume containing a fixed amount of gas. As discussed in the preceding sections, the amount of gas does change over time due to the gas exchange mechanisms with the blood in the mucosa, but over short periods of time the amount of gas enclosed in the ME cavity can be regarded as constant. When the ambient pressure changes, the pressure in the ear canal, P_{EC}, will change in the same way, resulting in a pressure gradient $P_{ME} - P_{EC}$ over the TM, where P_{ME} is the pressure in the ME cleft. The pressure load will displace the TM, thus compressing or expanding the gas in the ME. This compression or expansion increases or decreases ME pressure, thus reducing the pressure gradient over the TM. Displacements of the TM may therefore act as a regulative mechanism of ME pressure. Of course, the displacement itself also results in a stress on the TM. Ever since Shrapnell's observations on the pars flaccida, the idea has existed that this part of the TM could act as a pressure regulator by changing ME volume, and thus reducing the pressure load on the pars tensa (Shrapnell 1832).

Elner et al. (1971a) have presented clinical experiments in which ear canal air flow was measured to calculate the TM volume displacement as a function of pressure variations (volume–pressure relationship). In similar clinical experiments, Gaihede and Kabel (2000) presented measurements of ear canal pressure as a function of TM volume displacements (pressure–volume relationship). In both of these studies the volume displacements involved the entire TM. A more detailed approach has been reported by Dirckx and Decraemer (1992), who used Moiré profilometry, an optical method, on human cadaver temporal bones to measure the actual displacement of the TM over its entire surface. From such optical measurements TM volume displacements can be calculated also for the pars flaccida separately. Figure 5.5 shows the results of these experiments, and it is clear that the methods agree well for the entire TM. The graph also shows the results obtained separately for the pars flaccida, and it is immediately clear that its volume displacement is about 10 times smaller than the displacement of the entire TM. This observation suggests that the regulatory capacity of the pars flaccida, at least in humans, is limited.

For higher pressure loads, volume displacements of both the entire TM and the pars flaccida show a considerable asymmetry as a function of positive and negative pressure. Larger displacements were observed as the membrane was pushed laterally, as compared to medial displacements at the same pressure. Thus, at $P_{ME} - P_{EC} = +\ 1.5$ kPa the TM volume displacement was about 20 μL, while at $P_{ME} - P_{EC} = -\ 1.5$ kPa the displacement was only 15 μL. This asymmetry is probably connected to the conical and bent surface shape of the TM. For an outward motion, the TM can bulge and the manubrium can move rather easily. For an inward motion the shape of the slightly curved TM can only change into a straight-walled

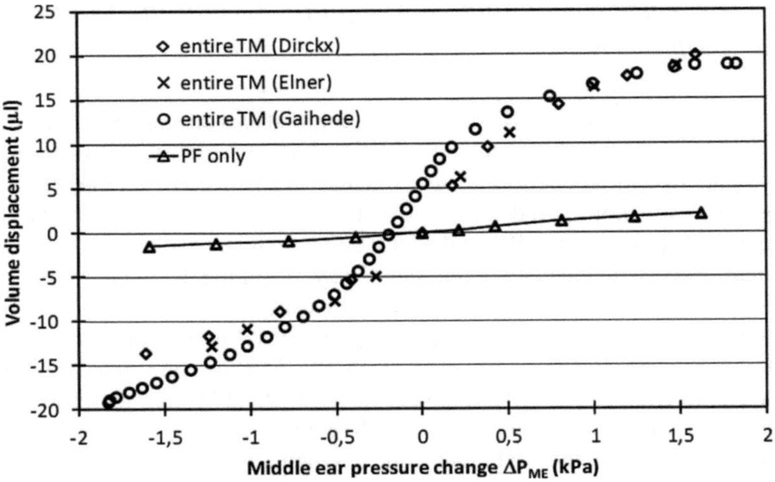

Fig. 5.5 Volume displacement of entire human TM from previous studies (*diamonds, crosses,* and *dots*) and of the TM pars flaccida PF separately (*triangles*). Data measured by Dirckx et al. (1998) originate from a human temporal bone, whereas data from Elner et al. (1971a) and Gaihede and Kabel (2000) are based on clinical experiments. (Data reproduced from Dirckx and Decraemer 1998, with kind permission of S. Karger AG; data reproduced from Gaihede and Kabel 2000, with kind permission from Kugler Publications, Amsterdam)

cone. It needs to be stretched for the manubrium to move further inward. The shape changes of the TM under pressure have been measured in full field using moiré profilometry (Dirckx and Decraemer 1992), and the conical shape of the TM plays an important role in the highly asymmetric motion of the manubrium for over- and underpressures.

The actual pressure load on the TM depends not only on volume displacement, but also on the amount of gas in the ME cleft. Using estimations of ME cleft volume, the relative pressure regulative capacity of TM and pars flaccida can be calculated. The volume of the ME in humans is rather constant over individuals, whereas the volume of the mastoid has much larger variability. Flisberg et al. (1963) calculated volume displacements of the intact TM in normal ears using manometers. One manometer was connected to the ear canal and another was connected to the mastoid (through a cannula). Pressure changes were produced using a syringe that was connected to the system between the manometer and the mastoid cannula. The effects of these changes were measured by the first manometer situated in the ear canal, and were then converted to TM volume changes. The maximal values for TM displacement as calculated by Flisberg et al. (1963) were 25–40 μL, which is in fair agreement with the 20 μL found in direct optical measurements on temporal bones (Dirckx and Decraemer 1992).

5.5.2 Measurements on Human Temporal Bones

Relative pressure compensation can be calculated in the following way. When the TM is in its resting or neutral position, when ear canal pressure P_{EC} is equal to ME pressure P_{ME}, the pressure in the ME is related to ME cleft volume V_{ME} and temperature T of the ME by the ideal gas law, or Boyle's law:

$$P_{ME} \times V_{ME} = \text{constant} \tag{5.1}$$

As stated before, it is a fair assumption to regard the amount of gas being constant over short periods of time. If the TM is now displaced, it will cause a volume change ΔV of the ME cleft. According to the gas law, this will in turn cause a change in ME pressure ΔP_{ME}. Assuming that the compression or expansion of the ME gas is isothermic, ME pressure will change to a value $P'_{ME} = (P_{ME} + \Delta P_{ME})$ determined by:

$$(P_{ME} + \Delta P_{ME}) = P_{ME} \times V_{ME}/(V_{ME} + \Delta V_{ME}) \tag{5.2}$$

The relationship between ΔV_{ME} and ΔP_{ME} is given by the experimental results shown in Fig. 5.5. These results were obtained from optical deformation measurements on pars tensa and pars flaccida in a normal human temporal bone (Dirckx and Decraemer 1991). In the experiments that yielded Fig. 5.5, ME pressure was changed by injecting gas into the ME and allowing the TM to move freely. In reality, it is the pressure in the ear canal that changes, thus displacing the TM and compressing (or expanding) the fixed amount of gas in the ME cavity. So the ear canal pressure will be equal to the sum of the ME pressure and the additional pressure needed to displace the TM and compress the gas present in the ME. The change in ear canal pressure ΔP_{EC}, which will cause a TM volume displacement equal to ΔV_{ME}, is calculated in the following way:

$$\Delta P_{EC} = \Delta P_{ME} + (P_0 \times \Delta V_{ME})/(V_{ME} + \Delta V_{ME}) \tag{5.3}$$

Here P_0 is the ambient pressure in the ear canal for the TM in its neutral position. Using this equation, Fig. 5.5 is reworked to a graph with ear canal pressure on the horizontal axis.

Next, a definition for the pressure regulative capacity of the membrane is needed. The relative pressure compensation due to TM displacement can be given as the difference between the pressure change ΔP_{ME} in the ME divided by the difference change ΔP_{EC} in the ear canal:

$$\Delta P_{ME}/\Delta P_{EC} \tag{5.4}$$

Using these equations and the experimental data shown in Fig. 5.5, one can now calculate the relative pressure compensation capacity of the TM and the pars

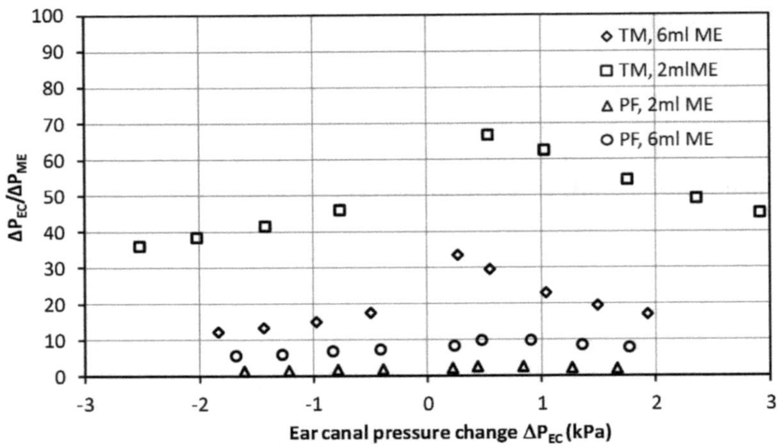

Fig. 5.6 Pressure regulation caused by displacement of the entire TM (*squares* and *diamonds*) and of the TM pars flaccida (*PF*) only (*crosses* and *triangles*) calculated for a ME volume of 2 mL (*squares* and *crosses*) and of 6 mL (*diamonds* and *triangles*). The pressure regulative capacity $\Delta P_{ME}/\Delta P_{EC}$ is significantly larger in ears with small ME volume. The contribution of the pars flaccida is marginal as compared to the entire TM (Data reproduced from Decraemer and Dirckx 1998, with kind permission of S. Karger AG)

flaccida under different circumstances. Figure 5.6 shows the result obtained at normal body temperature (37 °C or 310 K), for a ME cleft volume of 6 and 2 mL, respectively (6 mL is regarded as a normal ME cleft volume) (Elner et al. 1971a; Cinamon and Sadé 2003).

If the TM would be completely rigid, no pressure compensation would take place and $\Delta P_{ME}/\Delta P_{EC}$ equals 0 %. For a perfectly flaccid TM, ME pressure would always be exactly equal to ear canal pressure and $\Delta P_{ME}/\Delta P_{EC}$ would be 100 %. Figure 5.6 shows that in an ear with a normal large volume, TM displacement compensates some 10–35 % of the external pressure changes, while in an ear with reduced gas content the compensation effect reaches 30–65 %. As the displacement of the TM is a nonlinear function of pressure, the compensation effect is strongest for small pressure differences. Figure 5.6 also shows that the compensating capacity of the pars flaccida is marginal. In an ear of 6 mL volume, the pressure compensation due to the pars flaccida displacement is not larger than a mere 3 %.

From these results it is clear that volume displacement of the TM does indeed have some compensation effect on pressure differences between the ME and the ear canal, but the effect is rather limited and present mainly at very small pressures. The pars flaccida provides only a marginal contribution to pressure compensation. Reducing ME cleft volume from 6 to 2 mL doubles the pressure compensation effect, which might have useful implications on management procedures for ears with recurrent problems of impaired pressure regulation (see Sect. 5.7.1).

According to clinical observations reported by Luntz et al. (1997), pars flaccida retractions without retractions of the pars tensa are more frequent than the opposite situation. However, degeneration of the pars tensa discussed earlier, resulting in atrophy, may result in weaker spots compared with the pars flaccida. Therefore, retraction of the pars tensa may sometimes occur without simultaneous retraction of the pars flaccida (Luntz et al. 1997).

5.5.3 Animal Experiments

In gerbils, pars flaccida deformation and volume displacement have also been measured as a function of pressure (Dirckx et al. 1997, 1998). The major part of the displacement is reached within a pressure range of just 0.2 kPa, and beyond 0.4 kPa hardly any change is observed. The contribution of the pars flaccida to ME pressure regulation in gerbils is thus limited to a mere 200 Pa range. In three animals, maximal volume displacement at negative ME pressure varied between −0.476 and −0.301 μL between ears, and maximal volume displacement at positive ME pressure ranged from 0.239 to 0.358 μL. The volumes of the MEs were also measured varying between 250 and 270 μL. In any case, maximal volume displacement of the pars flaccida was smaller than 0.2 % of ME cleft volume. From these observations it is clear that even if the pars flaccida were perfectly flaccid, it cannot compensate for changes of more than 0.2 % of ambient pressure or 20 Pa.

The results obtained in these studies do not, of course, contradict a possible pressure regulative role of the pars flaccida at the very lowest pressures. To investigate this, it is necessary to see how the trans-tympanic pressure changes under very small pressure variations in the ear canal. Of course, in gerbil displacement of the pars tensa will also contribute to pressure buffering, but this has not been measured in detail. The main result of animal measurements is that the pars flaccida certainly is not the main pressure regulator, as was suggested in the past.

5.5.4 Summary and Future Research

Displacements of the TM due to pressure differences between the ear canal and the ME have some compensating effect on these pressure differences, but the effects are limited. There is no evidence that the pars flaccida has a significant function in ME pressure compensation, except possibly in the extremely small pressure range of a few tens of Pa's. In that pressure range the pars flaccida may react very quickly to sudden pressure changes, and could certainly also have a function in shunting ultralow- and low-frequency sound. Moreover, the pars flaccida and the pars tensa have been shown to contain nerve endings, so in addition of being a pressure regulator, the TM and especially the pars flaccida may act as a pressure detector involved in overall ME pressure regulation. Further research on TM innervation and on central control of ME pressure is needed to investigate this possibly very important role of the pars flaccida.

5.6 The Eustachian Tube

For a long time, the ET has been considered to be the prime active component in ME pressure regulation, and its functioning has been the subject of research for many years. The problem is that the clinical course of retraction pockets does not correlate at all with results of ET function tests. In this section a brief overview is given of findings with respect to ET function.

5.6.1 Anatomy

The ET connects the ME with the nasopharynx. The length of the ET is about 31–38 mm (Bluestone and Doyle 1988). It consists of a bony part, of about one third of its total length, as well as a cartilaginous part, of around two thirds of its length (Prades et al. 1998). The open bony part is connected at one end to the ME, and at its other end to the cartilaginous part through a narrow (~ 2–3 mm high, ~ 1–1.5 mm wide) isthmus (Magnuson and Falk 1988). The growth pattern of the ET lumen has been described by Luntz and Sadé (1988). Results obtained in these studies contradicted the previously held notion that the ET is wider in children and infants than in adults. Thus, other than describing the general growth of the ET lumen with age, Luntz and Sadé (1988) also indicated that the cartilaginous part grew to a greater extent as compared to the bony part.

The main peritubal muscles are the tensor veli palatini and the levator veli palatini. The tensor veli palatini is connected to the sphenoid, the soft palate, and the lateral lamina of the cartilaginous part of the ET. The levator veli palatini is connected to the petrous apex, the soft palate, and the ET (Bluestone and Doyle 1988; Prades et al. 1998).

5.6.2 Function

Bylander (1986) referred to three functions of the ET: equilibration of pressure, drainage or clearance of the ME, and protection of the ME. A good review of the clearance as well as the pressure equilibration functions was given by Sadé and Ar (1997). This section focuses mainly on the ET as a pressure equilibrating organ.

Several methods to test ET function have been proposed. Elner et al. (1971b) cited previous authors who used several different methods to test the ET function by recording TM displacement or ME pressure. In addition to methods involving intra-ET sound conduction, contrast media (radiographic studies) or tracer solutions have been used (Ingelstedt and Örtegren 1963). Recently, Swarts et al. (2011) reported normative values of ET function tests (force response test, inflation and deflation tests, “sniff” test, and Valsalva) obtained from healthy adults. Bylander et al. (1981)

studied the function of the ET in normal children and adults using tympanometry and compared their findings to those obtained by Elner et al. (1971b) in normal adults. Their findings indicate that the muscular opening function of the ET improves with age until adulthood (Bylander et al. 1981). Takahashi et al. (1987b) calculated a characteristic parameter called tubal compliance index. For this parameter they did not find any difference between normal children and adults. However, when subjects who had otitis media with effusion were examined by them, it seemed that the tubal compliance index was higher in children with otitis media with effusion than in normal subjects. On the other hand, the tubal compliance index was lower in adults, who suffered from this pathology, than in normal subjects. Thus, according to Takahashi et al. (1987b), it seems that children who suffer from this pathology have compliant ETs, in contrast to adults with otitis media with effusion who have rigid ETs.

Finally, the effect of deglutition events has been investigated. Different studies indicate that the amount of gas that passes between the nasopharynx and the ME during effective deglutition events (events in which the ET opens) is very small, as compared to the volume of the ME cleft. According to Elner (1977) the amount of gas that enters the ME in a deglutition event is ~1 μL. A more recent study found this amount to range between 0.79 and 2.79 μL (Mover-Lev et al. 1998). These findings indicate that deglutition cannot significantly change ME gas composition.

Harell et al. (1996) and Hergils and Magnuson (1998) measured the gas composition of the nose and the nasopharynx in order to estimate the composition of the gas that enters the ME during a deglutition event. They found that it was similar to the gas composition of expired air. Hence, it was closer to ME gas composition than to air. In a study performed by Mondain et al. (1997) on normal subjects, it was found that the ET was open for about 430 ms during effective deglutition events. The opening frequency of the ET was 1–2 times/min. In summary, the composition of the gas entering the ME during deglutition events is rather close to the composition of the gas present in ME, and the amount of injected gas is very small.

5.6.3 The Role of the Eustachian Tube in ME Aeration

Holmquist (1978) suggested that an active "pumping" mechanism may exist, whereby gas passes from the nasopharynx into the ME through the ET. This notion was further supported by Sadé et al. (2005). According to their findings, gas can enter the ME while pressure at both ends is equal. As this may indicate that an active mechanism exists in that direction, it was suggested as an explanation to the occurence of hyperectasis (Sadé 2001).

Honjo et al. (1983) recorded simultaneously electromyograms (EMGs) of the tensor veli palatini and the levator veli palatini as well as cineroentgenograms of the ET. They found that the distal and the proximal parts of the ET opened simultaneously, whereas a time lag was noted between the closures of each of these two parts. The proximal part (close to the ME) closed first, during the contraction of the tensor veli palatini muscle. Afterwards, the distal part closed, during the contraction of the levator veli palatini muscle. Hence, their results suggest that an active mechanism, whereby gas is pumped out of the ME through the ET, may exist. Apart from this possible active function at small pressure differences, the ET also has a passive and protective regulation function, opening spontaneously at high positive ME pressures.

5.6.4 Effects of Sleep and Body Position

Mover-Lev et al. (1998) found that the swallowing rate decreased from about 30 events per hour to about seven events per hour in patients who fell asleep during their experiments. In these patients, who were seated, a decrease in ME pressure was recorded. They mentioned previous authors who reported a positive ME pressure during and after sleep in a horizontal position. Apparently body position has an important effect on ME pressure during sleep.

Tideholm et al. (1999a) who measured ME pressure continuously (for 24 h) and found that the mean number of ET openings were as follows: 9.4 per hour in the (awake) erect position, 8.4 per hour in the (awake) horizontal position, and, significantly lower, 3.2, per hour during sleep in the horizontal position. In their study, a higher ME pressure was recorded when subjects were asleep in the horizontal position, as compared to the other two conditions in which the subjects were awake (erect and horizontal positions) (Tideholm et al. 1999a). They suggested that this was due to entrance of CO_2 into the ME resulting from depression of respiratory function during sleep.

Bonding and Tos (1981) reported that in control group patients, who stayed in a horizontal position for 12–24 h after their operation, no significant change in ME pressure was observed. On the other hand, according to previous studies cited by them, increased hydrostatic pressure in the ET vessels and thickening of the mucosa led to reduced ventilation of the ME in this position. The role of the ME mucosal volume was emphasized by Gaihede and Kjær (1998) in a study performed on healthy (adult) subjects. According to them, an increase of mucosal volume was the reason for the observed increase in ME pressure in the horizontal position, as compared to the erect position. In summary, the pressure regulative role of the ET is rather limited and its effect is dependent on the interplay between several factors such as posture and sleep.

5.6.5 Eustachian Tube Dysfunction/Occlusion

Bonding and Tos (1981) discussed various situations that can lead to negative ME pressure. These included insufficient deglutitions (for instance, after tonsillectomy), obstruction of the ET orifice, and mucosal inflammation. In a study performed by

Kindermann et al. (2008) on children (2–12 years old), obstruction of the ET orifice was usually associated with negative ME pressure. In other cases in which the ET orifice was not occluded, normal ME pressures were usually demonstrated. The question of whether ET dysfunction/occlusion plays a role in ME pathologies was addressed by Sadé and Ar (1997) and in Sect. 5.2 of this chapter.

As noted by previous authors, the ET can also be permanently open or patulous (Sadé and Ar 1997). In some patients who suffer from this pathology, "fullness" of the ear and autophony are reported, although it is usually an asymptomatic condition (Pulec and Hahn 1970; Sadé and Ar 1997). In patients who have a patulous ET, the TM is usually normal, but sometimes it is atelectatic (Sadé and Ar 1997). A possible explanation for this interesting phenomenon has been suggested by Sadé and Ar (1997): It might be that a Bernoulli effect could lead to negative ME pressure in patients who have a patulous ET as the velocity of the air passing by an opening will cause underpressure.

Tideholm et al. (1999b) performed continuous ME pressure measurements (for 24 h) in subjects with patulous ETs. Interestingly, they found more pressure variability within the patulous ET group than within a normal group, which they studied previously. The pattern of pressure changes recorded in these experiments was significantly different between the groups. In the patulous ET group, an overall negative mean pressure was observed while subjects were sleeping in the horizontal position. The pressure recorded was variable between subjects (in some of them it increased, in other subjects it decreased). In addition, no significant pressure difference was found between the erect and the horizontal position at night in subjects with patulous ET. In contrast to these findings, increased pressure was recorded in most of the normal subjects during night. Hence, Tideholm et al. (1999b) concluded that a diagnosis of patulous ET is not necessarily associated with an ET that is constantly open.

5.7 The Mastoid Air Cell System

The exact role of the mastoid air cell system in pressure regulation of the ME remains unknown. The mastoid contains clusters of a large number of connecting air-filled cells. These cells lie behind the ME cavity and the ear canal, and extend toward the inferior tip of the mastoid (Fig. 5.7). Thus, the mastoid is relatively inaccessible, which may have limited basic research. The structure of the air cells can have an acoustic function, as the many surfaces help to break up acoustic vibration modes, prevent sharp resonances, and increase hearing sensitivity for low frequencies (Fleischer 2010). However, there are numerous clinical and structural observations, as well as more recent experiments, indicating an important and active role for the mastoid in pressure regulation.

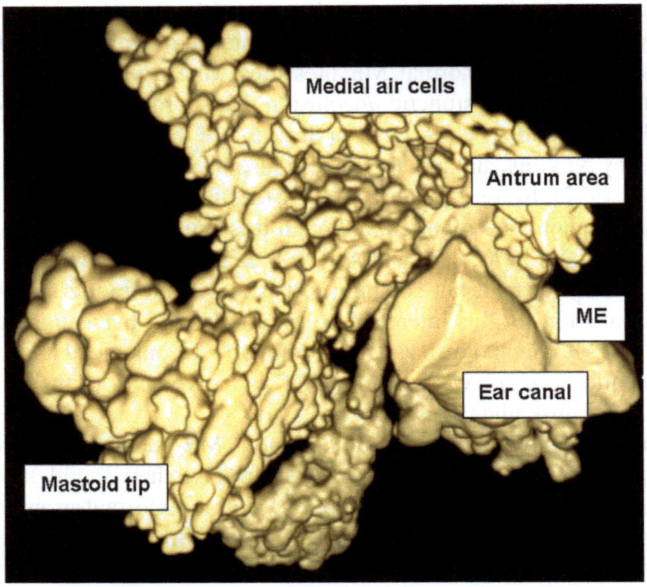

Fig. 5.7 The normal mastoid with clusters of air cells emerging from the posterior part of the *ME*. The ear canal is seen at the *right side* in the *middle* of the image. The image is based on a 3D reconstruction of a high-resolution clinical CT scanning, where the surface compact bone has been made transparent (Courtesy of Olivier Cros)

5.7.1 Pneumatization of the Mastoid and Clinical Observations

The air cell system is absent at birth, but develops during childhood by expansion of air-filled cells, mainly from the antrum; the normal extent of air-containing cells (pneumatization) of the mature system is reached at puberty (reviewed in Cinamon 2009). These developmental aspects are important, because there is a close clinical correlation between sclerotic changes with a decreased pneumatization of the mastoid, and chronic otitis media with formation of retraction pockets and cholesteatoma. Decreased pneumatization has been attributed to hereditary factors that increase the risk of development of chronic otitis media (Diamant et al. 1958); however, chronic otitis media itself may affect its normal development during childhood, leading to a decreased pneumatization (Tos and Stangerup 1984).

Further clinical evidence for the mastoid role in pressure regulation relates to the success of postoperative courses of surgical reconstructions; in patients with smaller and diseased sclerotic mastoids the recurrence risks of retractions and cholesteatoma are higher. Based on these observations, many centers apply surgical obliteration of the mastoid. The basic idea is that the diseased mastoid contributes to an impaired pressure regulation, and by its occlusion the formation of new underpressures can be prevented. Moreover, the pressure regulation capacity of the TM is favored by a smaller ME cleft (please refer to Sect. 5.5.2 in this chapter). Hence, mastoid obliteration has significantly reduced the risks of recurrence of cholesteatoma (Takahashi et al. 2007; Vercruysse et al. 2008).

5.7.2 Structural Properties of the Mastoid

The structure of the mastoid bears implications for its function. The volume of the ME cavity itself is rather small, around 0.5–1 mL (Sadé 1997; Alper et al. 2011), whereas the volume of the mastoid is highly variable. However, the gross anatomy is also dominated by the large numbers of air cells, which lead to an increased surface area such that its area-to-volume ratio (AV ratio) is enhanced. The mastoid size has traditionally been investigated by conventional X-ray projections (Schüller projection), wherein the extent of pneumatization can be measured by planimetry. This measure correlates with the mastoid volume, so that the ratio of volume to planimetric area is 0.7 (Cinamon 2009). In clinical work, Schüller projections are still used by some otosurgeons; they are simple to obtain, yet gives a good impression of the pneumatization.

More detailed analysis has emerged during the last decade based on computed tomography (CT) scans. Park et al. (2000) reported an average surface area for normal adult mastoids to be 167 cm^2, while the average volume was 10.4 cm^3; that is, the AV ratio was 16 cm^{-1}. These findings were supported by Alper et al. (2011). The gas exchange function previously discussed can no doubt be enhanced by the larger surface area if the mucosa is well vascularized. Consequently, it is important to determine both the surface area and volume, and to include the AV ratio in the analysis (Park et al. 2000; Alper et al. 2011). The literature is sparse on diseased ears, but recently Csakanyi et al. (2010) have reported on the mastoid area and volume in children with otitis media with effusion (2–18 years). They have found the AV ratio is higher in diseased ears, which seems unexpected and needs further investigations (Csakanyi et al. 2010).

The histological structure of the mastoid mucosa may also be adapted to gas exchange. According to some authors, the mucosa is relatively thin with an epithelium of flat cells and an underlying loose connective tissue with a rich vascularization (Hentzer 1970; Ars et al. 1997; Takahashi 2001). In comparison with the ME mucosa, the epithelium of the mastoid is lower and the distance from the surface to the underlying capillaries is significantly smaller (Ars et al. 1997). These features also favor gas exchange by a smaller diffusion distance.

5.7.3 Mastoid Passive Pressure Buffering

According to Sadé (1992), the mastoid may function as a passive pressure buffer, so that in a larger mastoid any ME pressure change induced by pressure change in the ear canal will be buffered or absorbed in the ME cleft, whereas in a smaller (non-pneumatized) mastoid this will not be feasible to the same extent. In agreement, smaller non-pneumatized sclerotic mastoids are correlated to retractions and atelectasis of the TM (Sadé 1992; Cinamon and Sadé 2003). This theory, which is based on clinical observations, has been contradicted by experimental research

(Doyle 2000). However, more recent work tried to explore this issue further: Doyle (2007) noted that large mastoid volumes are associated with smaller changes in pressure, which may actually support the notion that the mastoid is indeed a gas reserve. Swarts et al. (2010) concluded that the mastoid may function as a gas reserve only if its perfusion/surface area ratio is much lower than that of the ME itself.

5.7.4 Mastoid Mucosal Perfusion

Whereas ME cleft gas exchange has been held responsible for changes in ME pressure, an alternative hypothesis was proposed by Magnuson (2003). According to him, changes in perfusion may alter the congestion of the mucosa, and hence its thickness, which ultimately affects the ME pressure. A calculation based on the AV ratio reported by Park et al. (2000) shows that a small change in mucosal thickness of only 6 μm can alter the ME pressure by 1 kPa (Magnuson 2003). A similar mechanism has been found in diving mammals adapting to pressures at high depths in water; these animals have rich submucosal cavernous sinuses in their ME mucosa (Stenfors et al. 2001).

5.7.5 Recent Clinical Experiments on Mastoid Regulatory Function

Recent clinical experiments demonstrated that the mastoid can function as an active counter-regulator of small experimental deviations of ME pressures in both negative and positive directions without the involvement of the ET (Gaihede et al 2010). Fig. 5.8 depicts an experiment in which a positive ME pressure was induced by injection of a small amount of air, and the resulting counter-regulation of pressure was measured via the mastoid over 10 min. The counter-regulation was gradual and continued into negative pressures. Similar gradual counter-regulation was found for negative pressures. These responses are independent of the ET and can be related only to the mastoid; they may be explained both by changes in mucosa volume, as well as gas exchange (Gaihede et al. 2010).

At larger pressure deviations the counter-regulation also included openings of the ET, depicted as steeper pressure changes with a step-like pattern. Thus, the overall active regulation may consist of a complementary system including both the mastoid and the ET (Gaihede et al. 2010). Whereas the ET acts intermittently, the mastoid regulation of ME pressure acts continuously, and thus may play an important role in the long-term pressure loading of the TM, which in turn can be related to the pathological changes and clinical problems discussed in the preceding text.

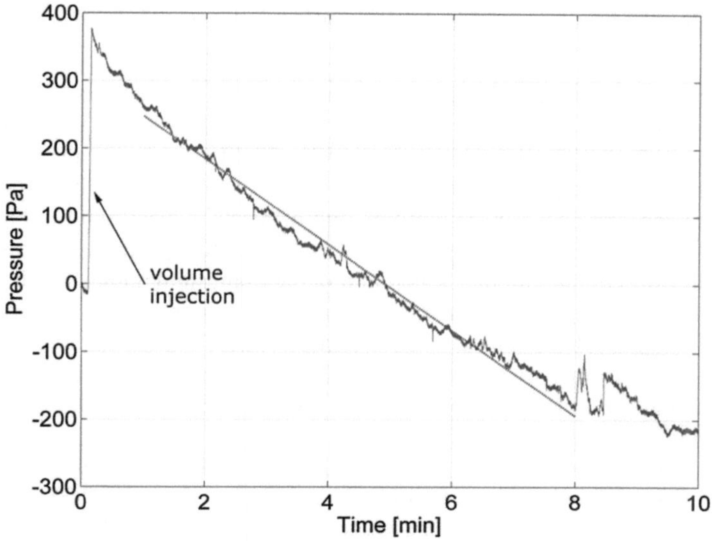

Fig. 5.8 The ME pressure is measured via a catheter inserted into the mastoid tip, and a volume injection of +50 mm^3 of air results in a pressure peak of 378 Pa. The resulting counter-regulation is reflected by a gradual decreasing pressure around – 63 Pa/min. The pressure decrease continues across 0 Pa at around 5 min; at 8 min, a swallow is seen (*M-configuration on curve*), but without any pressure equilibration, that is, no ET opening. After 10 min the pressure is around –200 Pa

5.7.6 Summary and Future Research

The mastoid air cell system has structural properties both in its gross anatomy and histology which favor gas exchange. However, perfusion may also play a role by altering the volume of the mucosa itself, and thus affect the pressure of the ME cleft. Studies on these properties are relatively few and mostly include normal mastoids. Moreover, systematic histological studies of the ME and mastoid mucosa are absent because current data are either sporadic or do not include the entire mastoid but rather the antrum. Such studies should focus on the vascularization of the mucosa. Structural analysis of the mastoid may be improved by applying micro-CT scanning on temporal bone specimens, so that higher resolution may yield more detailed information on volume and surface area.

5.8 Central Neural Feedback Control of ME Pressure

The structure of the mastoid has been described in the previous section. With its numerous air cells and its histological properties it bears many similarities to lung and alveolar structure. Further, in terms of functionality, the overall control of ME pressure regulation may also bear resemblances to the well-known neural feedback control of respiration (Eden 1981).

5.8.1 Basic Studies of Neural Feedback Control

The evidence for a neural feedback control in ME pressure regulation was originally based on experiments in rabbit MEs, in which a neural tracer (horse-radish peroxidase) was applied at two sites: (1) the ME mucosa at the promontory around the tympanic plexus and (2) the muscles of the ET and the palate (Eden 1981). For the first site, subsequent labeling of neurons was found in brain stem areas of respiratory control in the nucleus of the solitary tract, whereas for the second site subsequent labeling was found in the brain stem nucleus ambiguous as well as in the trigeminal motor nucleus. In respiratory control, neural projections, from the nucleus of the solitary tract to the nucleus ambiguous and trigenimal motor nucleus, form part of the neural reflex arch, and consequently, a similar reflex arch has been proposed for the aeration of the ME, and thus for ME pressure regulation (Eden 1981).

Based on his observations, Eden proposed that afferent stimulation of the nucleus of the solitary tract is attained by the tympanic nerve (NIX) containing information about ME aeration from oxygen sensitive glomus cells of the tympanic plexus. Further, projections from the nucleus of the solitary tract to the nucleus ambiguous and the trigenimal motor nucleus activate these areas which form the efferent parts by activation of the ET muscles via the trigeminal (NV) and the vagal nerve (NX). Figure 5.1b illustrates the principle and includes additional aspects discussed later.

This hypothesis was considered controversial at the time but has been substantiated in further studies in primates, where similar experiments have shown the same results (Eden and Gannon 1987). Moreover, in important neurophysiologic experiments performed in primates, electrical stimulation of the tympanic nerve resulted in activation of the ET muscles, as recorded by EMG with latencies of 9–28 ms, similarly to other brain stem reflexes (Eden et al. 1990). In addition, the concept of chemoreceptors of the tympanic plexus was extended to include also baroreceptors (Eden et al. 1990).

Similar ideas on reflex control were proposed by Nagai et al. (1989), who demonstrated a decreased ability to equilibrate positive ME pressures by the ET in clinical experiments after anesthetizing the TM by iontophoresis. They also demonstrated modified Vater-Pacinian corpuscles at the periphery of the pars tensa using electron microscopy of the TM, and suggested a mechanoreceptor function (Nagai and Tono 1989). In the same area smooth muscle fibers have been detected, and it has been speculated that these can influence TM tension, but no relation to ME pressure has been demonstrated (Henson and Henson 2000). Rockley and Hawke (1992) reported increasing thresholds of pressure sensation in response to similar TM anesthesia, when they pressurized the ear canal with an experimental tympanometer. In particular, thresholds were significantly increased in patients with TM pathological changes such as atrophy, myringosclerosis, and retractions. Thus, they concluded that an impaired pressure regulation could be attributed to depletion of neural receptors of the TM.

The actual site where receptors needed for ME pressure regulation are situated is unknown. However, due to its higher elastic properties, the pars flaccida of the TM seems to be an obvious position for detection of pressure changes (Hellström and Stenfors 1983). In fact, the pars flaccida contains numerous myelinated and unmyelinated nerve fibers, whereas the pars tensa contains fewer, mostly unmyelinated ones (Lim 1970). Specialized nerve endings serving as mechanoreceptors in the TM were not identified by Lim (1970). However, structures similar to Vater-Pacinian corpuscles were described in the mucosal strands of the ME and the mastoid of normal temporal bones (Lim et al. 1975). In conclusion, stretch-receptors as well as baroreceptors (mechanoreceptors) may be situated in the TM as well as in the ME cleft. The tympanic nerve forms the major part of the tympanic plexus and innervates the mucosa of the ME, the mastoid, and the ET, and thus it seems to be the afferent pathway for any of these receptors (Özveren et al. 2003).

5.8.2 Recent Evidence of Neural Feedback Control

Several more recent studies seem to support the hypothesis discussed in the preceding text. Ceylan et al. (2007) conducted an animal experiment in rabbits, where sectioning of the tympanic nerve was performed. They found that subsequent retraction pockets evolved over 3 months in 48 % of these animals compared to 4 % in controls. In addition, ME effusion evolved in 56 % compared to 12 % in controls. They concluded that tympanic glomus cells, innervated by the tympanic nerve, were involved in ME aeration, and, consequently, disruption of the nerve resulted in atelectasis. Moreover, it seemed that the ET capacity for clearance of mucus had been impaired, as reflected by the frequent effusion found in the study group (Ceylan et al. 2007).

Later, Songu et al. (2009) conducted clinical experiments in which lidocaine was administered: (1) to the promontory in subjects with a dry TM perforation, (2) to the entire ME cavity through a puncture in the TM, and (3) onto the lateral surface of the intact TM. Changes in ET function were subsequently determined using tympanometry, and automated Williams testing reflecting the opening ability of the ET. In groups 1 and 2, the ET function was significantly impaired, whereas in group 3 normal function was demonstrated. Their study compares with the methods applied by Nagai et al. (1989), though they found impaired ET function in cases with an intact TM. However, their anesthesia protocol was more effective, as they used iontophoresis to paralyze any neural activity, which was not employed by Songu et al. (2009). Hence, results reported by Songu et al. (2009) do not exclude the possible role of mechanoreceptors in the TM. The experiments conducted in groups 1 and 2 show that paralyzing the tympanic plexus results in an impaired ET function. Thus, a functional connection between ME afferents and ET efferents has been demonstrated.

Advances in neurophysiologic techniques include measurements of evoked brain potentials from various stimuli and the application of multichannel EEG recordings. This method includes up to 128 electrodes, where neural activities can be determined in a three-dimensional system, and combined with source localization analysis. Such a technique has been employed in clinical experiments in which evoked brain potentials have been measured in response to static pressure stimulation (3 kPa at 1 Hz) of the TM. Distinct activation of the brain stem was demonstrated with latencies around 5 ms, and with subsequent extension of neural activation to the cerebellum (Sami et al. 2009). Although the exact localization of the brain stem cannot be located to the nucleus of the solitary tract (Eden 1981; Eden and Gannon 1987), in similar control experiments, where acoustic stimulation (white noise) was employed, a separate brain stem activation was found with subsequent extension superiorly into the brain. Subsequent activation of the cerebellum was not possible to demonstrate using previous neural tracer techniques, as these tracers do not extend to second-order neurons (Eden 1981; Eden and Gannon 1987). However, in view of the static pressure experiments, this activation most likely relates to cerebellar nuclei (the fastigial nucleus). This nucleus shows connections to the trigenimal motor nucleus of the brain stem, which relates to activation of the m. tensor veli palatini involved in openings of the ET (Hecht et al. 1993).

Further evidence from the same experiments has been achieved by wavelet analysis, which describes the frequency contents of neural activity. Because the frequency content of different neural systems usually displays distinct characteristics, wavelet analysis can be used to distinguish between them (Darvishi and Al-Ani 2007). Static pressure stimulation of the TM has resulted mainly in θ-band activity (0–4 Hz), while acoustic pressure stimulation results in α-band activity (7–10 Hz). Thus, more distinct neural activation patterns exist for static and acoustic stimulation of the TM. This supports the idea that separate afferent pathways are related to static pressure stimulation (Gaihede et al. 2008; Sami et al. 2009).

Cortical activation in normal humans was recently demonstrated in a functional magnetic resonance imaging (fMRI) study in response to alternate static pressure stimulation of the TM up to 40 daPa (Job et al. 2011). The activation included the postcentral gyrus in Brodmann area 43, which is also involved in pharyngeal activities. Hence, it might represent a link between the ME and the ET activation (Job et al. 2011).

5.8.3 Summary and Future Research

In summary, various clinical, histological, and physiological experiments, as well as fMRI studies, have been employed in search for an overall mechanism of ME pressure regulation. This mechanism seems as obvious as respiratory control, and accordingly it would constitute a continuous process; thus, monitoring MEP directly with an intact TM over longer periods of time seems very important to understand its temporal variation and development of abnormal regulation

(Gaihede et al. 2010). The experiments performed by Eden and Gannon (1987) are important because they linked stimulation of the tympanic nerve directly to ET muscular activity. However, proper stimulation by static pressures seems a prerequisite to confirm the hypothesis, and improved experiments in this line will be valuable.

The role of the peripheral receptors is also important. In the study performed by Rockley and Hawke (1992) subjects were unable to distinguish positive from negative ear canal pressures. Hence, it may be hypothesized that pressure sensation depends on (1) highly sensitive stretch-receptors related to the highly flexible pars flaccida or less probably the pars tensa of the TM; these receptors may not be able to detect direction of pressure and (2) direction-sensitive baroreceptors in the ME cavity and the mastoid. Further research is needed to describe the locations and functions of such receptor components.

The studies reviewed in this section all focus on the role that the ET plays in pressure regulation, while the possible role of the ME and the mastoid mucosa is not mentioned at all. Whether perfusion of the ME and mastoid is involved still remains uncertain, but in principle this may be accomplished by vascular reflexes (Fig. 5.1b). These may constitute efferent autonomic innervation of precapillary arterioles of the mucosa or by local neurotransmitters. It constitutes a separate challenge for future research to demonstrate any link between static pressure changes and the mucosal perfusion.

5.9 Summary

In normal circumstances, the ET is closed and the ME cleft forms a closed gas volume. Because the volume is closed, pressure differences between the ME and the ambient pressures can easily develop on time scales from fractions of seconds to many hours. When such a pressure difference is present, the TM is subject to a mechanical load. If the load is large, it affects hearing ability, but even small pressure loads may have an effect on the TM in the long run. Long-term pressure loads on the TM are involved in the formation of a series of clinical ME conditions, which include otitis media with effusion, retraction, atelectasis, and cholesteatomas. These conditions can lead to degeneration of ME structures and to decreased hearing.

The balance between ME pressure and ambient pressure can be maintained either by changing the amount of gas in the ME cavity, or by changing ME volume. The only ways the volume of the cavity can change is by displacement of the TM, by changes in the congestion of the mucosa, or by fluid secretions. In older literature, it was suggested that the pars flaccida has an important pressure regulative function, but more recent quantitative measurements have shown that this pressure regulative role is marginal. Still, it may be important in buffering very small, rapid pressure changes. Displacement of the entire TM can reduce pressure changes in the order of 20–70 %, depending on the total volume of the ME cleft. Pressure compensation is much larger in a small ME cleft as opposed to a large one.

This finding may be important in the management of ears that show clinical problems related to impaired pressure regulation. Changes in thickness of mucosa also can alter the ME cleft volume. Such changes can happen over long time periods, for instance in case of inflammation, but changes in blood perfusion can in principle alter the mucosa thickness on a short time scale.

Gas can enter or leave the ME cavity in two ways: through the ET or by gas exchange. During actions of swallowing, the ET transfers very small amounts of gas from the nasopharynx to the ME or vice versa. When very large pressure differences between the ME and the nasopharynx develop, perhaps due to a sudden change in ambient pressure, the ET opens spontaneously to equilibrate the difference. Thus, the ET plays a role in pressure regulation, and much research has focused on this aspect. Hitherto, ET function tests do not correlate well with clinical problems; thus the behavior of the ET alone cannot be used to explain them. One possible explanation lies in the additional role of the mastoid, which has recently attracted much interest among researchers.

Gas exchange is an important factor in ME pressure regulation. When there is an imbalance between the partial pressure of the gas in the ME, and its partial pressure in the blood, it will enter or leave the ME cleft through the mucosa. Experiments have shown that this exchange process can go rather fast, and that significant amounts of gas can leave or enter the ME in this way. Whether the gas exchange process of each gas is limited mainly by diffusion or by perfusion is still a point of debate, but it is a debate of high relevance: if the process is limited by the diffusion through the ME mucosa, the gas exchange rate cannot change very quickly. If, however, blood perfusion of the mucosa is the main limiting factor, quick changes in perfusion rate can alter the gas exchange rate on a short time scale. If this is the case, gas exchange could form an essential part in an active pressure regulation mechanism. Experimental evidence indicates that at least CO_2 exchange is limited by perfusion.

Recent results indicate that the overall regulation of ME pressure includes an active neural feedback control based on peripheral mechano-receptors, and, mainly, respiratory brain stem centers. These centers may control the ET activity and also perfusion properties of the ME cleft. If such neural control exists, further research will bring insights into its connection with impaired ME pressure regulation. This may open up entirely new directions for treatment. Differences between ME pressure and ambient pressure are an essential part of everyday life. New insights in ME pressure regulation include future experiments on gas exchange, TM and ME biomechanics, the action of the ET and the mastoid, as well as any neural components. Because there is a clear connection between impaired ME pressure regulation and pathology, ongoing and future fundamental research is very important, as several questions still remain unanswered. This work will bring better insight into the underlying mechanisms, so that the development of pathologic changes can be prevented and treatment strategies improved.

References

Alper, C. M., Kitsko, D. J., Swarts, J. D., Martin, B., Yuksel, S., Cullen Doyle, B. M., Villardo, R. J., & Doyle, W. J. (2011). Role of the mastoid in middle ear pressure regulation. *The Laryngoscope*, 121, 404–408.

Aoki, K., Mitani, Y., Tuji, T., Hamada, Y., Utahashi, H., & Moriyama, H. (1998). Relationship between middle ear pressure, mucosal lesion and mastoid pneumatization. *The Laryngoscope*, 108, 1840–1845.

Ars, B., Decraemer, W., & Ars-Piret, N. (1989). The lamina propria and cholesteatoma. *Clinical Otolaryngology & Allied Sciences*, 14, 471–475.

Ars, B., Wuyts, F., Van de Heyning, P., Miled, I., Bogers, J., & Van Marck, E. (1997). Histomorphometric study of the normal middle ear mucosa. Preliminary results supporting the gas-exchange function in the postero-superior part of the middle ear cleft. *Acta Oto-Laryngologica*, 117, 704–707.

Bluestone, C. D., & Doyle, W. J. (1988). Anatomy and physiology of Eustachian tube and middle ear related to otitis media. *The Journal of Allergy and Clinical Immunology*, 81, 997–1003.

Bonding, P., & Tos, M. (1981). Middle ear pressure during brief pathological conditions of the nose and throat. *Acta Oto-Laryngologica*, 92, 63–69.

Buckingham, R. A., & Ferrer, J. L. (1973). Middle ear pressures in Eustachian tube malfunction: Manometric studies. *The Laryngoscope*, 83, 1585–1593.

Bylander, A. (1986). Pathophysiological aspects on Eustachian tube function and SOM. *Scandinavian Audiology, Supplementum*, 26, 59–63.

Bylander, A., Ivarsson, A., & Tjernström, Ö. (1981). Eustachian tube function in normal children and adults. *Acta Oto-Laryngologica*, 92, 481–491.

Bylander, A. K., Ivarsson, A., Tjernström, Ö., & Andréasson, L. (1985). Middle ear pressure variations during 24 hours in children. *Annals of Otology, Rhinology & Laryngology Supplement*, 120, 33–35.

Ceylan, A., Göksu, N., Kemaloğlu, Y. K., Uğur, B., Akyürek, N., & Bayazit, Y. A. (2007). Impact of Jacobson's (tympanic) nerve sectioning on middle ear functions. *Otology & Neurotology*, 28, 341–344.

Cinamon, U. (2009). The growth rate and size of the mastoid air cell system and mastoid bone: A review and reference. *European Archives of Oto-Rhino-Laryngology*, 266, 781–786.

Cinamon, U., & Sadé, J. (2003). Mastoid and tympanic membrane as pressure buffers: A quantitative study in a middle ear cleft model. *Otology & Neurotology*, 24, 839–842.

Csakanyi, Z., Katona, G. Josvai, E., Mohos, F, & Sziklai, I. (2010). Volume and surface of the mastoid cell system in otitis media with effusion in children: A case-control study by three-dimensional reconstruction of computed tomographic images. *Otology & Neurotology*, 32, 64–70.

Darvishi, S., & Al-Ani, A. (2007). Brain-computer interface analysis using continuous wavelet transform and adaptive neuro-fuzzy classifier. *Conference Proceedings—IEEE Engineering in Medicine and Biology Society*, 1, 3220–3223.

Decraemer W. F., & Dirckx, J. J. J. (1998). Pressure regulation due to the tympanic membrane pars flaccida and pars tensa displacements. *Acta Otorhinolaryngolica Nova*, 8, 277–281.

Decraemer, W., & Funnell, W. (2008). Anatomical and mechanical properties of the tympanic membrane. In B. Ars (Ed.) *Chronic otitis media: Pathogenesis-oriented therapeutic management* (pp. 51–84). The Hague, The Netherlands: Kugler Publications.

Decraemer, W. F., Creten, W. L., & Van Camp, K. J. (1984). Tympanometric middle ear pressure determination with two-component admittance meters. *Scandinavian Audiology*, 13, 165–172.

Diamant, M., Rubensohn, G., & Wallander, A. (1958). Otosalpingitis and mastoid pneumatization. *Acta Oto-Laryngologica*, 49, 381–389.

Dirckx, J. J. J., & Decraemer, W. F. (1991). Human tympanic membrane deformation under static pressure. *Hearing Research*, 51, 93–106.

Dirckx, J. J. J., & Decraemer, W. F. (1992). Area change and volume displacement of the human tympanic membrane under static pressure. *Hearing Research*, 62, 99–104.

Dirckx, J. J. J., & Decaemer, W. F. (1998). Contribution of pars flaccida volume displacement to middle ear pressure regulation. *Acta Otorhinolaryngolica Nova*, 8, 269–272.

Dirckx, J. J. J., Decraemer, W. F., von Unge, M., & Larsson, C. (1997). Measurement and modeling of boundary shape and surface deformation of the Mongolian gerbil pars flaccida. *Hearing Research*, 111, 153–164.

Dirckx, J. J. J., Decraemer, W. F., von Unge, M., & Larsson, C. (1998). Volume displacement of the gerbil eardrum pars flaccida as a function of middle ear pressure. *Hearing Research*, 118, 35–46.

Doyle, W. J. (2000). Experimental results do not support a gas reserve function for the mastoid. *International Journal of Pediatric Otorhinolaryngology*, 52, 229–238.

Doyle, W. J. (2007). The mastoid as a functional rate-limiter of middle ear pressure change. *International Journal of Pediatric Otorhinolaryngology*, 71, 393–402.

Doyle, W. J., & Seroky, J. T. (1994). Middle ear gas exchange in Rhesus monkeys. *The Annals of Otology, Rhinology & Laryngology*, 103(8), 636–645.

Doyle, W. J., Seroky, J. T., & Alper, C. M. (1995). Gas exchange across the middle ear mucosa in monkeys: Estimation of exchange rate. *Archives of Otolaryngology – Head & Neck Surgery*, 121, 887–892.

Doyle, W. J., Alper, C. M., & Seroky, J. T. (1999). Trans-mucosal inert gas exchange constants for the monkey middle ear. *Auris Nasus Larynx*, 26, 5–12.

Edelstein, D. R., & Parisier, S. C. (1989). Surgical techniques and recidivism in cholesteatoma. *Otolaryngologic Clinics of North America*, 22, 1029–1040.

Eden, A. R. (1981). Neural connections between the middle ear, Eustachian tube and brain: Implications for the reflex control of middle ear aeration. *Annals of Otology, Rhinology & Laryngology*, 90, 566–569.

Eden, A. R., & Gannon, P. J. (1987). Neural control of middle ear aeration. *Archives of Otolaryngology – Head & Neck Surg*ery, 113, 133–137.

Eden, A. R., Laitman, J. T., & Gannon, P. J. (1990). Mechanisms of middle ear aeration: Anatomic and physiologic evidence in primates. *The Laryngoscope*, 100, 67–75.

Elner, Å. (1970). Gas diffusion through the tympanic membrane: A model study in the diffusion chamber. *Acta Oto-Laryngologica*, 69, 185–191.

Elner, Å. (1972). Indirect determination of gas absorption from the middle ear. *Acta Oto-Laryngologica*, 74, 191–196.

Elner, Å. (1977). Quantitative studies of gas absorption from the normal middle ear. *Acta Oto-Laryngologica*, 83, 25–28.

Elner, Å., Ingelstedt, S., & Ivarsson, A. (1971a). A method for studies of the middle ear mechanics. *Acta Oto-Laryngologica*, 72, 191–200.

Elner, Å., Ingelstedt, S., & Ivarsson, A. (1971b). The normal function of the Eustachian tube. *Acta Oto-Laryngologica*, 72, 320–328.

Fay, J. P., Puria, S, & Steele, C. R. (2006). The discordant eardrum. *Proceedings of the National Academy of Sciences of the USA*, 103, 19743–19748.

Felding, J. U. (1998). Middle ear gas – its composition in the normal and in the tubulated ear. *Acta Oto-Laryngologica, Supplementum*, 536, 1–57.

Fleischer, G. (2010). Components for protecting and optimizing good hearing. In G. Theile (Ed.), *A contribution to the international convention of sound designers. Proceedings of the VDT 26th International Convention* (pp. 250–265). Bergisch-Gladbach, Germany.

Flisberg, K., Ingelstedt, S., & Örtegren, U. (1963). On middle ear pressure. *Acta Oto-Laryngologica, Supplementum*, 182, 43–56.

Gaihede, M. (2000). Middle ear volume and pressure effects on tympanometric middle ear pressure determination: Model experiments with special reference to secretory otitis media. *Auris Nasus Larynx*, 27, 231–239.

Gaihede, M., & Kjær, D. (1998). Positional changes and stabilization of middle ear pressure. *Auris Nasus Larynx*, 25, 255–259.
Gaihede, M., & Kabel, J. (2000). The normal pressure-volume relationship of the middle ear system and its biological variation. In J. J. Rosowski, S. N. Merchant (Eds.), *The function and mechanics of normal, diseased and reconstructed middle ears. Proceedings of the 2nd International Symposium on Middle Ear Mechanics in Research and Otosurgery* (pp. 59–70). Amsterdam: Kugler Publications.
Gaihede, M., Lildholdt, T., & Lunding, J. (1997). Sequelae of secretory otitis media: Changes in middle ear mechanics. *Acta Oto-Laryngologica*, 117, 382–389.
Gaihede, M., Bramstoft, M., Thomsen, L. T., & Fogh, A. (2005). Accuracy of tympanometric middle ear pressure determination in otitis media with effusion: Dose dependent overestimation related to the viscosity and amount of effusion. *Otology & Neurotology*, 26, 5–11.
Gaihede, M., Hald, K., Nørgaard, M., Wogelius, P., Buck, D., & Tveterås, K. (2007). Epidemiology of pressure regulation: Incidence of ventilation tube treatments and its preliminary correlation to subsequent ear surgery. In A. Huber & A. Eiber (Eds.), *Proceedings of the 4th Symposium on Middle Ear Mechanics in Research and Otology* (pp. 314–321). Singapore: World Scientific.
Gaihede, M., Sami, S., & Dirckx J. J. J. (2008). Middle ear cleft pressure regulation – the role of a central control system. In B. Ars (Ed.), *Chronic otitis media: Pathogenesis-oriented therapeutic management* (pp. 227–240). Amsterdam: Kugler Publications.
Gaihede, M., Dirckx, J. J. J., Jacobsen, H., Aernouts, J. E. F., Søvsø, M., & Tveterås, K. (2010). Middle ear pressure regulation – Complementary active actions of the mastoid and the Eustachian tube. *Otology & Neurotology*, 31, 603–611.
Grøntved, A., Krogh, H. J., Christensen, P. H., Jensen, P. O., Schousboe, H. H., & Hentzer, E. (1989). Monitoring middle ear pressure by tympanometry: A study of middle ear pressure variation through seven hours. *Acta Oto-Laryngologica*, 108, 101–106.
Harell, M., Mover-Lev, H., Levy, D., & Sadé, J. (1996). Gas composition of the human nose and nasopharyngeal space. *Acta Oto-Laryngologica*, 116, 82–84.
Hecht, C. S., Gannon, P. J., & Eden, A. R. (1993). Motor innervation of the Eustachian tube muscles in the guinea pig. *The Laryngoscope*, 103, 1218–1226.
Hellström, S., & Stenfors, L. E. (1983). The pressure equilibrating function of the pars flaccida in middle ear mechanics. *Acta Physiologica Scandinavica*, 118, 337–341.
Henson, O. W., & Henson, M. M. (2000). The tympanic membrane: Highly developed smooth muscle arrays in the annulus fibrosus of mustached bats. *Journal of the Association for Research in Otololaryngology*, 1, 1–25.
Hentzer, E. (1970). Histologic studies of the normal mucosa in the middle ear, mastoid cavities and Eustachian tube. *Annals of Otology, Rhinology & Laryngology*, 79, 825–833.
Hergils, L., & Magnuson, B. (1990). Human middle ear gas composition studied by mass spectrometry. *Acta Oto-Laryngologica*, 110, 92–99.
Hergils, L., & Magnuson, B. (1998). Nasal gas composition in humans and its implication on middle ear pressure. *Acta Oto-Laryngologica*, 118, 697–700.
Hergils, L. G., Magnuson, B., & Falk, B. (1990). Different tympanometric procedures compared with direct pressure measurements in healthy ears. *Scandinavian Audiology*, 19, 183–186.
Holmquist, J. (1978). Aeration in chronic otitis media. *Clinical Otolaryngology*, 3, 279–284.
Honjo, I., Ushiro, K., Nozoe, T., & Okazaki, N. (1983). Cineroentgenographic and electromyographic studies of Eustachian tube function. *Archives of Oto-Rhino-Laryngology*, 238, 63–67.
Ingelstedt, S., & Örtegren, U. (1963). Qualitative testing of the Eustachian tube function. *Acta Oto-Laryngologica, Supplementum*, 182, 7–23.
Ingelstedt, S., & Jonson, B. (1966). Mechanisms of the gas exchange in the normal human middle ear. *Acta Oto-Laryngologica, Supplementum*, 224, 452–461.
Ingelstedt, S., Ivarsson, A., & Jonson, B. (1967). Mechanics of the human middle ear. *Acta Oto-Laryngologica, Supplementum*, 228, 1–58.

Jacobsen, H., Dirckx, J. J. J., Gaihede, M., & Tveterås, K. (2007). Direct measurements and monitoring of middle ear pressure. In A. Eiber & A. Huber (Eds.), *Middle ear mechanics in research and otology* (pp. 26–35). Singapore: World Scientific.

Job, A., Paucod, J.-C., O'Beirne, G. A., & Delon-Martin, C. (2011). Cortical representation of tympanic membrane movements due to pressure variation: An fMRI study. *Human Brain Mapping*, 32, 744–749.

Kania, R., Portier, F., Lecain, E., Marcusohn, Y., Ar, A., Herman, P., & Tran Ba Huy, P. (2004). Experimental model for investigating trans-mucosal gas exchanges in the middle ear of the rat. *Acta Oto-Laryngologica*, 124, 408–410.

Kania, R. E., Herman, P., Tran Ba Huy, P., & Ar, A. (2006). Role of nitrogen in transmucosal gas exchange rate in the rat middle ear. *Journal of Applied Physiology*, 101, 1281–1287.

Kindermann, C. A., Roithmann, R., & Lubianca Neto, J. F. (2008). Obstruction of the Eustachian tube orifice and pressure changes in the middle ear: Are they correlated? *Annals of Otology, Rhinology & Laryngology*, 117(6), 425–429.

Lim, D. J. (1970). Human tympanic membrane: An ultrastructural observation. *Acta Oto-laryngologica*, 70, 176–186.

Lim, D., Jackson, D., & Bennett, J. (1975). Human middle ear corpuscles: A light and electron microscopic study. *The Laryngoscope*, 85, 1725–1737.

Loring, S. H., & Butler, J. P. (1987). Gas exchange in body cavities. In A. P. Fishman, L. E. Farhi, S. M. Tenny, & S. R. Geiger (Eds.), *Handbook of physiology*. Section 3: *The respiratory system*, Vol. IV. *Gas exchange* (pp. 283–295). Baltimore: American Physiological Society.

Luntz, M., & Sadé, J. (1988). Growth of the Eustachian tube lumen with age. *American Journal of Otolaryngology*, 9, 195–198.

Luntz, M., Fuchs, C. & Sadé, J. (1997). Correlation between retractions of the pars flaccida and the pars tensa. *The Journal of Laryngology and Otology*, 111, 322–324.

Magnuson, B. (2003). Functions of the mastoid cell system: Auto-regulation of temperature and gas pressure. *The Journal of Laryngology and Otology*, 117, 99–103.

Magnuson, B., & Falk, B. (1988). Physiology of the Eustachian tube and middle ear pressure regulation. In A. F. Jahn & J. Santos-Sacchi (Eds.), *Physiology of the ear* (pp. 81–100). New York: Raven Press.

Marcusohn, Y., Dirckx, J. J. J., & Ar, A. (2006). High-resolution measurements of middle ear gas volume changes in the rabbit enables estimation of its mucosal CO_2 conductance. *Journal of the Association for Research in Otolaryngology*, 7, 236–245.

Marcusohn, Y., Ar, A., & Dirckx, J. J. J. (2010). Perfusion and diffusion limitations in middle ear gas exchange: The exchange of CO_2 as a test case. *Hearing Research*, 265, 11–14.

Metz, O. (1946). *The acoustic impedance measured on normal and pathological ears*. Copenhagen: Einar Munksgaard.

Mondain, M., Vidal, D., Bouhanna, S., & Uziel, A. (1997). Monitoring eustachian tube opening: Preliminary results in normal subjects. *The Laryngoscope*, 107(10), 1414–1419.

Mover-Lev, H., Priner-Barenholz, R., Ar, A., & Sadé, J. (1998). Quantitative analysis of gas losses and gains in the middle ear. *Respiration Physiology*, 114, 143–151.

Nagai, T., & Tono, T. (1989). Encapsulated nerve corpuscles in the human tympanic membrane. *Archives of Oto-Rhino-Laryngology*, 246, 169–172.

Nagai, T., Nagai, M., Nagata, Y., & Morimitsu, T. (1989). The effects of anesthesia of the tympanic membrane on the Eustachian tube function. *Archives of Oto-Rhino-Laryngology*, 246, 210–212.

Ostfeld, E., Blonder, J., Crispin, M., & Szeinberg, A. (1980). The middle ear gas composition in air-ventilated dogs. *Acta Oto-Laryngologica*, 89, 105–108.

Özveren, M. F., Türe, U., Özek, M. M., & Pamir, M. N. (2003). Anatomic landmarks of the glossopharyngeal nerve: A microsurgical anatomic study. *Neurosurgery*, 52, 1400– 1410.

Park, M. S., Yoo, S. H., & Hoon, D. H. (2000). Measurement of surface area in the human mastoid air cell system. *The Journal of Laryngology and Otology*, 114, 93–96.

Piiper, J. (1965). Physiological equilibria of gas cavities in the body. In W. O. Fenn & H. Rahn (Eds.), *Handbook of physiology*. Section 3: *Respiration*, Vol. II (pp. 1205–1218). Washington, DC: American Physiological Society.

Piiper, J., Canfield, R. E., & Rahn, H. (1962). Absorption of various inert gases from subcutaneous gas pockets in rats. *Journal of Applied Physiology*, 17(2), 268–274.

Politzer, A. (1867). Diagnose und Therapie der Ansammlung seröser Flüssigkeit in der Trommelhöhle. *Wien Medicinishe Wochenschrift*, 17, 244–247.

Prades, J. M., Dumollard, J. M., Calloc'h, F., Merzougui, N., Veyret, C., & Martin, C. (1998). Descriptive anatomy of the human auditory tube. *Surgical and Radiologic Anatomy*, 20(5), 335–340.

Pulec, J. L., & Hahn, F. W., Jr. (1970). The abnormally patulous eustachian tube. *Otolaryngologic Clinics of North America*, 3(1), 131–140.

Riu, R., Flottes, L., Bouche, J., & Le Den, R. (1966). *La physiologie de la trompe d'Eustache*. Paris : Librairie Arnette.

Rockley, T. J., & Hawke, W. M. (1992). The middle ear as a baroreceptor. *Acta Oto-Laryngologica*, 112, 816–823.

Sadé, J. (1992). The correlation of middle ear aeration with mastoid pneumatization. The mastoid as a pressure buffer. *European Archives of Oto-Rhino-Laryngology*, 249, 301–304.

Sadé, J. (1993). Atelectatic tympanic membrane: Histologic study. *The Annals of Otology, Rhinology & Laryngology*, 102, 712–716.

Sadé, J. (1997). On the function of the pars flaccida: Retraction of the pars flaccida and buffering of negative middle ear pressure. *Acta Oto-Laryngologica*, 117, 289–292.

Sadé, J. (2001). Hyperectasis: The hyperinflated tympanic membrane: The middle ear as an actively controlled system. *Otology & Neurotology*, 22, 133–139.

Sadé, J., & Luntz, M. (1993). Dynamic measurement of gas composition in the middle ear. II Steady state values. *Acta Oto-Laryngologica*, 113, 353–357.

Sadé, J., & Ar, A. (1997). Middle ear and auditory tube: Middle ear clearance, gas exchange, and pressure regulation. *Otolaryngology – Head & Neck Surgery*, 116, 499–524.

Sadé, J., Halevy, A., & Hadas, E. (1976). Clearance of middle ear effusions and middle ear pressures. *Annals of Otology, Rhinology & Laryngology, Supplement*, 25, 58–62.

Sadé, J., Luntz, M., & Levy, D. (1995). Middle ear gas composition and middle ear aeration. *Annals of Otology, Rhinology & Laryngology*, 104(5), 369–373.

Sadé, J., Russo, E., Fuchs, C., & Cohen, D. (2003). Is secretory otitis media a single disease entity? *Annals of Otology, Rhinology & Laryngology*, 112, 342–347.

Sadé, J., Ar, A., & Marcusohn, Y. (2005). The hyperectatic outwardly-ballooned tympanic membrane as an indicator of an active mechanism of middle ear aeration. In J. J. J. Dirckx & J. Sadé (Eds.), *Middle ear pressure regulation: Basic research and clinical observations. Otology & Neurotology*, 26, 300–309.

Sami, S. A. K., Gaihede, M., Nielsen, L. G., & Drewes, A. M. (2009). Early static pressure related evoked brain potentials. Indications of central middle ear pressure control in humans. *Otology & Neurotology*, 30, 649–656.

Shanks, J., & Shelton, C. (1991). Basic principles and clinical applications of tympanometry. *Otolaryngologic Clinics of North America*, 24, 299–328.

Shrapnell, H. J. (1832). On the form and structure of the membrane tympani. *London Medical Gazette*, 10, 120–124.

Songu, M., Aslan, A., Unlu, H. H., & Celik, O. (2009). Neural control of Eustachian tube function. *The Laryngoscope*, 119, 1198–1202.

Stenfors, L. E., Stalen, B., & Windblad, B. (1979). The role of pars flaccida in the mechanics of the middle ear. *Acta Oto-Laryngologica*, 88, 395–400.

Stenfors, L. E., Sadé, J., Helstrom, S., & Anniko, M. (2001). How can the hooded seal dive to a depth of 100 m without rupturing his tympanic membrane? A morphological and functional study. *Acta Oto-Laryngologica*, 121, 689–695.

Strachan, D., Hope, G., & Hussain, M. (1996). Long-term follow-up of children inserted with T-tubes as a primary procedure for otitis media with effusion. *Clinical Otolaryngology & Allied Sciences*, 21, 537–541.

Sudhoff, H., & Tos, M. (2000). Pathogenesis of attic cholesteatoma: Clinical and immunohistochemical support for combination of the retraction and proliferation theory. *The American Journal of Otology*, 21, 786–782.

Swarts, J. D., Doyle, B. M. C., Alper, C. M., & Doyle, W. J. (2010). Surface area-volume relationships for the mastoid air cell system and tympanum in adult humans: Implications for mastoid function. *Acta Oto-Laryngologica*, 130(11), 1230–1236.

Swarts, J. D., Alper, C. M., Mandel, E. M., Villardo, R., & Doyle, W. J. (2011). Eustachian tube function in adults without middle ear disease. *Annals of Otology, Rhinology and Laryngology*, 120(4), 220–225.

Takahashi, H. (2001). *The middle ear. The role of ventilation in disease and surgery*. Tokyo: Springer.

Takahashi, H., Hayashi, M., & Honjo, I. (1987a). Direct measurement of middle ear pressure through the Eustachian tube. *Archives of Oto-Rhino-Laryngology*, 243, 378–381.

Takahashi, H., Hayashi, M., & Honjo, I. (1987b). Compliance of the Eustachian tube in patients with otitis media with effusion. *American Journal of Otolaryngology*, 8(3), 154–156.

Takahashi, H., Iwanaga, T., Kaieda, S., Fukuda, T., Kumagami, H., Takasaki, K., Hasebe, S., & Funabiki, K. (2007). Mastoid obliteration combined with soft-wall reconstruction of posterior ear canal. *European Archives of Oto-Rhino-Laryngology*, 264, 867–871.

Therkildsen, A. G., & Gaihede, M. (2005). Accuracy of tympanometric middle ear pressure Determination: The role of direction and rate of pressure change with a fast modern tympanometer. *Otology & Neurotology*, 26, 252–256.

Thomsen, K. A. (1960). Investigations on the tubal function and measurements of the middle ear pressure in pressure chamber. *Acta Oto-Laryngologica, Supplementum*, 140, 269–278.

Tideholm, B., Jönsson, S., Carlborg, B., Welinder, R., & Grenner, J. (1996). Continuous 24–hour measurement of middle ear pressure. *Acta Oto-Laryngologica*, 116, 581–588.

Tideholm, B., Carlborg, B., Jönsson, S., & Bylander-Groth, A. (1998). Continuous long-term measurements of the middle ear pressure in subjects without a history of ear disease. *Acta Oto-Laryngologica*, 118, 369–374.

Tideholm, B., Brattmo, M., & Carlborg, B. (1999a). Middle ear pressure: Effect of body position and sleep. *Acta Oto-Laryngologica*, 119, 880–885.

Tideholm, B., Carlborg, B., & Brattmo, M. (1999b). Continuous long-term measurements of the middle ear pressure in subjects with symptoms of patulous Eustachian tube. *Acta Oto-Laryngologica*, 119, 809–815.

Tos, M., & Stangerup, S. E. (1984). Mastoid pneumatization in secretory otitis. Further support for the environmental theory. *Acta Oto-Laryngologica*, 98, 110–118.

Tos, M., Stangerup, S. E., Holm-Jensen, S., & Sørensen, C. H. (1984). Spontaneous course of secretory otitis and changes of the eardrum. *Archives of Otolaryngology*, 110, 281–289.

Uchimizu, H., Utahashi, H., Hamada, Y., & Aoki, K. (2005). Middle ear total pressure measurement as a useful parameter for outcome prediction in pediatric otitis media with effusion. *International Journal of Pediatric Otorhinolaryngology*, 69(12), 1659–1665.

Van Liew, H. D. (1962). Oxygen and carbon dioxide permeability of subcutaneous pockets. *The American Journal of Physiology*, 202(1), 53–58.

Vercruysse, J. P., De Foer, B., Somers, T., Casselman, J. W., & Offeciers, E. (2008). Mastoid and epitympanic bony obliteration in pediatric cholesteatoma. *Otology & Neurotology*, 29, 953–960.

von Unge, M., Decraemer, W. F., Dirckx, J. J. J., & Bagger-Sjöbäck, D. (1999). Tympanic membrane displacement patterns in experimental cholasteatoma. *Hearing Research*, 128, 1–15.

Yuksel, S., Swarts, J. D., Banks, J., Seroky, J. T., & Doyle, W. J. (2009). In vivo measurement of O_2 and CO_2 gas exchange across the human tympanic membrane. *Acta Oto-Laryngologica*, 129, 716–725.

Zahnert, T., Hüttenbrink, K. B., Mürbe, D., & Bornitz, M. (2000). Experimental investigations of the use of cartilage in tympanic membrane reconstruction. *The American Journal of Otology*, 21, 322–328.

Zielhuis, G. A., Rach, G. H., & van den Broek, P. (1990). The occurrence of otitis media with effusion in Dutch pre-school children. *Clinical Otolaryngology & Allied Sciences*, 15, 147–153.

Chapter 6
Bone Conduction and the Middle Ear

Stefan Stenfelt

Keywords Audiometry • Bone conduction • Bone conduction hearing aid • Carhart notch • Fluid • Inertia • Middle ear • Occlusion effect • Otosclerosis • Own voice • Skull vibration • Third window • Transcranial attenuation • Wave transmission

6.1 Introduction to the Field of Bone Conduction

There is no general definition of what is meant by bone conducted (BC) sound, but it is often understood as the way vibration of the skull bones can result in a sound percept. However, BC sound usually involves transmission in cartilage and soft tissues, for example, in normal BC audiometry in which the BC transducer is positioned by a static force at the skin-covered mastoid. Even if the BC vibration is transmitted from the transducer to the skull bone, the soft tissue is involved in the transmission. It therefore has been argued sometimes that "body conduction" describes the sound transmission better, or that BC and soft tissue conduction should be separated. It is not clear how to separate the different modes of transmission and how fluid conduction should be included. Here, BC (bone conduction) is used to describe sound energy that is transmitted through the body (in bone or soft tissue) and that it involves the outer, middle, or inner ear to finally produce a perception of sound.

Even before the concept of BC sound was understood, BC sound was used to separate a sensory impairment from a conduction impairment. With the introduction of tuning forks, tests were developed during the nineteenth century to diagnose a hearing impairment as either sensory or conductive, for example, the Weber test

S. Stenfelt (✉)
Department of Clinical and Experimental Medicine/Technical Audiology, Linköping University, 58185 Linköping, Sweden
e-mail: stefan.stenfelt@liu.se

S. Puria et al. (eds.), *The Middle Ear: Science, Otosurgery, and Technology*, Springer Handbook of Auditory Research 46, DOI 10.1007/978-1-4614-6591-1_6,

or Bing test. After the electric audiometer was introduced in the twentieth century, these tests were superfluous because they gave unreliable results versus a comparison of air-conducted (AC) and BC hearing thresholds (Miltenburg 1994; Behn et al. 2007). The increased usage of BC hearing aids in recent times has also increased the need for understanding the processes underlying perception of BC sound.

6.1.1 Basic Inner Ear Processes with Bone-Conducted Sound

One of the quintessential questions about BC hearing is the end organ for transforming BC vibration in the skull to neural code. The first to show BC vibration leading to a basilar membrane motion in the cochlea was von Békésy when he cancelled the perception of a 400-Hz BC tone with an AC tone of the same frequency in a human subject (von Békésy 1932, 1960). Similar experiments in human subjects were later conducted at several stimulation levels (Khanna et al. 1976) and using multiple tones simultaneously (Stenfelt 2007). It has also been used over a broad frequency range in animal experiments cancelling the cochlear microphonics when stimulating with an AC and BC tone at the same frequency (Lowy 1942; Wever and Lawrence 1954).

More direct measures of basilar membrane motion were performed in human temporal bone specimens in which the relative motion between a position on the basilar membrane and the surrounding bone was measured while shaking the whole specimen (Stenfelt et al. 2003b). This is not conclusive evidence that BC sound has the same end organ as AC sound but the study indicated that the maximum vibration amplitude for a specific position on the basilar membrane appeared at the same frequency regardless if stimulation was provided as AC or BC. Moreover, a modeling approach for understanding basilar membrane response with BC stimulation showed results similar to those in the temporal bone specimen study (Stenfelt and Puria 2010; Kim et al. 2011).

Other evidence for the cochlea and basilar membrane motion as the mechanism for BC sound perception is electrophysiological measures of AC and BC sound. It has been shown that BC electrocochleography (BC-ECoG) correlates well with behavioral hearing thresholds (Kylen et al. 1982). Moreover, BC-evoked brain stem response (BC-ABR) disappears subsequent to masking by an AC source showing that the evoked potentials from the BC stimulation are purely auditory (Collet et al. 1989). However, there is a difference in the latency-intensity function for Jewett wave V between click evoked BC-ABR and AC-ABR (Beattie 1998). This difference can be explained by the difference in spectral content of the stimuli due to the filtering effect of the AC and BC transducers and is not an inherent difference related to AC and BC sound perception in the human (Schwartz et al. 1985). A further indication of the same end organ for AC and BC sound is the ability to produce distortion-product otoacoustic emissions (DPOAE) with AC and BC stimulation (Purcell et al. 1998; Watanabe et al. 2008).

One documented difference between AC and BC sound perception is the ability to perceive BC ultrasonic sound (20–120 kHz); when modulated it can be used for speech detection in profoundly deaf individuals (Lenhardt et al. 1991; Hosoi et al. 1998). The mechanisms for ultrasonic BC perception are not clarified and a number of possible explanations have been provided. One such possibility is the demodulation of the ultrasound due to nonlinearities of BC transmission in the skull itself (Haeff and Knox 1963) or to nonlinear processing in the cochlea, for example, due to processing of the inner hair cells (Nishimura et al. 2011).

6.1.2 Human and Animal Studies

Although animal studies are important for understanding processes involved in hearing, BC data are obtained mainly in humans. One reason for this is that perception of BC sound is strongly dependent on the specific anatomy of the skull and ear. For example, most animal heads differ in composition and geometry compared with the human head. Large interspecies differences also exist. Moreover, the cochlea is positioned in hard dense bone in the skull base in the human whereas it is protruding in the air-filled bulla in some animals (e.g., guinea pigs). Such differences make it difficult to extrapolate findings in animal models to human BC sound perception. In addition, it is far easier to perform psychoacoustic experiments using BC stimulation on humans than it is on animals.

There are only a few studies on animal hearing thresholds when stimulation is by BC. One such study is in dogs, in which brain stem responses indicated that the BC hearing thresholds were similar to those for humans (Munro et al. 1997). However, the measurements were not conducted in a controlled environment, and surrounding noise may have affected the measurement negatively. Tonndorf used primarily cats for his experiments, explaining underlying mechanisms for BC sound perception (Tonndorf 1966). Animals have also been used for artificial middle ear manipulations (stapes or tympanic membrane fixation, mass loading, removal of the ossicular chain) to explore the importance of the middle and outer ear for BC sound perception or BC sensitivity alterations in middle ear disease (Legouix and Tarab 1959; Tonndorf 1966; Irvine et al. 1979).

6.2 Bone Conduction Wave Transmission in the Skull Bone and Soft Tissue

One aspect for understanding BC sound perception in the human skull is how BC sound is transmitted in the skull bone. The vibration response of the human head is complex, involving the thin, sphere-like cranial vault, as well as the more dense bone in the skull base, both types loaded with soft tissue and fluids. Moreover, the thin bone in the cranial vault is not homogeneous but comprises hard shell-like structures with fluid-filled matrix-like bone structures (diploae) in between. Also, the skull consists of many bones connected by sutures. Consequently, such complicated structures, both in terms of geometry and composition, make analytical approaches difficult. Even so, researchers have attempted to formulate analytical or finite element computations to achieve the vibration pattern of the human skull (Advani and Lee 1970; Khalil and Hubbard 1977; Young 2002, 2003). These theoretical approaches were intended primarily for head injury protection and not for the transmission and hearing of BC sound.

An early attempt to analyze the response pattern of the human skull during BC stimulation consisted of approximating the vibration mode to that of a vibrating thin-shell sphere (von Békésy 1932). Another approach to investigate the human skull vibration pattern is modal analyses in which the skull resonance frequencies are extracted. Such a method was used in dry skulls (Franke 1956; Khalil et al. 1979; Stenfelt et al. 2000) and living humans (Håkansson et al. 1994). From these studies it is clear that dry skulls do not represent the true response in living humans. For the dry skulls, the first resonance frequency was in one skull at 1.2 kHz (Stenfelt et al. 2000) whereas in another study it was at 1.4 kHz for a male skull and 1.6 kHz for a female skull (Khalil et al. 1979). In six living human skulls, 14–19 resonance frequencies were identified at frequencies below 7.5 kHz; the average of the lowest two were 0.97 and 1.23 kHz (Håkansson et al. 1994). There were no obvious relations between resonance frequencies and head size; other parameters such as stiffness and thickness of the bone may influence the frequencies of the resonances (Håkansson et al. 1986). However, even if the resonance frequencies are important to determine the mechanical characteristic of the human skull, their effect on BC hearing is minor owing to high damping (Håkansson et al. 1994, 1996; Stenfelt and Goode 2005b).

Skull bone transmission of BC sound has been suggested to produce nonlinear distortion (Khanna et al. 1976; Arlinger et al. 1978). In a study to investigate nonlinearities of BC sound in the living human skull, transcranial BC transmission was measured in subjects with skin-penetrating fixtures ensuring rigid connection to the skull bone (Håkansson et al. 1996): they report skull bone transmission of BC sound to be linear up to at least 77 dB HL at frequencies between 0.1 and 10 kHz. In an investigation of skull vibration in cadaver heads, no indications of nonlinear distortion caused by the skull bone were detected at levels corresponding to 80–100 dB HL (Stenfelt and Goode 2005b). Investigations of the human mechanical point impedance have also suggested skull bone transmission to be linear at hearing frequencies and levels (Flottorp and Solberg 1976; Khalil et al. 1979; Håkansson et al. 1986).

6.2.1 Vibration Transmission to the Cochlea

The earlier investigations on BC sound transmission in the human skull focused on the vibration pattern of the cranial vault, either as whole head vibrations (von Békésy 1932; Ogura et al. 1979; Hoyer and Dörheide 1983) or as transcranial transmission measurements (Håkansson et al. 1994). This is not the same as the vibration pattern of the cochlea, and more recent investigations have studied the three-dimensional cochlear vibration during BC stimulation at the skull surface in a damped dry skull (Stenfelt et al. 2000) and in cadaver heads (Stenfelt and Goode 2005b), or as one-dimensional cochlear vibration in cadaver heads (Eeg-Olofsson et al. 2008, 2011) or living humans (Eeg-Olofsson et al. 2013). These studies indicated that the vibration pattern for the human skull can be categorized into four regions (for frequencies below 10 kHz, see Fig. 6.1). At the lowest frequencies, below the resonance frequency of the mechanical point impedance (150–400 Hz; Stenfelt and Goode 2005b), the skull moves as a rigid body (Fig. 6.1a) and above this resonance frequency and up to approximately 1 kHz where the first free resonance of the skull appears (Håkansson et al. 1994), the motion can be described as a mass-spring system wherein large parts of the skull move in phase and in the direction of the stimulation (Fig. 6.1b). This also means that at these frequencies, bilateral stimulation is primarily added in-phase or out-of-phase depending on the stimulation direction (Deas et al. 2010; Eeg-Olofsson et al. 2011).

At frequencies above 1 kHz, the wavelength of the BC sound is short enough to facilitate wave transmission in the skull bone. Between 1 and 2 kHz, the skull transitions from a mass-spring system to a system dominated by wave transmission, and at frequencies above 2 kHz, wave transmission dominates the skull vibration pattern of the skull (Fig. 6.1c). At these high frequencies, the vibration response at the cochlea is more or less independent of the stimulation direction; the response vibration is in all three space dimensions without any dominating direction (Stenfelt and Goode 2005b). The types of vibration in the cranial vault have been suggested to occur as plate waves constituting both longitudinal and transverse components (Tonndorf and Jahn 1981). More recently, the vibration in the cranial vault was separated from the skull base, enabling separate analysis of the two. At frequencies above 2 kHz, the phase velocity at the skull base was almost constant at 400 m/s whereas it increased with frequency at the cranial vault (250 m/s at 2 kHz to 300 m/s at 10 kHz). This suggests that the sound transmission at the thicker skull base is dominated by longitudinal wave motion whereas a mixture of modes including bending wave motion is present in the thinner cranial vault (Stenfelt and Goode 2005b) (Fig. 6.1c). Others who have investigated the group and phase velocity of BC sound in the human head have reported it to be between 260 and 540 m/s (von Békésy 1948; Zwislocki 1953; Franke 1956; Tonndorf and Jahn 1981).

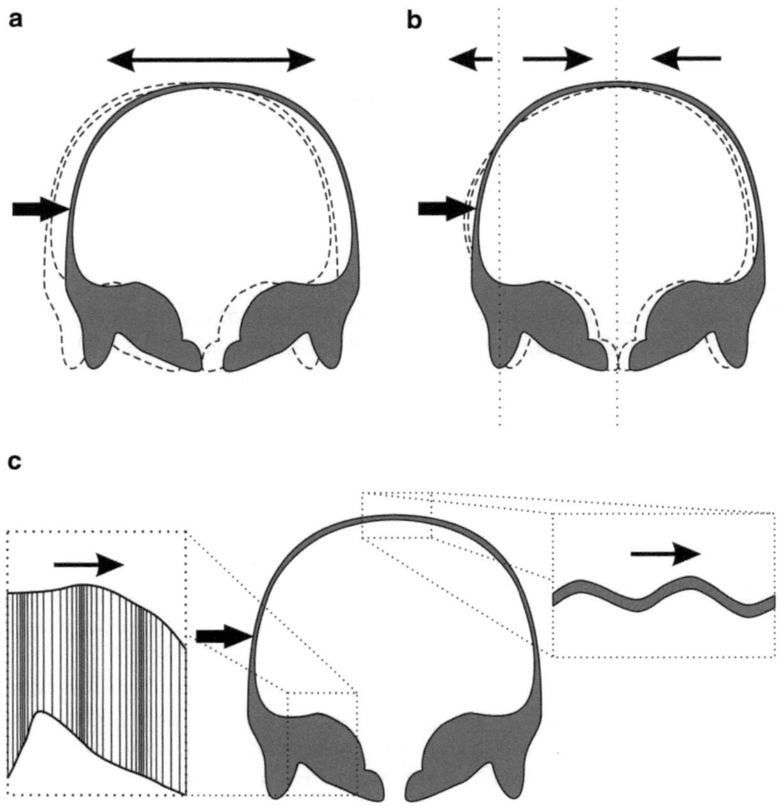

Fig. 6.1 Two-dimensional illustration of the vibration modes of the human skull at frequencies between 0.1 and 10 kHz. The *thick arrows* indicate the stimulation position and the *thin arrows* indicate the response directions. The rigid body response at the lowest frequencies is illustrated in (**a**) while the response at frequencies between approximately 0.3–1.0 kHz that is similar to a mass-spring system is shown in (**b**) where three sections of the skull move sequentially in opposite directions. In (**c**) the vibration responses for frequencies above 2 kHz is illustrated differently for the skull base and the cranial vault: at the skull base longitudinal wave propagation dominates the response while a mixture of vibration modes including bending waves is present at the cranial vault (From Stenfelt 2011)

6.2.2 Influence of Skin and Soft Tissue

The most frequently used mode of BC stimulation is by a transducer pressed on the skin-covered bone at the mastoid or forehead. This means that the skin and soft tissue are interposed between the transducer and the skull bone, affecting BC sound transmission. This is often suggested to interact negatively on BC sound transmission, wherein thicker soft tissue between transducer and bone results in greater attenuation. Except at the very low frequencies, below 250 Hz, the mechanical parameters of the skin and soft tissue are the primary mechanical load (mechanical point impedance) for a transducer on the skin-covered skull. At frequencies below 3 kHz the impedance is stiffness controlled whereas at the higher frequencies, the mass of the skin and soft tissues determine the impedance (Flottorp and Solberg 1976; Håkansson et al. 1986). However, the transmission depends on the interaction between the mechanical parameters of the transducer and the skin and soft tissues. For example, the acceleration transmission is attenuated (Håkansson et al. 1985a) whereas the force transmission is affected only at the higher frequencies (Carlsson et al. 1995; Stenfelt and Håkansson 1999); the effect seen at the electrical input to the transducer is usually between the two (Håkansson et al. 1984; Stenfelt and Håkansson 1999). Moreover, when investigating attenuation of the soft tissue, no relation to the thickness of the soft tissue was found (Mylanus et al. 1994).

Because the mechanical parameters of the skin and soft tissue in the transmission path affect the BC sound transmission, all manipulations that affect the mechanical parameters also affect the transmission. One such parameter is the compression of the skin. It has been shown that increased static force influences BC hearing sensitivity at 1 kHz and below; an increase from about 1.5 N to 6–10 N can improve sensitivity of up to 10 dB (Nilo 1968; Khanna et al. 1976). However, recently it was shown that hearing thresholds improved only 1.5 dB when the static force increased from 2.4 to 5.4 N (Toll et al. 2011); such a difference is of small clinical significance. Another parameter is the size of the vibration interface. A larger size of the vibration interface improves the BC hearing sensitivity primarily at the higher frequencies (above 1 kHz); however, there is an interaction between size and static force of the transducer (Nilo 1968; Khanna et al. 1976).

6.2.3 *Influence of Stimulation Position*

Generally speaking, the closer the stimulation is to the cochlea, the better the sensitivity of BC sound. As a consequence, mastoid placement is preferred over forehead placement of the transducer due to 11 dB improved sensitivity (Richter and Brinkmann 1981). It has been suggested that the forehead is less sensitive to variation in the stimulation position (von Békésy 1960) but the forehead has been shown to be sensitive to small changes in stimulation position with up to a 25 dB difference between adjacent positions (Khanna et al. 1976). There are also no differences in test–retest results or intersubject variability between the mastoid and forehead (Studebaker 1962; Dirks 1964).

Also, when the stimulation position is at the mastoid and adjacent bone structures, a stimulation position closer to the cochlea compared with a position farther away results in greater response, either as cochlear vibration (Stenfelt and Goode 2005b; Eeg-Olofsson et al. 2008) or as improved hearing thresholds (Eeg-Olofsson et al. 2013; Stenfelt 2012b). The reason for the improved sensitivity at the

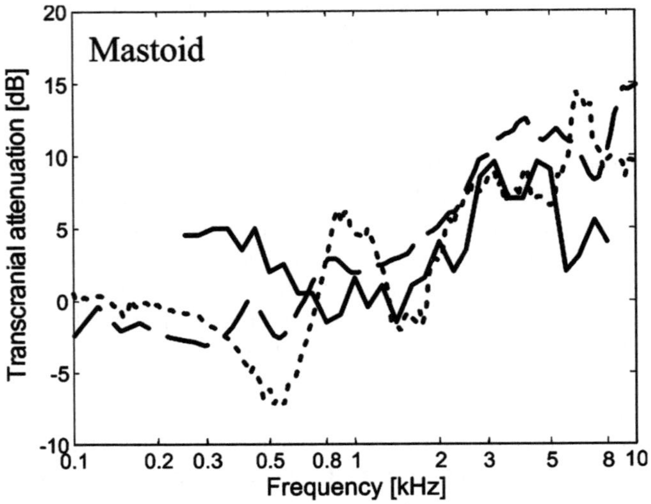

Fig. 6.2 Transcranial attenuation measured in three ways: (1) by hearing thresholds in unilaterally deaf subjects (*solid line*), (2) by one-dimensional vibration response of the cochlea (*dotted line*), and (3) by three-dimensional vibration response of the cochlea (*dashed line*) (From Stenfelt 2012b)

mastoid may be that the sound transmission from the skull surface to the petrous bone encapsulating the cochlea is more efficient when stimulation is positioned directly in line with this bone structure (Eeg-Olofsson et al. 2008). However, it could also be a result of not involving any of the skull bone sutures (Eeg-Olofsson et al. 2008).

Another issue often debated is the amount of BC transcranial attenuation in the human head, that is, how much less the stimulation is at the contralateral cochlea than at the ipsilateral cochlea. When measured as differences in hearing thresholds with ipsilateral and contralateral BC stimulation, it is between 0 and 15 dB, with large individual variability (Hurley and Berger 1970; Snyder 1973; Nolan and Lyon 1981; Reinfeldt et al. 2007a; Stenfelt 2012b). It has also been estimated from cochlear vibration with ipsilateral and contralateral stimulation; it shows almost no attenuation up to 1 kHz, where it increases and becomes close to 20 dB at 10 kHz (Stenfelt and Goode 2005b; Eeg-Olofsson et al. 2011, 2013). A comparison of threshold-based and cochlear vibration-based transcranial attenuation is shown in Fig. 6.2. Similar results were obtained when the transcranial attenuation was assessed by ear canal sound pressures caused by the BC sound (Reinfeldt et al. 2007a). The transcranial attenuations obtained using thresholds are, on average, similar to average vibration measurements at the cochlea at frequencies between 0.8 and 6 kHz (Reinfeldt et al. 2007a; Stenfelt 2012b); at frequencies below 0.8 kHz and above 6 kHz, there are discrepancies in the results from the different methods (Fig. 6.2).

6.3 Perception of Bone-Conducted Sound: Influences from the Outer, Middle, and Inner Ear

Several theories of how skull vibrations ultimately result in a hearing perception have been proposed. Early theories often suggested one or two pathways to dominate the BC perception (Allen and Fernandez 1960; Brinkman et al. 1965) whereas more recent literature suggests multiple pathways that contribute to BC sound perception (Tonndorf 1966; Stenfelt and Goode 2005a; Stenfelt 2011). However, there is no obvious way to distinguish between them because they are interconnected. One often used categorization is the anatomical division inspired by early investigators (von Békésy 1960). This categorization does not differentiate between the different physical mechanisms involved in the transformation from skull vibration to sound pressure differences between the scala vestibuli and scala tympani setting up a traveling wave on the basilar membrane. Recent literature has presented five components as being important for BC sound perception in normal and impaired ears, as indicated in Fig. 6.3 (Stenfelt and Goode 2005a; Stenfelt 2011); these are presented here divided according to ear anatomy.

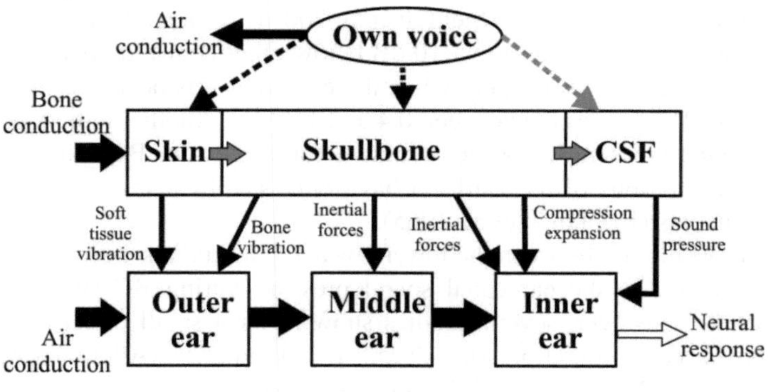

Fig. 6.3 A model of the multiple pathways for hearing BC sounds. A BC vibration onto the compressed skin of the skull bone causes vibrations of the skull and also produces a sound pressure in the skull interior. The vibration of the skin and bone produces a sound pressure in the ear canal while inertial forces cause relative vibration between the ossicles and the surrounding bone. The sound is transmitted to the inner ear from the outer and middle ear, but also directly through inertial forces acting in the cochlear fluids, through compression and expansion of the cochlear space, and, to some extent, through sound pressure transmission from the skull interior. The own sound production is transmitted to the inner ear by both airborne sound and BC (From Stenfelt 2011)

6.3.1 Outer Ear

The human ear canal is approximately 30–35 mm deep with two sharp bends; roughly half is surrounded by cartilage and half by bone. During BC stimulation a sound pressure is produced in the ear canal, primarily due to motions of the bony and cartilaginous ear canal walls. This sound pressure is then transmitted via the eardrum and ossicular chain to the cochlea in a way similar to AC sound. This also means that this pathway is affected by the status of the middle ear. Because the ear canal is easily accessible, several manipulations of the ear canal have been made and results reported. However, many of these involve occluding the ear canal and thereby changing the sound pressure in the ear canal; the occlusion effect is discussed in Sect. 6.6. It has been suggested that the cartilage part rather than the bony part of the ear canal contributes to the low-frequency sound; at higher frequencies the bony part become dominant (Naunton 1963; Stenfelt et al. 2003a).

One way to elucidate the importance of the ear canal sound pressure for BC perception is to compare ear canal sound pressure during AC and BC stimulation for the same sensation. To achieve the result for the normal ear, such comparison should be made for an open ear canal. This was tested, and greater sound pressure with AC stimulation than with BC stimulation at frequencies above 0.5 kHz was found, indicating that ear canal sound pressure is not the most important pathway for BC perception at frequencies above 0.5 kHz (Huizing 1960). A similar approach was to examine the umbo motion in relation to the ear canal sound pressure when stimulation was by AC and BC (Stenfelt et al. 2003a). That study indicated that the ear canal sound pressure was about 10 dB below other contributors to BC perception when the ear canal was open; when the ear canal was occluded the outer ear dominated the BC perception between 0.4 and 1.2 kHz. Another indication that the ear canal sound pressure is not the dominant contributor for BC sound is the fact that large perforations of the eardrum have only small effects on the BC sound perception in cats (Brinkman et al. 1965).

The relative motion between the lower jaw and the skull has been suggested as a major contributor to the ear canal sound pressure during BC stimulation (von Békésy 1932). However, several studies show no or a small effect of the lower jaw on the ear canal sound pressure during BC stimulation (Allen and Fernandez 1960; Howell and Williams 1989; Stenfelt et al. 2003a).

6.3.2 Middle Ear

The middle ear can contribute to BC sound perception primarily by two components: ossicular inertia and sound pressure in the middle ear cavity. The latter is deemed insignificant because middle ear sound pressure measured in sealed human temporal bones did not indicate any significant sound pressure level during BC stimulation (Stenfelt et al. 2002). The ossicular inertia effect relies on the mass

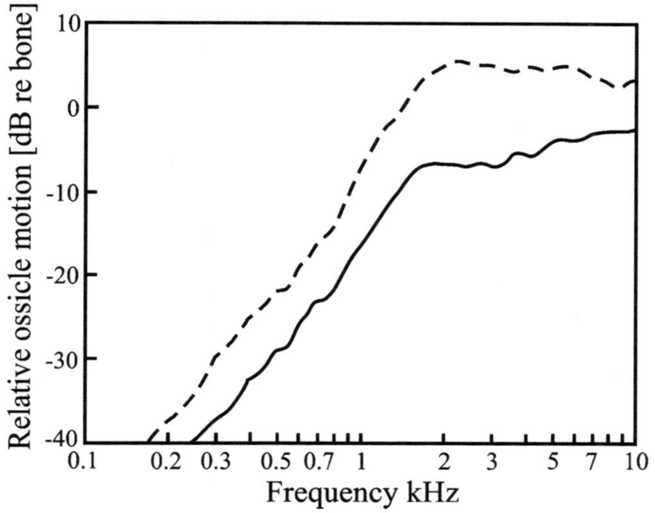

Fig. 6.4 The relative motion between the stapes footplate and the promontory bone (*solid line*) and the malleus umbo and the promontory bone (*dashed line*) in temporal bone specimens when stimulation is in line with the ossicles. The results are averages from 26 ears (Data from Stenfelt et al. 2002)

of the ossicles suspended by ligaments and tendons in the middle ear cavity producing inertial forces when the skull bone vibrates during BC stimulation. This effect is low at low frequencies, at which the stiffness of the suspensory ligaments forces the ossicles to move in phase with the surrounding bone whereas the ossicles become mechanically decoupled from the skull bone at higher frequencies, resulting in a large relative motion between ossicles and skull bone. This behavior is shown in Fig. 6.4 for temporal bone specimens where the relative motion between ossicles and skull bone increases at 40 dB/decade at frequencies below the resonance frequency above which the relative motion flattens out; it stays between 5 and 10 dB re bone motion for the umbo and −5 to 0 dB re bone motion for the stapes (Stenfelt et al. 2002; Homma et al. 2009). The resonance frequency was found to be 1.5–1.7 kHz in the aforementioned studies.

The influence of the middle ear ossicles on BC has been studied extensively; experimental and pathological findings are described in Sect. 6.5. When comparing ossicle motion at AC and BC hearing threshold levels, the vibration of the ossicles is approximately 10 dB below other contributors for BC sound at frequencies below the resonance frequency (1.5–1.7 kHz) (Stenfelt 2006). That analysis also indicated that the ossicles may contribute to BC perception at the resonance frequency and up to approximately 3 kHz. Another experimental and computational modeling study of middle ear ossicle vibration also suggested a contribution at the ossicle resonance frequency of about 1.7 kHz due to a “pivoting” mode that is dominantly excited by BC stimulation (Homma et al. 2009).

6.3.3 *Cochlea*

In audiology, BC thresholds are compared with AC thresholds to diagnose a conductive impairment. This is based on the notion that BC thresholds are minimally affected by the impairment (situated in the outer or middle ear) while the AC thresholds are affected more significantly. This means that even if the middle or outer ears are involved in the BC sound perception, there is very little effect on the BC sensitivity. Consequently, the BC stimulation of the cochlea can be seen as the dominating part for BC sound perception. However, the processes in the cochlea that result in the perception of sound are to date still debated. The major theories are described in sections "Compression," "Inertia," "Third-Window Theory," "Dynamic Pressure Transmission" and "Pathological Third Window of the Inner Ear."

6.3.3.1 Compression

Due to the wave motion in the skull bone, the bone itself compresses and expands. Such deformation forces displacement of the inner ear fluids and creates a sound pressure in the cochlea. This mode of BC was first termed "inner ear compression" (von Békésy 1932) and later renamed as the "distortional component" (Tonndorf 1966). A better name is "cochlear space alteration" because it is based on the idea that fluid is incompressible during the expansion and compression phase of the cochlear space. During the compression phase the space is reduced and the excess fluid displaced at the compliant oval and round windows. The round window is more compliant (lower impedance) than the oval window and can displace more volume, forcing fluid from the scala vestibuli toward the scala tympani, thus exciting the basilar membrane in the process. Further, the greater volume in the scala vestibuli than in the scala tympani, combined with the higher impedance of the oval window, also forces excess fluid in the direction of the scala tympani adding to the process. During expansion of the cochlear space the process is the opposite.

The significance of this mode of BC perception is disputed and there are clinical findings with obstruction of the cochlear windows and reopening them (fenestration), indicating it not to be important at lower frequencies. Moreover, the cochlea is coiled and its dimensions in the bone can be approximated with that of a sphere with a diameter of 10 mm. If the limit for effective compression response is set to a wavelength that is less than 10 times the size of the cochlea, the lowest frequency for an effective compressive response of the cochlea would be 4 kHz. This is in line with other estimations of the importance of the compression response in the human (Stenfelt and Goode 2005a). Also, based on finite element simulations, the compression component of the bone encapsulating the cochlea is some 25 dB below the rigid body motion of the same up to around 5 kHz (Hudde 2005).

6.3.3.2 Inertia

Similar to the middle ear ossicles, inertial forces also act on the cochlear fluid during vibration of the cochlea. The result of such forces is a sound pressure gradient across the basilar membrane. Consequently, the greater this pressure gradient is, the more efficient is the fluid inertia as a contributor for BC sound. If the fluid is considered incompressible, the fluid flow would require a compliant inlet and a compliant outlet, one on each side of the basilar membrane. In the healthy ear the oval and round windows accomplish this. However, BC sound is relatively unaffected by closing, for example, the oval window in otosclerosis. This can still be explained as an inertial response in the cochlea due to other compliant pathways, known collectively as the "third window" (see section "Third-Window Theory") (Ranke et al. 1952). This means that as long as there is a pressure gradient between the two scalae, there will be fluid flow acting on the basilar membrane resulting in a traveling wave.

Another way to explain the contribution of fluid inertia was suggested by Kim et al. (2011), who stimulated basilar membrane response to BC excitation in a 3D tapered box model of the cochlea. In their model, they decomposed the volume velocities at the oval and round windows into antisymmetric (slow wave) and symmetric (fast wave) volume velocities similar to Peterson and Bogert (1950) for AC. Kim et al. found that the basilar membrane vibration correlated to the antisymmetric volume velocities of the round and oval windows with no dependence on stimulation direction. They therefore argue that manipulations of the middle ear and fixation of the oval and/or the round windows directly influence the antisymmetric volume velocity input to the cochlea (i.e., the difference in volume displacement between the oval and the round window). The caveat is that the model simulations are made in a simplified geometry of the cochlea and the results need experimental validation.

It should be realized that only a small fluid flow is required to produce a hearing sensation based on inertial fluid flow in the cochlea. An estimation of the flow with BC stimulation at 80–100 dB HL yields a displacement that is less than one-millionth of the total fluid volume in the cochlea (Stenfelt and Goode 2005a). It should be pointed out that even if the compressibility of both the fluid and the cochlear bone is small, it is nonzero (Shera and Zweig 1992), and part of the displacement may be attributed to the small compliance of these entities. Because the BC response is nearly unaffected at low frequencies with removal of outer and middle ear components (immobility or removal of middle ear ossicles; Sect. 6.5), the response is directly due to the cochlea. Also, at low frequencies compression is unlikely to be a major contributor (Hudde 2005; Stenfelt and Goode 2005a), indicating that fluid inertia is likely the most important contributor to BC perception at frequencies below 4–5 kHz (Stenfelt and Goode 2005a; Taschke and Hudde 2006; Kim et al. 2011). However, it may be less important at higher frequencies.

6.3.3.3 Third-Window Theory

As presented in the previous section, the major compliant pathways of the inner ear are the oval and round windows. But besides these there are several other compliant pathways that may serve as compliant inlets and outlets for fluid motion. These include the cochlear and vestibular aqueducts (Gopen et al. 1997), as well as nerve fibers, blood vessels, and microchannels entering the cochlea (Kucuk et al. 1991). Also, the compliance of the fluid itself and the bone encapsulating the cochlea yields a general compliance for fluid displacement. These structures provide a combined compliant pathway collectively known as the third window (Ranke et al. 1952). Such a compliant pathway facilitates two possible excitation modes for BC sound. One is the displacing of fluid due to inertial forces of the fluid, described in section "Inertia." The other is by providing a channel for sound pressure transmission from the cranial space to the cochlea, described in section "Dynamic Pressure Transmission."

For AC transmission in the normal cochlea, the volume displacements at the oval and round windows are equal but with opposite phases (Kringlebotn 1995; Voss et al. 1996; Stenfelt et al. 2004). However, this was not found when stimulation is by BC, where the volume displacement between the two windows may differ by up to 10 dB (Stenfelt et al. 2004); this was also found in a model of inertial BC in the cochlea (Kim et al. 2011). This indicates that fluid in the cochlea is displaced at places other than the oval and round windows during BC stimulation, or that the cochlear space is deforming and being "squeezed." However, the difference of volume displacement at the oval and round window is seen at low as well as high frequencies. Because cochlear space alteration is not believed to be present at lower frequencies owing to the nature of wave transmission in the skull bone at these frequencies, the third window effect is considered as important for BC sound. In experiments in cats, it was found that the third window effect was important for BC but not for AC stimulation (Tonndorf 1966). However, in diseased cochleae with round window atresia, AC thresholds are elevated but not absent (Linder et al. 2003). This suggests that the third window may also be important in for AC stimulation in pathological ears.

6.3.3.4 Dynamic Pressure Transmission

It has recently been shown that a sound percept can be evoked in the cochlea with stimulation on the body without involving the skull bone, for example, by applying a vibration stimulus to the eye (Perez et al. 2011). This type of soft tissue transmission is hypothesized to rely on sound pressure transmission from the cerebrospinal fluid through compliant pathways to the cochlea (Sohmer et al. 2000). However, it is not clarified how this sound pressure transmission occurs. Also, several clinical findings such as BC sensitivity change due to a semicircular canal dehiscence indicate that this mode is not the most important for BC sound perception (see section "Pathological Third Window of the Inner Ear"). Also, similarities between cochlear vibration pattern and BC perception suggest that that the vibration of the cochlea itself is responsible for BC sound perception (Stenfelt 2012b).

6.3.3.5 Pathological Third Window of the Inner Ear

As described in section "Third-Window Theory," the third window provides compliant inlets and outlets of the cochlea. The impedance of these pathways depend on their diameter and length; they are normally long and thin (relative to the oval and round window), resulting in a high impedance not affecting normal AC stimulation (Gopen et al. 1997). However, when they become wider (pathological) their impedance decreases and they affect both AC and BC sensitivity when situated on the scala vestibule side. Two such pathologies are semicircular dehiscence and large vestibular aqueduct syndrome (Merchant and Rosowski 2008).

Both types of pathologies have, from the AC and BC sound perspective, similar explanations. In semicircular dehiscence, AC hearing sensitivity is decreased at frequencies below 2 kHz while BC sensitivity is improved in the same frequency region (Mikulec et al. 2004; Merchant and Rosowski 2008). Large vestibular aqueduct syndrome also shows low-frequency pathological differences between AC and BC thresholds (known as air–bone gaps) (Merchant et al. 2007; Sato et al. 2007). The explanation for the low-frequency air–bone gaps is the reduced impedance for fluid flow at the scala vestibuli side: for the AC sound, part of the sound energy is rerouted to the enlarged canal or dehiscence instead of the round window, reducing the stimulation of the basilar membrane, whereas for BC sound the reduced impedance parallels the oval window, facilitating larger fluid flow between the scalae and increased basilar membrane stimulation (Songer and Rosowski 2007, 2010).

6.4 Bone Conduction Audiometry

BC hearing thresholds are together with AC thresholds the fundamental measure of a person's hearing ability. The configuration of the absolute thresholds as well as the difference between AC and BC thresholds guide the clinician categorizing a hearing impairment as sensorineural, conductive, or mixed. This relies on the notion that BC thresholds are minimally affected by the outer and middle ear status, while the AC thresholds are highly influenced. The maximum possible difference between AC and BC thresholds depends on the specific test method and equipment; at a certain level the AC stimulation induces a BC vibration in the skull that is audible. For a sound field, this level is between 40 and 60 dB (Reinfeldt et al. 2007b).

Throughout history, different transducers have been used for BC testing but today the Radioear B71 is most frequently used. This transducer is limited in useable frequency (approximately 250–4,000 Hz) and dynamic range, and there have been suggestions for other designs to overcome some of the current limitations (Håkansson 2003; Popelka et al. 2010a). Test–retest variability is another problem associated with BC threshold testing; however, with careful positioning of the transducer, the standard deviation of test–retest is in the 3–5 dB range (Laukli and Fjermedal 1990); this variability does not improve with smaller step size (Jervall and Arlinger 1986). Besides conventional pure tone threshold estimations, BC stimulation can be used in brain stem response audiometry (Collet et al. 1989; Beattie 1998), auditory steady-state response audiometry (Ishida et al. 2011), otoacoustic emissions (Kandzia et al. 2011), and speech testing (Beattie and Smiarowski 1981).

6.4.1 Factors Affecting Bone Conduction Threshold Estimations

Variables that have a particular influence on the reliability of BC testing are the specific type of BC transducer, the static force, presence or absence of contralateral masking (see Sect. 6.4.2), and location of the transducer (Dirks 1964). Adding to the list of uncertainties are the functional state of the middle ear (see Sect. 6.5), the position of the lower jaw, and the large amount of distortion produced at low frequencies (Salomon and Elberling 1988). Most of these variability problems can be avoided by carefully following standardized testing procedures (ISO:8253-1 2010).

As stated in Sect. 6.2.3, the forehead has an overall lower sensitivity than the mastoid with a difference of approximately 11 dB at the normal test frequencies (Richter and Brinkmann 1981). This sensitivity difference is important because nonlinear distortion affects the maximum level available from the BC transducer at low frequencies: hence the maximum hearing loss testable is less at the forehead than at the mastoid. Another problem is vibrotactile sensitivity: At the mastoid the vibrotactile thresholds for BC testing are 43 and 55 dB HL at 250 and 500 Hz, respectively (Brinkmann and Richter 1983). That is another limit for the maximum testable hearing loss at low frequencies.

A problem with BC testing at the higher frequencies, at 3–4 kHz, is that the vibration of the transducer couples to the air causing airborne sound at the same level as the BC stimulation (Shipton et al. 1980; Frank and Crandell 1986). It is therefore sometimes advised that BC threshold testing should be done with ear plugs. The caveat is the occlusion effect at low frequencies caused by the ear plug giving erroneous BC threshold data (see Sect. 6.6).

6.4.2 *Masking*

To ensure BC testing of a specific ear, the nontest ear requires masking. The level and frequency of the masking noise is important. Inadequate masking allows the nontest ear to participate while excessive masking falsely makes the BC threshold worse. Optimum masking is produced by narrow band noise centered at the test frequency. The optimal level is difficult to predict beforehand. One issue is the attenuation of the BC sound to the contralateral side. It is assumed to be, on average, 0–15 dB (see Fig. 6.2), but the large variability may cause the BC stimulation at the opposite ear to be 20 dB higher than at the test ear at certain frequencies (Stenfelt 2012b). The masking is usually provided by circumaural or insert earphones that causes an occlusion effect at low frequencies that can amount to 20 dB (Elpern and Naunton 1963). Consequently, the BC stimulation at the nontest ear may be 40 dB greater than at the test ear.

Because the exact masking level is difficult to predict according to the foregoing, an adaptive masking procedure is often used (Studebaker 1964). This is also known as the plateau technique, wherein the unmasked threshold is elevated by increasing masking level in the nontest ear. Above a certain level, a further increase of the masking noise in the nontest ear causes no further threshold elevation in the test ear. The threshold at this plateau is considered as the true masked threshold of the test ear.

6.5 Bone Conduction Thresholds Influenced by the Status of the Middle Ear

The role of the middle ear in BC was explained in Sect. 6.3.2. To reveal the importance of the middle ear ossicles in BC, different types of artificial lesions have been made in living humans, for example, mass loading of the eardrum and ossicles or the addition of a static pressure in the ear canal, and also by manipulation of the ossicles in research animals. Moreover, the alteration of the BC sensitivity in pathological middle ears adds to the understanding of the middle ear ossicles' role in perception of BC sound.

6.5.1 *Experimental Conditions*

Because the middle ear ossicles are not accessible in the normal ear, most investigations have used mass loading of the eardrum to increase the mass of the ossicular system (Bárány 1938; Legouix and Tarab 1959; Huizing 1960; Stenfelt et al. 2002) or by increasing the static pressure in the ear canal and thereby increasing the stiffness of the ossicular chain (Bárány 1938; Huizing 1960; Aazh et al. 2005; Homma et al. 2010). Most studies agree that adding mass improves BC thresholds at frequencies below 2 kHz; direct measurement of the ossicle vibration shows that adding a mass to either malleus umbo or stapes footplate decreases the resonance frequency for the ossicular vibration with BC stimulation and thereby improves sensitivity below the normal resonance frequency at approximately 1.5 kHz (Stenfelt et al. 2002). The lowering of the resonance frequency is an effect of the increased mass while the stiffness is the same; this result in greater velocity at low frequencies owing to the increased inertial force caused by the greater mass. The evidence that increased static pressure decreases BC sensitivity is not equally conclusive; there are studies that show decreased threshold sensitivity (Huizing 1960; Humes 1979; Nolan et al. 1985) but also show improved sensitivity (Aazh et al. 2005). A caveat is that achieving the static pressure in humans also creates an occlusion effect known to improve low-frequency sensitivity of BC sound (see Sect. 6.6). The static pressure does increase the stiffness and by that the resonance frequency of the ossicles with BC stimulation, leading to decreased vibration velocity of the ossicles below the new resonance frequency (Homma et al. 2010); however, its effect on BC sensitivity for the normal human ear is not clear.

In a thorough study of the vibration response of the middle ear ossicles during BC stimulation in temporal bone specimens, the effects of several manipulations were studied (Stenfelt et al. 2002). Gluing the stapes or the malleus to the surrounding bone reduced the ossicle vibration, more so for gluing the stapes than for gluing the malleus; almost no effect was seen at frequencies above 3 kHz for gluing the malleus on the stapes vibration. This is attributed to the ossicular joints, primarily the incudo–stapedial joint but the incudo–malleolar joint may also contribute. It was also shown that severing the incudo–stapedial joint affected the vibration of the stapes only at 1.5–2 kHz, where it decreased by approximately 10 dB; this is the frequency region of the middle ear ossicle resonance where the inertia of the ossicles may contribute to the perceived BC sound (see Sect. 6.3.2). Another interesting finding was that when the cochlea was drained of the fluids, the low-frequency response of the ossicles decreased while it increased at around 2 kHz; yet another indication of a middle ear ossicle inertia contribution at this frequency.

6.5.2 Pathological Conditions

There are indications that the status of the middle ear affects BC thresholds more when the stimulation is at the mastoid than at the forehead. Approximately 5 dB worse BC thresholds at frequencies between 0.5 and 4 kHz are obtained with the stimulation at mastoid compared with at the forehead for several different middle ear lesions (Studebaker 1962; Dirks and Malmquist 1969; Goodhill et al. 1970). This may suggest that the middle ear is more influential for BC perception when stimulation is in line with the ossicles than when directed perpendicularly.

However, there is only a minor vibration response difference for the ossicles when stimulation is in line with the ossicles than when perpendicular (Stenfelt et al. 2002). It is also argued that the difference for BC sensitivity at mastoid and forehead does not hold for all middle ear impairments (Dirks and Malmquist 1969).

6.5.2.1 Otosclerosis of Stapes and/or Malleus

One well accepted change in the BC sensitivity with middle ear lesions is the so-called Carhart notch that manifests itself in otosclerosis of the stapes and oval window. The depressed BC threshold in otosclerosis is approximately 20 dB at 2 kHz, with lesser losses at frequencies below and above this frequency (Carhart 1950, 1971). The depressed BC sensitivity has been explained as a lack of contribution from the resonating middle ear ossicles (Tonndorf 1966). However, it may also be caused by the impedance change seen for the fluid flow in the cochlea.

It has been suggested that the Carhart notch may be used to diagnose a stapedial otosclerosis based on the depressed BC threshold at and around 2 kHz when the same depression is not seen in the AC thresholds. However, there are a few caveats in doing so. One is that not just conductive lesions affect otosclerotic ears. When comparing nonoperated ears with proven otosclerosis with a control group, it was shown that the otosclerotic ears had greater BC deterioration than the normal ear even after correction for the Carhart effect, indicating additional cochlear impairment due to the otosclerosis (Browning and Gatehouse 1984). Such an effect obscures the Carhart notch, making the diagnosis uncertain. Another caveat is that the Carhart notch is not specific to otosclerosis or congenital absence of the oval window. To a lesser degree it also exists in cases of otitis media with effusion, tympanosclerosis, and congenital ossicular anomalies (Ysan2007). However, it is only when the incudo–stapedial joint has become part of the ossicular fixation that a fixed malleus produces BC thresholds comparable to those seen in otosclerosis of the stapes (Goodhill 1966).

6.5.2.2 Fluid in the Middle Ear Cavity

There are several reports indicating that fluid in the middle ear, for example, associated with serous or adhesive otitis media, temporally worsen the BC thresholds, and after incision of the eardrum and insertion of a ventilation tube, the BC thresholds recover (Palva and Ojala 1955; Huizing 1960; Milner et al. 1983). As pointed out in section "Otosclerosis of Stapes and/or Malleus," the alteration of the BC thresholds in patients with chronic ear diseases such as chronic suppurative otitis media, cholesteatoma, and adhesive otitis media show BC thresholds similar to those of patients with otosclerosis (Lindstrom et al. 2001). After tympanoplasty and ossicular reconstruction, the BC thresholds improved between 4 and 10 dB at frequencies between 0.25 and 4 kHz. Also, children suffering from otitis media with effusion showed BC threshold depression similar to that found in otosclerosis (Carhart notch) (Ahmad and Pahor 2002; Shishegar et al. 2009). However, a study of children with suppurative otitis media without cholesteatoma reported no alteration of the BC thresholds (Kaplan et al. 1996).

6.5.2.3 Ossicular Discontinuity

The literature on the effect of ossicular discontinuity on BC thresholds is less conclusive than for other middle ear pathologies. However, most studies report no or insignificant alteration of the BC threshold as a result of ossicular discontinuity (Møller 2000). In a group of patients with chronic otitis media who underwent tympanomastoidectemy without ossicular reconstruction, no change in the BC thresholds was seen. However, in a group who underwent ossicular reconstruction, the gain in BC thresholds were on average 2.3–3.9 dB at frequencies between 0.5 and 4 kHz, with the greatest improvement at 2 kHz (Lee et al. 2008). Another study indicated no significant difference in BC thresholds in a group of patients after ossiculoplasty subsequent to traumatic ossicular dislocation (Yetiser et al. 2008). However, in patients with radical mastoidectomy removing the major part of the ossicular chain, the greatest reduction was around 2 kHz, where the ossicles normally resonate (Dirks and Malmquist 1969). This finding is in line with artificial manipulations of the ossicles in which severing the incudo–stapedial joint did reduce vibration of the stapes at 2 kHz (Stenfelt et al. 2002).

6.5.2.4 Oval-Window and Round-Window Occlusion

The most common reason for oval-window and/or round-window occlusion is otosclerosis, but there exist cases with congenital absence of one or both inner ear windows (House 1959). Early on, the common treatment for otosclerosis was a fenestration of the vestibule. With this treatment the BC thresholds almost returned to normal (Walsh 1962). The improvement of the BC thresholds was, as a pure tone average (PTA) for the frequencies 0.5, 1, and 2 kHz, on the order of 6.5–12 dB, reflecting the depression of the Carhart notch (Miyamoto and House 1978; Brooks 1985). It was reported that a patient with congenital absence of the oval window due to malformation showed BC thresholds close to normal (Everberg 1968), but most report that congenital absence of the oval window shows BC results similar to that in otosclerosis of the stapes (Yi et al. 2003).

Today, more common treatments in otosclerosis are stapedectomy or stapedotomy. With total footplate stapedectomy, the BC thresholds generally improve by more than 5 dB at frequencies between 0.5 and 2 kHz (Awengen 1993). Stapedectomy (total or partial) gives significantly larger improvement in BC thresholds (in otosclerotic ears) of 10–12 dB at 1 and 2 kHz compared with stapedotomy, giving 3–6 dB improvement at the same frequencies (Persson et al. 1997). Another study reports mean BC threshold improvement of 8 dB at 2 kHz after a piston was inserted in a series of otosclerotic ears (Tange et al. 2000).

There have also been studies investigating the BC improvement with relation to diameter of the piston in stapedotomy. One such study reported the improvement of the BC thresholds at 1 and 2 kHz after stapedotomy to increase with increased diameter of the stapedotomy piston (Teig and Lindeman 2000) whereas another study pointed in another direction, indicating that the increase of BC thresholds is greater for a 0.4-mm piston compared with a 0.6-mm piston, PTA (0.5, 1k, 2k, 4k) 4.5 and 2 dB, respectively (Shabana et al. 1999). A meta-analysis of the literature indicated no difference between 0.4- and 0.6-mm pistons on BC sensitivity (Laske et al. 2011).

An occlusion of the round window affects the BC thresholds more than occlusion of the oval window. In a case of otosclerosis primarily of the round window, the audiogram showed BC thresholds that decreased by approximately 16 dB per octave (Groen and Hoogland 1958). The BC improvement after surgery was 4 dB per octave, with a gain of 5 dB at 250 Hz and 20 dB at 8 kHz. In other cases with absence of the round window, the BC thresholds are reported as approximately 20 dB worse than normal (Martin et al. 2002; Linder et al. 2003).

6.5.3 Influence from the Stapedius Muscle

The effect of the stapedius muscle on AC transmission in humans have been reported to increase with decreased frequency at frequencies below 2 kHz, reaching 40 dB attenuation (Morgan and Dirks 1975). There are no similar reports on the effect of the stapedius muscle on BC transmission in the human. A study on the effect of the stapedius muscle on BC transmission in the cat suggests it to be similar to the effect of AC: The BC attenuation was approximately 10 dB at frequencies below 1.5 kHz (Irvine 1976). The effect may be similar to that of static pressure in the ear canal increasing the stiffness of the ossicular chain. If so, the effect of the stapedius muscle on BC transmission in humans is expected to be similar to that obtained in cats.

6.6 The Occlusion Effect

The occlusion effect is the low-frequency BC sound increase subsequent to an occlusion of the ear canal. One common manifestation of the occlusion effect is the low-frequency emphasis of one's own voice while speaking with the ears occluded (see Sect. 6.7). Because the ear canal is easily accessible and the occlusion effect is important in several areas for BC sound, the occlusion effect has been investigated thoroughly, either as the perceptual change (alteration of hearing thresholds) (Klodd and Egerton 1977; Small and Stapells 2003), a change of the ear canal sound pressure (Howell et al. 1988; Stenfelt et al. 2003a), or both (Huizing 1960; Berger 1983; Stenfelt and Reinfeldt 2007). Several explanations for the occlusion

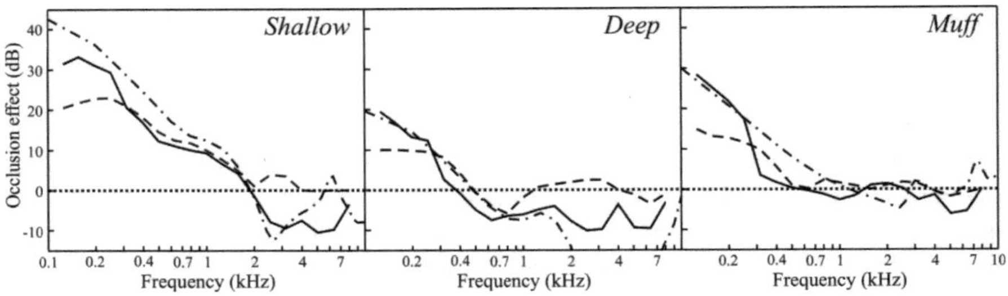

Fig. 6.5 The occlusion effect obtained as ear canal sound pressure changes (*solid line*), threshold differences (*dashed line*), and estimated according to an acoustic model (*dashed-dotted line*). Three conditions are displayed: (1) shallow occlusion (7 mm down the ear canal), (2) deep occlusion (22 mm down the ear canal), and (3) earmuff with 30 cm^3 internal volume (From Stenfelt and Reinfeldt 2007)

effect have been proposed: Huizing measured the ear-canal sound pressure and related threshold changes and explained the occlusion effect by the impedance change of the ear canal caused by the ear canal length and terminations, changing resonances and anti-resonances (Huizing 1960). Tonndorf presented a simple mass-spring model of the ear canal to explain the occlusion effect; the occlusion changes the filter parameters rerouting low-frequency BC sound to the cochlea (Tonndorf 1972). In an acoustic modeling effort, the ear canal sound pressure change with the occlusion effect could be well predicted by a transmission line model of the ear canal in which the length of the ear canal and radiation impedance were altered due to the occlusion device (Stenfelt and Reinfeldt 2007).

Figure 6.5 shows the occlusion effect measured as the changes of ear canal sound pressure, BC threshold change, and according to a model (Stenfelt and Reinfeldt 2007) for three different conditions: (1) shallow occlusion (7 mm down the ear canal), (2) deep occlusion (22 mm down the ear canal), and (3) earmuff with 30 cm^3 internal volume. The model verified that different positions in the ear canal gave different occlusion effects (Stenfelt and Reinfeldt 2007). It corroborated the assumption that deep occlusion causes no or negligible occlusion effect (von Békésy 1941). Also, large enough air volume inside ear phones or earmuffs removes the occlusion effect (Elpern and Naunton 1963; Khanna et al. 1976; Stenfelt and Reinfeldt 2007; Reinfeldt et al. 2010).

With BC stimulation at the mastoid, the perceived occlusion effect is often 10 dB lower than the change in ear canal sound pressure (Huizing 1960; Berger 1983; Stenfelt and Reinfeldt 2007). This is explained by the significance of the ear canal sound pressure for BC sound perception; the ear canal component is believed to be about 10 dB below other contributing parts for BC perception (see Sect. 6.3.1). However, when the stimulation is at the forehead the perceived occlusion effect is greater than when stimulation is at the mastoid (Klodd and Egerton 1977; Dean and Martin 2000; Stenfelt and Reinfeldt 2007). This may indicate that the contribution of pathways to BC sound depends on the stimulation position or the direction of the stimulation. It should be noted that not all studies show differences between mastoid- and forehead-stimulated occlusion effect (Goldstein and Hayes 1971).

The occlusion effect is most pronounced at the lowest frequencies, meaning that low-frequency hearing sensitivity is increased for BC sound. However, the sensitivity for other body sounds is also increased. When these body sounds are enhanced (breathing, heartbeat, and swallowing) they can mask the BC sound at the lowest frequencies (125 and 250 Hz). The masking due to this enhancement of body sound is as much as 5 dB (Berger and Kerivan 1983). This means that if hearing thresholds are used for estimating the occlusion effect, the occlusion effect at the lowest frequencies may be underestimated.

6.7 Own Voice Perception

One example in which the influence of BC sound is familiar to most people is in the act of hearing one's own voice. When listening to a recording of one's own voice people are often struck by the difference between the sound of the recording and the way that they normally perceive their voice. The reason for this difference is that a person hears his or her own voice through two routes, airborne (AC transmission) and via the skull bones (BC transmission), whereas the recording contains only the airborne sound.

There have been attempts at estimating the two components of one's own voice. One attempt was done by attaching tubes filled with cotton to the ears, removing the AC component without causing an occlusion effect (not affecting the BC component) (von Békésy 1949). The decrease in loudness of vocalization subsequent to attaching the tubes was about 6 dB, and this was assumed to be equivalent to the AC component of the subject's own voice; this indicated that the AC and BC components of one's own voice were similar in magnitude (von Békésy 1949). Another estimate used masked thresholds by one's own voice, in which the AC and BC components were manipulated. The estimation was performed for a voiced (/z/) and an unvoiced (/s/) sound: The results showed the BC components to be greater at frequencies between 0.7 and 1.2 kHz whereas the AC component dominated outside this frequency region (Pörschmann 2000). A third investigation of the two components were conducted with a large earmuff removing the AC component without adding the occlusion while measuring the ear canal sound pressures from the AC and BC components of the subject's own voice (Reinfeldt et al. 2010). That study used ten different utterances: four vowels and six consonants. It was found that the relative contribution of AC and BC for one's own voice perception depends on the utterance, but similar utterances gave approximately similar relative contributions. For example, /m/ and /n/ gave similar relative contributions but differed from, for example, /k/ and /t/, that were similar to each other. At frequencies below 2 kHz, there were different regions for the different utterances where the AC and BC components dominated. However, at frequencies above 2 kHz the AC component dominated the subject's own voice regardless of utterance (Reinfeldt et al. 2010).

Besides better understanding of one's own voice as provided by the aforementioned studies, the knowledge about the two components during vocalization is important for designing and fitting hearing aids because the perception of one's own voice is affected by a hearing aid (Killion et al. 1988; Carle et al. 2002; Stenfelt 2012a). Also, BC transmission of one's own voice is important for designing BC microphones that record the BC component of a subject's own voice (Ono 1977; Zheng et al. 2003).

6.8 Bone Conduction Hearing Aids

One area in which BC hearing is of importance is the use of BC hearing aids. The primary use of BC hearing aids is in patients needing amplification but in whom normal AC hearing aids are contraindicated. The typical BC hearing aid wearer has a large conductive hearing loss but BC hearing aids are also used in patients with draining ears or eczema in the ear canal and have recently been introduced as an alternative for contralateral routing of signal (CROS) hearing aids.

6.8.1 Conventional Bone Conduction Hearing Aids

The concept of BC hearing is old, as are BC hearing aids (Berger 1976; Mudry and Tjellström 2011). The general function of a BC hearing aid is the same as an AC hearing aid: A microphone picks up the sound and a sound processor shapes the sound according to a fitting rule for the hearing loss. The difference is that instead of a normal receiver as in the case of the AC hearing aid, the signal is supplied to a transducer vibrating the skull by pressing on the skin-covered mastoid. The typical BC hearing aid is either positioned by a headband or incorporated into spectacles for improved cosmetics (Banga et al. 2011). The requirement of a static force to facilitate sound transfer as well as positioning of the device causes discomfort when used for long periods of time, and the sound quality of these devices is often poor (Snik et al. 1995). These shortcomings have led to the development of the bone-anchored hearing aid (BAHA). Recently, a conventional BC hearing aid was introduced that uses implanted magnets in the mastoid to retain the transducer and achieve the necessary static force to transmit the BC sound (Siegert 2011).

6.8.2 Bone-Anchored Hearing Aids

The BAHA is attached to the skull using a skin-penetrating titanium implant (Håkansson et al. 1985b). Such a design avoids problems associated with the static pressure in the conventional BC hearing aid, and the direct coupling to the skull bone offered by the implant removes the high-frequency attenuation of the skin. As with conventional BC hearing aids, the degree of conductive loss is not important for BAHA users; it is the sensorineural hearing loss that is the limit. When used within the suggested limits of the different devices, the BAHA system is considered beneficial and is a well-accepted rehabilitation (Snik et al. 2005).

Because a BC sound is transmitted to the contralateral side with low attenuation (Stenfelt 2012b), the BAHA is usually fitted only unilaterally even though the hearing problem is bilateral. One reason is the unclear binaural effect with bilateral bone conduction (Stenfelt 2005). However, a systematic review of the literature of patients fitted bilaterally with BAHAs shows a clear benefit with bilateral fitting compared with unilateral fittings (Colquitt et al. 2011). It should be noted that due to the cross transmission with BC stimulation, the binaural effect is less for BC sound than for AC sound bilaterally stimulated at the ears (Stenfelt and Zeitooni 2013).

Another patient group fitted with the BAHA is one with unilateral profound deafness. In this case, the low attenuation of cross transmission for BC sound is used to transmit sound from the deaf side, picked up by the BAHA and transmitted to the skull bone at the deaf side, reaching the healthy cochlea by means of BC sound transmission (Stenfelt 2005). A review of the literature shows benefit, both subjective and objective, with this rehabilitation (Stewart et al. 2011). It should be noted that using BC hearing aids for treatment of unilateral deafness, or any other type of CROS hearing system, does not provide any binaural hearing but solely improves hearing sensitivity from the deaf side that is deteriorated by the head shadow.

A drawback of the current BAHA system is the skin penetration itself; such a solution requires special care and good hygiene. In a few percent of BAHA users, the skin problems become severe and, if untreated, can result is loss of the implant. Consequently, other alternatives to the BAHA are considered and two such alternatives are presented in Sects. 6.8.3 and 6.8.4.

6.8.3 Teeth-Applied Bone Conduction Hearing Aids

The teeth as a sensitive site for BC stimulation have been reported in the literature (Sabatino and Stromsta 1969; Dahlin et al. 1973; Stenfelt and Håkansson 1999). The teeth provide a more or less direct attachment to the skull bone (in the upper jaw). However, there is compliance in the teeth–bone interface reducing the high-frequency BC sound transmission (Stenfelt and Håkansson 1999). The teeth have been suggested as a site for attaching a BC hearing aid but the unfriendly environment in the oral cavity has hindered several attempts. Recently, a system was presented for unilaterally deaf patients in which the sound is picked up at the ear canal of the deaf side with a microphone and wirelessly transmitted to a removable bone transducer attached to the teeth in the oral cavity (Popelka et al. 2010b). This is yet another system using BC sound transmission for patients with unilateral deafness. One benefit compared to the BAHA system is that it avoids skin penetration and by that, surgery. But the use of the oral cavity for an active system is not yet proven in the marketplace.

6.8.4 Implanted Bone Conduction Hearing Aids

Another way to avoid skin penetration is to implant the BC hearing aid. Such systems are termed BCI: bone conduction implants (Håkansson et al. 2008, 2010). The idea of such a system is to have a microphone and sound processing unit on the outside and transmit the signal wirelessly to a receiver and transducer implanted in the mastoid of the skull, similar to that of cochlear implants or active middle ear implants. However, even if the skin penetration is avoided, surgery is still required. The concept of implanting the transducer facilitates a new transducer solution that may provide more effective stimulations modes (Adamson et al. 2010). Also, implanting the system enables a closer position of the transducer to the cochlea. A closer position provides greater BC sensitivity and separation between the cochleae beneficial for binaural hearing (Stenfelt and Goode 2005b; Eeg-Olofsson et al. 2008; Håkansson et al. 2010).

6.9 Summary

The research to understand the underlying processes of BC sound was intense in the mid-1900s. However, after Tonndorf's 1966 publication, the number of BC publications was meager for a period of time. With the advent of the BC hearing aids based on implantable techniques, there has been a significant increase in BC-related research. At the end of the twentieth century and beginning of the twenty-first century, BC research is once again flourishing. There are several reasons for this increase. The development of BC hearing aids requires new data to optimize behavior of these devices, but equally important are advances in measuring techniques (e.g., the laser Doppler vibrometer for contactless vibration measurements) and modeling and computational methods. Also, understanding the mechanisms for BC sound relies to a large extent on the understanding of normal AC hearing and the mechanical and physiological processes in the outer, middle, and inner ear. There remains a need to integrate knowledge of AC mechanisms and BC mechanisms into a common framework.

Most pathology in the outer and middle ear that severely affects the AC sound transmission affects the BC sensitivity only to a minor extent. Thus even if the changed BC sensitivity in a middle ear lesion is helpful for understanding underlying BC physiology, its clinical relevance is minor. Also, the use of BC thresholds for differential diagnosis of the specific middle ear lesion is risky; the Carhart notch is not always identifiable in cases of otosclerotic ears, and other lesions show BC depression similar to the Carhart notch. There are several pitfalls when conducting BC testing. The most common are occlusion of the ear canal, airborne sound radiation from the transducers, and unmasked or over-masked nontest ear.

With more than a century of research in the field of BC hearing, the importance of the contributors for BC sound is not clarified and there is no consensus on the issues. However, the literature suggests that the inner ear fluid inertia is the most important mechanism for speech frequencies. But several other contributors are generally within 10 dB of the most important one.

References

Aazh, H., Moore, B., Peyvandi, A., & Stenfelt, S. (2005). Influence of ear canal occlusion and static pressure difference on bone conduction thresholds: Implications for mechanisms of bone conduction. *International Journal of Audiology*, 44, 302–306.

Adamson, R., Bance, M., & Brown, J. (2010). A piezoelectric bone-conduction bending hearing actuator. *The Journal of the Acoustical Society of America*, 128(4), 2003–2008.

Advani, S. H., & Lee, Y.-C. (1970). Free vibrations of fluid-filled spherical shells. *Journal of Sound and Vibration*, 12(4), 453–462.

Ahmad, I., & Pahor, A. (2002). Carhart's notch: A finding in otitis media with effusion. *International Journal of Pediatric Otorhinolaryngology*, 64, 165–170.

Allen, G., & Fernandez, C. (1960). The mechanism of bone conduction. *Annals of Otology, Rhinology and Laryngology*, 69(1), 5–28.

Arlinger, S. D., Kylén, P., & Hellqvist, H. (1978). Skull distortion of bone conducted signals. *Acta Otolaryngologica*, 85, 318–323.

Awengen, D. (1993). Change of bone conduction thresholds by total footplate stapedectomy in relation to age. *American Journal of Otolaryngology*, 14(2), 105–110.

Banga, R., Lawrence, R., Reid, A., & McDermott, A. (2011). Bone-anchored hearing aids versus conventional hearing aids. *Advances in Oto-Rhino-Laryngology*, 71, 132–139.

Bárány, E. (1938). A contribution to the physiology of bone conduction. *Acta Oto-Laryngologica, Supplementum* 26, 1–223.

Beattie, R. (1998). Normative wave V latency-intensity functions using the EARTONE 3A insert earphone and the Radioear B-71 bone vibrator. *Scandinavian Audiology*, 27, 120–126.

Beattie, R., & Smiarowski, R. (1981). Bone-conducted speech: Intelligibility functions and threshold force levels for spondees. *The American Journal of Otology*, 3(2), 109–115.

Behn, A., Westerberg, B., Zhang, H., Riding, K., Ludemann, J., & Kozak, F. (2007). Accuracy of the Weber and Rhinne tuning fork tests in evaluation of children with otitis media with effusion. *Journal of Otolaryngology*, 36(4), 197–202.

Berger, E. H. (1983). Laboratory attenuation of earmuffs and earplugs both singly and in combination. *American Industrial Hygiene Association Journal*, 44(5), 321–329.

Berger, E. H., & Kerivan, J. E. (1983). Influence of physiological noise and the occlusion effect on the measurement of real-ear attenuation threshold. *The Journal of the Acoustical Society of America*, 74(1), 81–94.

Berger, K. W. (1976). Early bone conduction hearing aid devices. *Archives of Otolaryngology*, 102, 315–318.

Brinkmann, K., & Richter, U. (1983). Determination of the normal threshold of hearing by bone conduction using different types of bone vibrators. Part I. *Audiological Acustics* 22(3), 62–85.

Brinkman, W., Marres, E., & Tolk, J. (1965). The mechanism of bone conduction. *Acta Otolaryngol*, 59, 109–115.

Brooks, D. (1985). Fenestration: A twenty-five-year evaluation. *The Journal of Laryngology and Otology*, 99, 225–230.

Browning, G., & Gatehouse, S. (1984). Sensorineural hearing loss in stapedial otosclerosis. *Annals of Otology Rhinology and Laryngology*, 93, 13–16.

Carhart, R. (1950). Clinical application of bone conduction audiometry. *Archives of Otolaryngology*, 798–808.
Carhart, R. (1971). Effects of stapes fixation on bone-conductio response. In I. Ventry, J. Chaiklin & R. Dixon (Eds.), *Hearing measurement: A book of readings* (pp. 116–129). New York: Appleton-Century-Crofts.
Carle, R., Laugesen, S., & Nielsen, C. (2002). Observation on the relations among occlusion effect, compliance, and vent size. *Journal of American Academy of Audiology*, 10, 25–37.
Carlsson, P., Håkansson, B., & Ringdahl, A. (1995). Force threshold for hearing by direct bone conduction. *The Journal of the Acoustical Society of America*, 97(2), 1124–1129.
Collet, L., Chanal, J., Hellal, H., Gartner, M., & Morgon, A. (1989). Validity of bone conduction stimulated ABR, MLR and otoacoustic emissions. *Scandinavian Audiology*, 18(1), 43–46.
Colquitt, J., Loveman, E., Baguley, D., Mitchell, T., Sheehan, P., Harris, P., Proops, D., Jones, J., Clegg, A., & Welch, K. (2011). Bone-anchored hearing aids for people with bilateral hearing impairment: A systematic review. *Clinical Otolaryngology*, 36(5), 419–441.
Dahlin, G., Allen, F., & Collard, E. (1973). Bone-conduction thresholds of human teeth. *The Journal of the Acoustical Society of America*, 53(5), 1434–1437.
Dean, M. S., & Martin, F. N. (2000). Insert earphone depth and the occlusion effect. *American Journal of Audiology*, 9, 131–134.
Deas, R., Adamson, R., Curran, L., Makki, F., Bance, M., & Brown, J. (2010). Audiometric thresholds measured with single and dual BAHA transducers: The effect of phase inversion. *International Journal of Audiology*, 49(12), 933–999.
Dirks, D. (1964). Factors related to bone conduction reliability. *Archives of Otolaryngology*, 79, 551–558.
Dirks, D., & Malmquist, C. (1969). Comparison of frontal and mastoid bone-conduction thresholda in various conductive lesions. *Journal of Speech and Hearing Research*, 12, 725–746.
Eeg-Olofsson, M., Stenfelt, S., Tjellström, A., & Granström, G. (2008). Transmission of bone-conducted sound in the human skull measured by cochlear vibrations. *International Journal of Audiology*, 47(12), 761–769.
Eeg-Olofsson, M., Stenfelt, S., & Granström, G. (2011). Implications for contralateral bone conducted transmission as measured by cochlear vibrations. *Otology and Neurotology*, 32, 192–198.
Eeg-Olofsson, M., Stenfelt, S., Håkansson, B., Taghavi, H., & Reinfeldt, S. (2013). Transmission of bone conducted sound—correlation between hearing perception and cochlear vibration. In press.
Elpern, B., & Naunton, R. (1963). The stability of the occlusion effect. *Archives of Otolaryngology*, 77, 44–52.
Everberg, G. (1968). Congenital absence of the oval window. *Acta Oto-Laryngologica*, 66, 320–332.
Flottorp, G., & Solberg, S. (1976). Mechanical impedance of human headbones (forehead and mastoid portion of temporal bone) measured under ISO/IEC conditions. *The Journal of the Acoustical Society of America*, 59(4), 899–906.
Frank, T., & Crandell, C. (1986). Acoustic radiation produced by B-71, B-72, and KH 70 bone vibrators. *Ear and Hearing*, 7(5), 344–347.
Franke, E. (1956). Response of the human skull to mechanical vibrations. *The Journal of the Acoustical Society of America*, 28(6), 1277–1284.
Goldstein, D., & Hayes, C. (1971). The occlusion effect in bone-conduction hearing. In I. Ventry, J. Chaiklin & R. Dixon (Eds.), *Hearing measurement: A book of readings* (pp. 150–157). New York: Appleton-Century-Crofts.
Goodhill, V. (1966). External conductive hypacusis and the fixed malleus syndrome. *Acta Oto-Laryngologica, Supplementum* 217, 1–39.

Goodhill, V., Dirks, D., & Malmquist, C. (1970). Bone-conduction thresholds. Relationships of frontal and mastoid measurement in conductive hypacusis. *Archives Otolaryngology*, 91, 250–256.

Gopen, Q., Rosowski, J., & Merchant, S. (1997). Anatomy of the normal human cochlear aqueduct with functional implications. *Hearing Research*, 107, 9–22.

Groen, J., & Hoogland, G. (1958). Bone conduction and otosclerosis of the round window. *Acta Oto-Laryngologica*, 49, 206–212.

Håkansson, B. (2003). The balanced electromagnetic separationtransducer: Anew bone conduction transducer. *The Journal of the Acoustical Society of America*, 113(2), 818–825.

Håkansson, B., Tjellström, A., & Rosenhall, U. (1984). Hearing thresholds with direct bone conduction versus conventional bone conduction. *Scandinavian Audiology*, 13, 3–13.

Håkansson, B., Tjellström, A., & Rosenhall, U. (1985a). Acceleration levels at hearing threshold with direct bone conduction versus conventional bone conduction. *Acta Oto-Laryngologica*, 100, 240–252.

Håkansson, B., Tjellström, A., Rosenhall, U., & Carlsson, P. (1985b). The bone-anchored hearing aid. Principal design and psychoacoustical evaluation. *Acta Oto-Laryngologica*, 100, 229–239.

Håkansson, B., Carlsson, P., & Tjellström, A. (1986). The mechanical point impedance of the human head, with and without skin penetration. *The Journal of the Acoustical Society of America*, 80(4), 1065–1075.

Håkansson, B., Brandt, A., Carlsson, P., & Tjellström, A. (1994). Resonance frequency of the human skull in vivo. *The Journal of the Acoustical Society of America*, 95(3), 1474–1481.

Håkansson, B., Carlsson, P., Brandt, A., & Stenfelt, S. (1996). Linearity of sound transmission through the human skull in vivo. *The Journal of the Acoustical Society of America*, 99(4), 2239–2243.

Håkansson, B., Eeg-Olofsson, M., Reinfeldt, S., Stenfelt, S., & Granström, G. (2008). Percutaneous versus transcutaneous bone conduction implant system: A feasibility study on a cadaver head. *Otology and Neurotology*, 29(8), 1132–1139.

Håkansson, B., Reinfeldt, S., Eeg-Olofsson, M., Ostli, P., Taghavi, H., Adler, J., Gabrielsson, J., Stenfelt, S., & Granström, G. (2010). A novel bone conduction implant (BCI): Engineering aspects and pre-clinical studies. *International Journal of Audiology*, 49, 203–215.

Haeff, A., & Knox, C. (1963). Perception of ultrasound. *Science*, 139, 590–592.

Homma, K., Du, Y., Shimizu, Y., & Puria, S. (2009). Ossicular resonance modes of the human middle ear for bone and air conduction. *The Journal of the Acoustical Society of America*, 125(2), 968–979.

Homma, K., Shimizu, Y., Kim, N., Du, Y., & Puria, S. (2010). Effects of ear-canal pressurization on middle-ear bone- and air-conduction responses *Hearing Research*, 263, 204–215.

Hosoi, H., Imaizumi, S., Sakaguchi, T., Tonoike, M., & Murata, K. (1998). Activation of the auditory cortex by ultrasound. *The Lancet*, 351, 496–497.

House, W. (1959). Oval window and round window surgery in extensive otosclerosis, a preliminary report. *The Laryngoscope*, 69, 693–701.

Howell, P., & Williams, M. (1989). Jaw movement and bone-conduction in normal listeners and a unilateral hemi-mandibulectomee. *Scandinavian Audiology*, 18, 231–236.

Howell, P., Williams, M., & Dix, H. (1988). Assessment of sound in the ear canal caused by movement of the jaw relative to the skull. *Scandinavian Audiology*, 17, 93–98.

Hoyer, H.-E., & Dörheide, J. (1983). A study of human head vibrations using time-avaraged holography. *Journal of Neurosurgery*, 58, 729–733.

Hudde, H. (2005). A Functional View on the Peripheral Human Hearing Organ. In J. Blauert (Ed.), *Communication acoustics* (pp. 47–74). Berlin: Springer.

Huizing, E. (1960). Bone conduction—The influence of the middle ear. *Acta Oto-Laryngologica, Supplementum*, 155, 1–99.

Humes, L. (1979). The middle ear inertial component of bone conduction hearing in man. *Audiology*, 18, 24–35.

Hurley, R. M., & Berger, K. W. (1970). The relationship between vibrator placement and bone conduction measurements with monaurally deaf subjects. *Journal of Auditory Research*, 10, 147–150.

Irvine, D. (1976). Effects of reflex middle-ear muscle contractions on cochlear responses to bone-conducted sound. *Audiology*, 15, 433–444.

Irvine, D., Yates, G., & Johnstone, B. (1979). Bone conduction mechanisms: Mössbauer measurements on the role of ossicular inertia. *Hearing Research*, 1, 101–109.

Ishida, I., Cuthbert, B., & Stapells, D. (2011). Multiple auditory steady state response thresholds to bone conduction stimuli in adults with normal and elevated thresholds. *Ear and Hearing*, 32(3), 373–381.

ISO:8253-1. (2010). International Organization for Standardization *Acoustics – Audiometric test methods—*Part 1*: Pure-tone air and bone conduction audiometry*. Geneva.

Jervall, L., & Arlinger, S. (1986). A compairson of 2-dB and 5-dB step size in pure-tone audiometry. *Scandinavian Audiology*, 15, 51–56.

Kandzia, F., Oswald, J., & Janssen, T. (2011). Binaural measurement of bone conduction click evoked otoacoustic emissions in adults and infants. *The Journal of the Acoustical Society of America*, 129(3), 1464–1474.

Kaplan, D., Fliss, D., Kraus, M., Dagan, R., & Leiberman, A. (1996). Audiometric findings in children with chronic suppurative otitis media without cholesteatoma. *International Journal of Pediatric Otorhinolaryngology*, 35(2), 89–96.

Khalil, T. B., & Hubbard, R. P. (1977). Parametric study of head response by finite element modeling. *Journal of Biomechanics*, 10, 119–132.

Khalil, T. B., Viano, D. C., & Smith, D. L. (1979). Experimental analysis of the vibrational characteristics of the human skull. *Journal of Sound and Vibration*, 63(3), 351–376.

Khanna, S. M., Tonndorf, J., & Queller, J. (1976). Mechanical parameters of hearing by bone conduction. *The Journal of the Acoustical Society of America*, 60, 139–154.

Killion, M., Wilber, L., & Gudmundsen, G. (1988). Zwislocki was right... *Hearing Instruments*, 39, 14–18

Kim, N., Homma, K., & Puria, S. (2011). Inertial bone conduction: Symmetric and anti-symmetric components. *Journal of the Association for Research in Otolaryngology*, 12, 261–279.

Klodd, D., & Egerton, B. (1977). Occlusion effect: bone conduction speech audiometry using forehead and mastoid placement. *Audiology*, 16(6), 522–529.

Kringlebotn, M. (1995). The equality of volume displacement in the inner ear windows. *The Journal of the Acoustical Society of America*, 98(1), 192–196.

Kucuk, B., Abe, K., & Ushiki, T. (1991). Microstructures of the bony modiolus in the human cochlea: Scanning electron microscopic study. *Journal of Electron Microscopy*, 40, 193–197.

Kylen, P., Harder, H., Jerlvall, L., & Arlinger, S. (1982). Reliability of bone-conducted electrocochleography. A clinical study. *Scandinavian Audiology*, 11(4), 223–226.

Laske, R., Röösli, C., Chatzimichalis, M., Sim, J., & Huber, A. (2011). The influence of prosthesis diameter in stapes suregery: A meta analysis and systematic review of the literature *Otology and Neurotology*, 32(4), 520–528.

Laukli, E., & Fjermedal, O. (1990). Reproducibility of hearing threshold measurements. Supplementary data on bone-conduction and speech audiometry. *Scandinavian Audiology*, 19(3), 187–190.

Lee, H., Hong, S., Hong, S., Choi, Y., & Chung, W. (2008). Ossicular chain reconstruction improves bone conduction threshold in chronic otitis media. *The Journal of Laryngology and Otology*, 122, 351–356.

Legouix, J., & Tarab, S. (1959). Experimental study of bone conduction in ears with mechanical impairment of the ossicles. *The Journal of the Acoustical Society of America*, 31(11), 1453–1457.

Lenhardt, M., Skellett, R., Wang, P., & Clarke, A. (1991). Human ultrasonic speech perception. *Science*, 253, 82–85.

Linder, T., Ma, F., & Huber, A. (2003). Round window atresia and its effect on sound transmission. *Otology and Neurotology*, 24(2), 259–263.
Lindstrom, C., Rosen, A., Silverman, C., & Meiteles, L. (2001). Bone conduction impairment in chronic ear disease. *Annals of Otology, Rhinology and Laryngology*, 110, 437–441.
Lowy, K. (1942). Cancellation of the electrical cochlear response with air- and bone-conducted sound. *The Journal of the Acoustical Society of America*, 14(2), 156–158.
Martin, C., Tringali, S., Bertholon, P., Pouget, J.-F., & Prades, J.-M. (2002). Isolated congenital round window absence. *Annals of Otology Rhinology and Laryngology*, 111, 799–801.
Merchant, S., & Rosowski, J. (2008). Conductive hearing loss caused by third-window lesions of the inner ear. *Otology and Neurotology*, 29(3), 282–289.
Merchant, S., Nakajima, H., Halpin, C., Nadol, J. J., Lee, D., Innis, W., Curtin, H., & Rosowski, J. (2007). Clinical investigation and mechanism of air-bone gaps in large vestibular aqueduct syndrome. *Annals of Otology Rhinology and Laryngology*, 116(7), 532–541.
Mikulec, A., McKenna, M., Ramsey, M., Rosowski, J., Herrmann, B., Rauch, S., Curtin, H., & Merchant, S. (2004). Superior semicircular canal dehiscence presenting as conductive hearing loss without vertigo. *Otology and Neurotology*, 25, 121–129.
Milner, R., Weeller, C., & Breman, A. (1983). Elevated bone conduction thresholds associated with middle ear fluid in adults. *International Journal of Pediatric Otorhinolaryngology*, 6(2), 163–169.
Miltenburg, D. (1994). The validity of tuning fork tests in diagnosing hearing loss. *Journal of Otolaryngology*, 23(4), 254–259.
Miyamoto, R., & House, H. (1978). Cochlear reserve in otosclerosis. A long-term follow-up of fenestration cases. *Archives of Otolaryngology*, 104, 464–466.
Møller, A. (2000). *Hearing: Its physiology and pathophysiology*. San Diego: Academic Press.
Morgan, D., & Dirks, D. (1975). Influence of middle-ear muscle contraction on pure-tone suprathreshold loudness judgments. *The Journal of the Acoustical Society of America*, 57, 411–420.
Mudry, A., & Tjellström, A. (2011). Historical background of bone conduction hearing devices and bone conduction hearing aids. *Advances in Oto-Rhino-Laryngology*, 71, 1–9.
Munro, K. J., Paul, B., & Cox, C. L. (1997). Normative auditory brainstem response data for bone conduction in the dog. *Journal of Small Animal Practice*, 38, 353–356.
Mylanus, E., Snik, A., & Cremers, C. (1994). Influence of the thickness of the skin and subcutaneous tissue covering the mastoid on bone-conduction thresholds obtained transcutaneously versus percutaneously. *Scandinavian Audiology*, 23, 201–203.
Naunton, R. (1963). The measurement of hearing by bone conduction. In J. Jerger (Ed.), *Modern developments in audiology* (pp. 1–29). New York: Academic Press.
Nilo, E. (1968). The relation of vibrator surface area and static application force to the vibrator-to-head coupling. *Journal of Speech and Hearing Research*, 11(4), 805–810.
Nishimura, T., Okayasu, T., Uratani, Y., Fukuda, F., Saito, O., & Hosoi, H. (2011). Peripheral perception mechanism of ultrasonic hearing. *Hearing Research*, 277, 176–183.
Nolan, M., & Lyon, D. J. (1981). Transcranial attenuation in bone conduction audiometry. *The Journal of Laryngology and Otology*, 95, 597–608.
Nolan, M., Lyon, D., & Mok, C. (1985). Air pressure changes in the external auditory meatus: The influence on pure tone bone conduction thresholds. *The Journal of Laryngology and Otology*, 99, 315–326.
Ogura, Y., Masuda, Y., Miki, M., Takeda, T., Watanabe, S., Ogawara, T., Shibata, S., Uyemura, T., & Yamamoto, Y. (1979). Vibration analysis of the human skull and auditory ossicles by holographic interferometry. In G. v. Bally (Ed.), *Holography in medicine and biology*. Berlin: Springer-Verlag.
Ono, H. (1977). Improvement and evaluation of the vibration pick-up-type ear microphone and two-way communication device. *The Journal of the Acoustical Society of America*, 62(3), 760–768.

Palva, T., & Ojala, L. (1955). Middle ear conduction deafness and bone conduction. *Acta Oto-Laryngologica*, 45, 137–152.
Perez, R., Adelman, C., & Sohmer, H. (2011). Bone conduction activation through soft tissues following complete immobilization of the ossicular chain, stapes footplate and round window. *Hearing Research*, 280, 82–85.
Persson, P., Harder, H., & Magnuson, B. (1997). Hearing results in otosclerosis surgery after partial stapedectomy, total stapedectomy and stapedotomy. *Acta Oto-Laryngologica*, 117(1), 94–99.
Peterson, L., & Bogert, B. (1950). A dynamical theory of the cochlea. *The Journal of the Acoustical Society of America*, 22, 369–381.
Pörschmann, C. (2000). Influences of bone conduction and air conduction on the sound of one's own voice. *Acustica – Acta Acustica*, 86, 1038–1045.
Popelka, G., Telukuntla, G., & Puria, S. (2010a). Middle-ear function at high frequencies quantified with advanced bone-conduction measures. *Hearing Research*, 263, 85–92.
Popelka, G., Derebery, J., Blevins, N., Murray, M., Moore, B., Sweetow, R., Wu, B., & Katsis, M. (2010b). Preliminary evaluation of a novel bone-conduction device for single-sided deafness. *Otology and Neurotology*, 31(3), 492–497.
Purcell, D., Kunov, H., Madsen, P., & Cleghorn, W. (1998). Distortion product otoacoustic emissions stimulated through bone conduction. *Ear and Hearing*, 19(5), 362–370.
Ranke, O., Keidel, W., & Weschke, H. (1952). Des Hören beim Verschluss des runden Fensters. *Zeitschrift für Laryngologie*, 31, 467–475.
Reinfeldt, S., Stenfelt, S., & Håkansson, B. (2007a). Transcranial transmission of bone conducted sound measured acoustically and psychoacoustically. In A. Huber & A. Eiber (Eds.), *Middle ear mechanics in research and otology: Proceedings of the 4th international symposium* (pp. 276–281). Singapore: World Scientific.
Reinfeldt, S., Stenfelt, S., Good, T., & Håkansson, B. (2007b). Examination of bone-conducted transmission from sound field excitation measured by thresholds, ear-canal sound pressure, and skull vibrations. *The Journal of the Acoustical Society of America*, 121(3), 1576–1587.
Reinfeldt, S., Ostli, P., Håkansson, B., & Stenfelt, S. (2010). Hearing one's own voice during phoneme vocalization–transmission by air and bone conduction. *The Journal of the Acoustical Society of America*, 128, 751–762.
Richter, U., & Brinkmann, K. (1981). Threshold of hearing by bone conduction. *Scandinavian Audiology*, 10, 235–237.
Sabatino, D., & Stromsta, C. (1969). Bone conduction thresholds from three locations on the skull. *The Journal of Auditory Research*, 9, 194–198.
Salomon, G., & Elberling, C. (1988). Estimation of inner ear function and conductive hearing loss based on electrocochleography. *Advances in Audiology*, 5, 46–55.
Sato, E., Sugiura, M., Naganawa, S., Yoshino, T., Mizuno, T., Otake, H., Ishida, I., & Nakashima, T. (2007). Effect of an enlarged endolymphatic duct on bone conduction threshold. *Acta Oto-Laryngologica*, 128, 534–538.
Schwartz, D., Larson, V., & DeChicchis, A. (1985). Spectral characteristics of air and bone conduction transducers used to record the auditory brain stem response. *Ear and Hearing*, 6(5), 274–277.
Shabana, Y., Ghonim, M., & Pedersen, C. (1999). Stapedotomy, does prosthesis diameter affect outcome. *Clinical Otolaryngology*, 24, 91–94.
Shera, C., & Zweig, G. (1992). An empirical bound on the compressibility of the cochlea. *The Journal of the Acoustical Society of America*, 92(3), 1382–1388.
Shipton, M. S., John, A. J., & Robinson, D. W. (1980). Air-radiated sound from bone vibration transducers and its implications for bone conduction audiometry. *British Journal of Audiology*, 14, 86–99.
Shishegar, M., Faramarzi, A., Esmaili, N., & Heydari, S. (2009). Is Carhart notch an accurate predictor of otitis media with effusion. *International Journal of Pediatric Otorhinolaryngology*, 73, 1799–1802.

Siegert, R. (2011). Partially implantable bone conduction hearing aids without a percutaneous abutment (Otomag): Technique and preliminary clinical results. *Advances in Oto-Rhino-Laryngology*, 71, 41–46.

Small, S., & Stapells, D. (2003). Normal brief-tone bone-conduction behavioral thresholds using the B-71 transducer: Three occlusion conditions. *Journal of American Academy of Audiology*, 14(10), 556–562.

Snik, A., Mylanus, E., & Cremers, C. (1995). The bone-anchored hearing aid compared with conventional hearing aids. *Otolaryngyologic Clinics of North America*, 28(1), 73–83.

Snik, A. F., Mylanus, E. A. M., Proops, D. W., Wolfaardt, J. F., Hodgetts, W. E., Somers, T., Niparko, J. K., Wazen, J. J., Sterkers, O., Cremers, C. W. R. J., & Tjellström, A. (2005). Consensus statements on the BAHA system: Where do we stand at present? *Annals of Otology, Rhinology and Laryngology*, 114(12), *Supplementum* 195:191–112.

Snyder, J. (1973). Interaural attenuation characteristics in audiometry. *The Laryngoscope*, 83, 1847–1855.

Sohmer, H., Freeman, S., Geal-Dor, M., Adelman, C., & Savion, I. (2000). Bone conduction experiments in humans—a fluid pathway from bone to ear. *Hearing Research*, 146, 81–88.

Songer, J., & Rosowski, J. (2007). A mechano-acoustic model of the effect of superior canal dehiscence on hearing in chinchilla. *The Journal of the Acoustical Society of America*, 122(2), 943–951.

Songer, J., & Rosowski, J. (2010). A superior semicircular canal dehiscence-induced air-bone gap in chinchilla. *Hearing Research*, 269, 70–80.

Stenfelt, S. (2005). Bilateral fitting of BAHAs and BAHA fitted in unilateral deaf persons: Acoustical aspects. *International Journal of Audiology*, 44, 178–189.

Stenfelt, S. (2006). Middle ear ossicles motion at hearing thresholds with air conduction and bone conduction stimulation. *The Journal of the Acoustical Society of America*, 119(5), 2848–2858.

Stenfelt, S. (2007). Simultaneous cancellation of air and bone conduction tones at two frequencies: Extension of the famous experiment by von Békésy. *Hearing Research*, 225, 105–116.

Stenfelt, S. (2011). Acoustic and physiologic aspects of bone conduction hearing. *Advances in Oto-Rhino-Laryngology*, 71, 10–21.

Stenfelt, S. (2012a). A model for prediction of own voice alteration with hearing aids. In T. Dau, M.L. Jepsen, T. Poulsen, J.C. Dalsgaard (Eds.), *Speech Perception and Auditory Disorders* (pp. 323–330). The Danavox Jubilee Foundation

Stenfelt, S. (2012b). Transcranial attenuation of bone conducted sound when stimulation is at the mastoid and at the bone conduction hearing aid position. *Otology and Neurotology*, 33, 105–114.

Stenfelt, S., & Håkansson, B. (1999). Sensitivity to bone-conducted sound: Excitation of the mastoid vs the teeth. *Scandinavian Audiology*, 28(3), 190–198.

Stenfelt, S., & Goode, R. (2005a). Bone conducted sound: Physiological and clinical aspects. *Otology and Neurotology*, 26, 1245–1261.

Stenfelt, S., & Goode, R. L. (2005b). Transmission properties of bone conducted sound: Measurements in cadaver heads. *The Journal of the Acoustical Society of America*, 118(4), 2373–2391.

Stenfelt, S., & Reinfeldt, S. (2007). A model of the occlusion effect with bone-conducted stimulation. *International Journal of Audiology*, 46(10), 595–608.

Stenfelt, S., & Puria, S. (2010). Consider bone-conducted human hearing. In C. O'Connell-Rodwell (Ed.), *The use of vibrations in communication: Properties, mechanisms and function across taxa* (pp. 142–162). Kerala: Research Signpost.

Stenfelt, S., & Zeitooni, M. (2013). Binaural hearing with bone conduction stimulation. *In press*.

Stenfelt, S., Håkansson, B., & Tjellström, A. (2000). Vibration characteristics of bone conducted sound *in vitro*. *The Journal of the Acoustical Society of America*, 107(1), 422–431.

Stenfelt, S., Hato, N., & Goode, R. (2002). Factors contributing to bone conduction: The middle ear. *The Acoustical Society of America*, 111(2), 947–959.

Stenfelt, S., Wild, T., Hato, N., & Goode, R. L. (2003a). Factors contributing to bone conduction: The outer ear. *The Journal of the Acoustical Society of America*, 113(2), 902–912.

Stenfelt, S., Puria, S., Hato, N., & Goode, R. L. (2003b). Basilar membrane and osseous spiral lamina motion in human cadavers with air and bone conduction stimuli. *Hearing Research*, 181, 131–143.

Stenfelt, S., Hato, N., & Goode, R. L. (2004). Round window membrane motion with air conduction and bone conduction stimulation. *Hearing Research*, 198, 10–24.

Stewart, C., Clark, J., & Niparko, J. (2011). Bone-anchored devices in single-sided deafness. *Advances in Oto-Rhino-Laryngology*, 71, 92–102.

Studebaker, G. (1962). Placement of vibrator in bone-conduction testing. *Journal of Speech and Hearing Research*, 5(4), 321–331.

Studebaker, G. (1964). Clinical masking of air and bone conducted stimuli. *Journal of Speech and Hearing Disorders*, 29, 23–35.

Tange, R., Bruijn, A., & Dreschler, W. (2000). Gold and teflon in the oval window: A comparison of stapes prostheses. In J. Rosowski & S. Merchant (Eds.), *The function and mechanics of normal, diseased and reconstructed middle ears* (pp. 255–260). Amsterdam: Kugler Publications.

Taschke, H., & Hudde, H. (2006). A finite element model of the human head for auditory bone conduction simulation. *ORL; Journal for Oto-Rhino-Laryngology and Its Related Specialties*, 68(6), 319–323.

Teig, E., & Lindeman, H. (2000). Stapedotomy piston diameter: Is bigger better? In J. Rosowski & S. Merchant (Eds.), *The function and mechanics of normal, diseased and reconstructed middle ears* (pp. 281–287). Amsterdam: Kugler Publications.

Toll, L., Emanuel, D., & Letowski, T. (2011). Effect of static force on bone conduction hearing thresholds and comfort. *International Journal of Audiology*, 50(9), 632–635.

Tonndorf, J. (1966). Bone conduction: Studies in experimental Animals. *Acta Oto-Laryngologica, Supplementum* (213), 1–132.

Tonndorf, J. (1972). Bone conduction. In J. Tobias (Ed.), *Foundations of modern auditory theory* (Vol. II, pp. 197–237). New York: Academic Press.

Tonndorf, J., & Jahn, A. F. (1981). Velocity of propagation of bone-conducted sound in a human head. *The Journal of the Acoustical Society of America*, 70(5), 1294–1297.

Walsh, T. (1962). Fenestration: Results, indications, limitations. In H. Schuknecht (Ed.), *Otosclerosis* (pp. 245–250). Boston: Little, Brown and Company.

Watanabe, T., Bertoli, S., & Probst, R. (2008). Transmission pathways of vibratory stimulation as measured by subjective thresholds and distortion-product otoacoustic emissions. *Ear and Hearing*, 29, 667–673.

Wever, E. G., & Lawrence, M. (1954). *Physiological acoustics*. Princeton, NJ: Princeton University Press.

von Békésy, G. (1932). Zur Theorie des Hörens bei der Schallaufnahme durch Knochenleitung. *Annalen der Physik*, 13, 111–136.

von Békésy, G. (1941). Uber die Schallausbreitung bei Knochenleitung. *Zeitschrift für Hals-, Nasen- und Ohrenheilkunde*, 47, 430–442.

von Békésy, G. (1948). Vibration of the head in a sound field, and its role in hearing by bone conduction. *The Journal of the Acoustical Society of America*, 20, 727–748.

von Békésy, G. (1949). The structure of the middle ear and the hearing of one's own voice by bone conduction. *The Journal of the Acoustical Society of America*, 21(3), 217–232.

von Békésy, G. (1960). *Experiments in hearing*. New York: McGraw- Hill.

Voss, S., Rosowski, J., & Peake, W. (1996). Is the pressure difference between the oval and round windows the effective acoustic stimulus for the cochlea. *The Journal of the Acoustical Society of America*, 100(3), 1602–1616.

Yetiser, S., Hidir, Y., Birkent, H., Satar, B., & Durmaz, A. (2008). Traumatic ossicular dislocations: Etiology and management. *American Journal of Otolaryngology*, 29, 31–36.

Yi, Z., Yang, J., Li, Z., Zhou, A., & Lin, Y. (2003). Bilateral congenital absence of stapes and oval window in 2 members of a family: Etiology and management. *The Journal of Laryngology and Otology*, 121, 219–221.

Young, P. G. (2002). A parametric study on the axisymmetric modes of vibration of multi-layered spherical shells with liquid cores of relevance to head impact modelling. *Journal of Sound and Vibration*, 256(4), 665–680.

Young, P. G. (2003). An analytical model to predict the response of fluid-filled shells to impact—a mosell for blunt head impacts. *Journal of Sound and Vibration*, 267, 1107–1126.

Ysan, H. 2007. Predictive role of Carhart's notch in pre-operative assesment for middle ear surgery. *The Journal of Laryngology and Otology* 121, 219–221.

Zheng, Y., Liu, Z., Zhang, Z., Sinclair, M., Droppo, J., Deng, L., Acero, A., & Huang, X. (2003). Air and bone conductive integrated microphones for robust speech detection and enhancement. *IEEE Workshop on Automatic Speech Recognition and Understanding*, 249–254.

Zwislocki, J. (1953). Acoustic attenuation between the ears. *The Journal of the Acoustical Society of America*, 25(4), 752–759.

Chapter 7
Modeling of Middle Ear Mechanics

W. Robert J. Funnell, Nima Maftoon, and Willem F. Decraemer

Keywords Air cavities • Circuit models • Eardrum • External ear canal • Finite-element models • Image segmentation • Material properties • Mathematical models • Mesh generation • Ossicular chain • Parameter fitting • 3-D shape measurement • Tympanic membrane • Uncertainty analysis • Verification and validation

7.1 Introduction

Quantitative understanding of the mechanical behavior of the external and middle ear is important, not only in the quest for improved diagnosis and treatment of conductive hearing loss but also in relation to other aspects of hearing that depend on the conductive pathways. Mathematical modeling is useful in arriving at that understanding.

The middle ear is of course more than just a mechanical system: it has physiological aspects (e.g., muscle contraction, healing) and biochemical aspects (e.g., gas exchange) that directly affect its mechanical behavior. Even when it is studied only from a mechanical point of view, however, it presents considerable challenges.

W.R.J. Funnell (✉)
Departments of BioMedical Engineering and Otolaryngology – Head & Neck Surgery, McGill University, 3775, rue University, H3A 2B4 Montréal, QC, Canada
e-mail: robert.funnell@mcgill.ca

N. Maftoon
Department of BioMedical Engineering, McGill University, 3775, rue University, H3A 2B4 Montréal, QC, Canada
e-mail: nima.maftoon@mail.mcgill.ca

W.F. Decraemer
Laboratory of BioMedical Physics, University of Antwerp, Groenenborgercampus, Groenenborgerlaan 171, B-2020 Antwerp, Belgium
e-mail: wim.decraemer@ua.ac.be

S. Puria et al. (eds.), *The Middle Ear: Science, Otosurgery, and Technology*, Springer Handbook of Auditory Research 46, DOI 10.1007/978-1-4614-6591-1_7,

For one thing, it has a complicated and irregular geometry involving a number of distinct structures encompassing a wide range of sizes. Its overall dimensions are in the range of tens of millimeters but it has important dimensions measured in micrometers (e.g., the thickness of the eardrum). One can go even further down the scale and consider the dimensions of the collagen fibers that are mechanically important in the eardrum. The displacements that one must be able to measure to characterize middle ear mechanics are as small as nanometers in response to sound pressures but as large as millimeters in response to static pressures. The time scales for the mechanical responses of the middle ear range from tens of microseconds for high-frequency sounds to tens of seconds for changes of static pressure, and even millions of seconds for the mechanical changes involved in development and healing.

The challenge of the external and middle ear is increased by the many different tissue types involved with very different mechanical behaviors: bone; fibrous connective tissue, with its collagen, elastin, and ground substance; muscle, both striated and smooth; cartilage, both calcified and uncalcified; and synovial fluid. The mechanical properties of low-density air (in the canal and cavities) and high-density water (in the cochlea) are also involved.

This chapter starts by reviewing some background modeling topics: Sects. 7.2 and 7.3 discuss some general issues related to the modeling of geometry and of material properties as required for realistic models, while Sect. 7.4 is a discussion of model verification and validation, including the issues of uncertainty analysis and parameter fitting. (See Funnell et al. (2012) for a tutorial review of the underlying mechanical principles and modeling approaches.) Sect. 7.5 is a review of models that have been presented for the outer and middle ear, divided into canal, air cavities, eardrum, ossicular chain, and cochlea, followed by a very brief treatment of nonlinearity. The chapter ends with the discussion in Sect. 7.6.

7.2 Geometry Modeling

Realistic modeling requires more or less accurate three-dimensional (3-D) shapes. This section includes a brief and qualitative review of the sources of such shape data, and then a discussion of the processing required for preparation of the geometric meshes used in finite-element models.

7.2.1 Sources of Shape Data

This section discusses various sources of 3-D shape data. Most of the techniques involve cross-sectional images of some kind, but section "Surface-Shape Measurement" includes techniques that work directly from the surfaces of objects.

Decraemer et al. (2003) presented a brief overview of some of these methods as used in the middle ear. Clinical techniques are reviewed by Popelka and Hunter in Chap. 8.

7.2.1.1 Light Microscopy

Precise 3-D shape data may be obtained from serial-section histology. This technique is very time consuming and involves many processing steps: fixation, decalcification, dehydration, embedding, sectioning, staining, and mounting. Some of these processes can be automated (e.g., Odgaard et al. 1994). The embedding step may be replaced by freezing. Histology is particularly challenging for the ear because the petrosal part of the temporal bone is very dense and hard; because the eardrum is unsupported and extremely thin; and because the ossicles are suspended in air by small ligaments.

It is challenging to make 3-D reconstructions from histological sections because of the need to align the images to one another. The alignment problem is made worse by the fact that individual sections are typically stretched, folded, and torn in unpredictable ways. The processing also involves some degree of tissue shrinkage (e.g., Kuypers et al. 2005a for the eardrum). Alignment problems can be reduced by photographing the surface of the tissue block as each successive slice is removed (e.g., Sørensen et al. 2002; Jang et al. 2011), but the resolution is limited to that of the camera.

In confocal microscopy the sectioning is done optically by using pinhole apertures or very narrow slits (e.g., Koester et al. 1994). This technique is often combined with the use of fluorescent dyes. The effectiveness of optical sectioning can be greatly improved by the use of multiphoton microscopy. In second-harmonic generation (SHG) and third-harmonic generation (THG) microscopy, two or three photons are transformed into one photon with two or three times the energy (e.g., Sun 2005). Because the amount of energy emitted is the same as the amount of energy absorbed, there is no net energy absorption and the photo-bleaching and damage problems of conventional fluorescence do not occur. Jackson et al. (2008) used a combination of two-photon fluorescence and SHG to visualize collagen fibers in human eardrums. Lee et al. (2010) used SHG and THG on the rat eardrum; unlike previous users of confocal microscopy, they did not need to excise and flatten the eardrum, so they could observe the conical shape of the drum as well as its thickness and layered structure.

Optical coherence tomography (OCT) obtains the effect of optical sectioning by effectively measuring the different travel times of light reflected from different depths, either directly in the time domain or in the frequency domain (Fercher 2010; Wojtkowski 2010). OCT has shown promise in imaging the middle ear (Just et al. 2009) and can also be used for vibration measurements (Subhash et al. 2012).

These various types of optical sectioning can be used in vivo, and much effort is being put into minimizing the amount of light required and maximizing the speed with which changes can be tracked (e.g., De Mey et al. 2008; Carlton et al. 2010).

Another light-microscopy technique is orthogonal-plane fluorescence optical sectioning (OPFOS), in which the effect of sectioning is obtained by shining a thin sheet of laser light through the specimen from the side. Voie et al. (1993) introduced the use of this technique for the inner ear, and Buytaert et al. (2011) have developed a higher-resolution version and applied it to the middle ear. It requires tissue processing similar to that required for histological sections, but it provides very high resolution without any of the alignment problems associated with physical sections.

7.2.1.2 X-Ray Computed Tomography

X-rays can provide information about the interiors of solid objects because they penetrate further than visible light does. Computed tomography (CT) uses image-processing algorithms to combine multiple X-ray images, from many different angles around an object, to produce cross-sectional images of the interior. Compared with histology, this provides the dramatic advantages that it is not necessary to physically cut (and thus destroy) the object, and that there are no alignment problems at all.

The spatial resolution of current clinical CT scanners is such that the outer ear and the general form of the middle ear air cavities are fairly clear (e.g., Egolf et al. 1993), but few details of the ossicles can be seen and none of the ligaments (e.g., Lee et al. 2006). However, Vogel and Schmitt (1998) demonstrated the use of a "microfocus" X-ray tube for microtomography for the ear. At about the same time, Sasov and Van Dyck (1998) described a desktop microCT scanner built with commercially available components and demonstrated its use for the ear. That scanner was quickly used to support the analysis of middle ear vibration measurements (Decraemer and Khanna 1999). Several models of microCT scanner are now commercially available and further development continues (e.g., Salih et al. 2012). Resolutions down to a few micrometers can be obtained for small, dissected specimens; scan times tend to be tens of minutes.

X-ray absorption increases as bone density increases, and this can be used to estimate variations of Young's modulus within a bone. This is often done for large bones and has been attempted for the middle ear ossicles (Yoo et al. 2004).

One limitation of current microCT scanners is that their X-ray sources produce a fairly broad band of frequencies. The fact that softer (lower-frequency) X-rays are absorbed more than harder ones leads to a phenomenon known as beam hardening, which causes image artifacts that are difficult to avoid. It is possible to filter out some of the softer X-rays, but this greatly reduces the already limited intensity of the beam. Synchrotron radiation, however, although available only in a few centers, provides very bright and highly collimated X-ray beams, with a very narrow (practically monochromatic) frequency range. Vogel (1999) used it for the middle and inner ear. Both absorption-contrast and phase-contrast modes can be used. It is possible to use multiple beam energies and to combine the individual gray-scale images to produce false-color images. Synchrotron-radiation CT has not been much used for the middle ear, but the appearance of some recent papers (Neudert et al. 2010; Kanzaki et al. 2011) suggests that it may become more common.

The resolutions of clinical CT scanners will continue to improve, and hand-held X-ray scanners are possible (Webber et al. 2002).

7.2.1.3 Magnetic Resonance Imaging

The resolution of current clinical magnetic resonance (MR) imaging is even lower than that of clinical X-ray CT. However, Johnson et al. (1986) first reported on MR "microscopy" using a modified clinical MR scanner. The resolution that is obtained can be improved by attention to many factors, including smaller coils (and thus smaller specimens), more averaging (and thus longer acquisition time), higher magnetic-field gradients, and larger image-matrix size. In that year, three different groups reported resolutions of tens of micrometers in two axes but hundreds of micrometers in the third axis (e.g., Johnson et al. 1986). The resolutions that they achieved depended in part on what they chose to image. Within a few years, Henson et al. (1994) obtained an isotropic voxel size of $25 \times 25 \times 25$ μm in the ear.

MR imaging provides good contrast between different types of soft tissue. Although MR images are inherently monochromatic, multiple data-acquisition parameters can be used to emphasize different tissue types and the individual gray-scale images can be combined to produce false-color images. Given that MR depends on the presence of protons (e.g., Reiser et al. 2008), it provides practically no contrast between air and cortical bone because neither has many protons. This is a problem for imaging the middle ear air space and ossicles but it can be circumvented by filling the air cavities with a liquid gadolinium–based contrast agent (Wilson et al. 1996). The filling has to be done very carefully to avoid air bubbles, and it is likely to displace the eardrum significantly.

MR scanners, both clinical and microscopic, are more expensive and less widely available than their X-ray CT counterparts but are becoming more common.

7.2.1.4 Electron Microscopy

Electron microscopy (EM) can provide much higher spatial resolutions than light microscopy, but there is no possibility of using different stains to enhance contrast. Scanning EM is analogous to looking at solid objects under a microscope by reflected light, while transmission EM is analogous to looking at histological sections. EM is currently the method of choice for imaging details like the fibrous ultrastructure of the eardrum (e.g., Lim 1995). 3-D reconstructions have been performed using both physical sectioning with scanning EM (Denk and Horstmann 2004) and computed tomography with transmission EM (e.g., Koning and Koster 2009). In addition to electrons, various ions can be used for surface microscopy, and helium ions are particularly attractive (e.g., Bell 2009).

7.2.1.5 Ultrasound

Conventional ultrasound imaging has not had high enough resolution for use in the ear, but high-frequency ultrasound has recently shown promise (Brown et al. 2009). One disadvantage is that it does not work with structures in air.

7.2.1.6 Surface-Shape Measurement

In addition to the use of sequences of cross-sectional images, shape can also be measured from purely surface measurements. Surface shape can be measured optically by taking advantage of small depths of focus and varying the position of the focal plane (e.g., Danzl et al. 2011). With larger depths of focus it is possible to reconstruct the 3-D shape of an object from photographs taken from multiple orientations, even without knowledge of the camera positions (Snavely et al. 2008). This can also be done with tilted images from scanning electron microscopy.

Many other optical methods for surface-shape measurement exist, including moiré topography and laser range finding, which are mentioned in section "Finite-Element Models" as having been used for the eardrum, and estimation from silhouettes (e.g., Weistenhöfer and Hudde 1999 for the ossicles). The concepts of plenoptic functions and light fields, combined with the availability of microlens arrays, make 3-D photography and microscopy possible (e.g., Georgiev et al. 2011).

Techniques that have been used for the inside surfaces of cavities, such as the external ear canal, include the use of molds (Stinson and Lawton 1989), acoustical measurements (Hudde 1983), and fluorescence (Hart et al. 2010). Information about surfaces, such as texture, can be obtained using, for example, near-field optical techniques (e.g., Novotny 2011) and tactile techniques, including atomic force microscopy (e.g., Leach 2010).

7.2.2 *Model Creation*

7.2.2.1 Introduction

"Why is building 3D content so expensive and time-consuming?" Polys et al. (2008, p. 94) answer their own question largely in terms of the variety of approaches used and the lack of standards, but part of the answer is simply that building 3-D models is hard. The difficulty is especially great when dealing with complex dynamic natural structures (anatomical, biological, geological, etc.) composed of large numbers of irregular and inhomogeneous parts that are attached to multiple other parts at shared surfaces, which themselves have arbitrarily complex shapes.

Faithful models of natural structures must be created from experimental shape data, often in the form of sets of images of parallel sections, whether derived from physical cutting or from tomographic imaging. The process of creating 3-D models from such data may be considered to consist of four steps: (1) definition of relationships, (2) segmentation, (3) surface generation, and (4) volume mesh generation. These four steps are briefly discussed in sections "Relationships," "Segmentation," "Surface Generation," and "Volume Mesh Generation." In view of the considerable anatomical variability among individual ears, it is important to make the creation of models easier than it now is.

7.2.2.2 Relationships

Three types of relationships are relevant here. The first type involves an object hierarchy (e.g., manubrium *is part of* malleus). In making models from 3-D image data, this type of hierarchy has generally been dealt with on a voxel-by-voxel basis (e.g., Gehrmann et al. 2006) rather than with the more efficient surface or solid models used in computer-aided design (CAD). The second type of relationship involves a class hierarchy (e.g., cortical bone *is a kind of* bone). This type of hierarchy has not generally been made explicit—both a cause and an effect of the fact that models have been oversimplified. It is also complicated to simultaneously handle both types of relationships systematically (e.g., Cerveri and Pinciroli 2001). The third type of relationship concerns physical attachments (e.g., a surface shared between tendon and bone). These relationships have generally been ignored because they are not necessary for the visualization of static models, but they are very important for interactive and dynamic models. They also make it much easier to create variants of a model to represent, for example, anatomical variability or pathological cases.

7.2.2.3 Segmentation

Segmentation involves identifying the outlines of structures of interest within images. Considerable research has been and is being done on methods for automatic segmentation (e.g., Zhang 2006, Chap. 1). Most currently available systems represent individual structures either by filled regions or by closed contours. In neither case is it possible to explicitly represent the shared surface between two adjoining structures, often resulting in unwanted gaps or overlaps. The use of explicitly connected open contours can address this problem, as well as the representation of very thin structures like the eardrum (Decraemer et al. 2003).

Automatic techniques are fast but so far are successful only for relatively simple segmentation tasks (e.g., distinguishing between bone and non-bone). Even an image that seems very easy to segment may be very difficult for an automatic algorithm. The human visual system after all is very good at pattern recognition (e.g., von Ahn et al. 2008), sometimes too good (e.g., Lowell 1908).

It is generally accepted that manual intervention is often required, and attention is increasingly being given to integrating user interaction with powerful

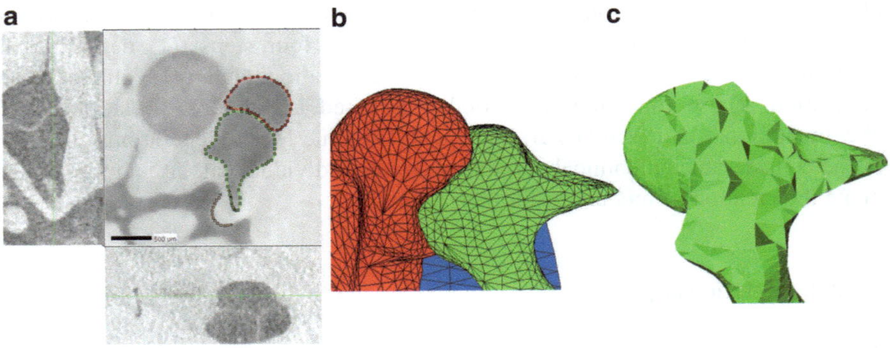

Fig. 7.1 Examples of stages in finite-element modeling of gerbil middle ear. (**a**) Example of semi-automatic 2-D segmentation of microCT data. *Red* = malleus, *green* = incus. The image in which the segmentation is done has been median filtered; the side-view images have not. (**b**) Surface meshes generated from results of segmentation. *Red* = malleus, *green* = incus, *blue* = pars flaccida. (**c**) Volume mesh of incus, with some elements removed to show that the interior is filled with tetrahedra

segmentation tools (e.g., Liang et al. 2006). Figure 7.1a shows an example of semi-automatic 2-D segmentation. Often what is needed is 3-D segmentation, and applying 2-D algorithms slice by slice often gives poor results. Fully 3-D algorithms are available but tend to be difficult to visualize and control.

7.2.2.4 Surface Generation

Surface generation often involves the generation of triangular meshes. Voxel-based algorithms such as "marching" cubes (Schroeder et al. 1996) or tetrahedra (Bourke 1997) must typically be followed by a step to greatly reduce the number of polygons, and automatic polygon-reduction algorithms often give unsatisfactory results, with many unnecessary polygons in some regions and/or excessive loss of detail in other regions. One also generally loses the original serial-section slice structure.

Alternatively, the surface can be formed by triangulating at the desired resolution between vector-based contours in different slices. Figure 7.1b shows surfaces created in this way. Determining which of the many possible triangulations to use can be done heuristically or by globally minimizing a cost function, but the choice of cost function can have a drastic effect on the quality of the triangulation (e.g., Funnell 1984). The quality is critical if the model is to be used for simulation and not just for visualization. For an extended, thin structure like the eardrum, the smoothness of the surface mesh is critical to its mechanical behavior; any local curvature will have a strong effect, as do ripples on a potato chip.

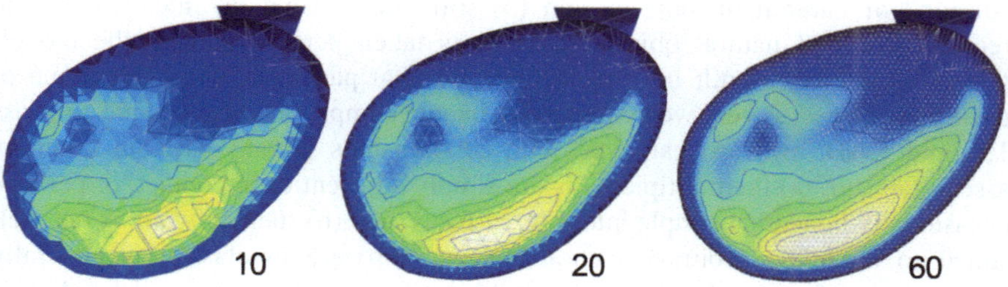

Fig. 7.2 Finite-element models of eardrum and corresponding low-frequency displacement patterns, with three different mesh resolutions: nominally 10, 20, and 60 elements/diameter. The elements (*triangles*) are indicated by *gray lines*. Displacements are color coded from *black* (zero) to *white* (maximum). The finer meshes produce more accurate displacement patterns

7.2.2.5 Volume Mesh Generation

Volume mesh generation involves creating a mesh of solid elements (e.g., hexahedra, tetrahedra) to fill the volume enclosed by a surface mesh, as shown in Fig. 7.1c. This is necessary for many applications, including finite-element modeling. A great deal of work has been done on methods for 3-D mesh generation but research continues, especially for image-based models (e.g., Young et al. 2008).

Meshes to be used for finite-element modeling must fulfill certain requirements. First, they must be topologically correct, that is, there must be no overlaps between neighboring elements and no unintended gaps. Second, the elements must not be too long and thin, because this leads to numerical problems when doing the finite-element calculations. Third, the mesh must be fine enough to avoid excessive discretization errors but not so fine as to require excessive computation. On the one hand, a mesh that is too coarse will usually tend to lead to model behavior that is too stiff. On the other hand, the computational requirements increase dramatically as the number of elements increases. Figure 7.2 shows an eardrum model with three different mesh resolutions, and the resulting low-frequency displacement patterns. Higher frequencies will lead to more complicated patterns that require finer meshes. The trade-off between accuracy and computational expense must be judged according to the requirements of the analysis, and it is often desirable to make the mesh finer in some parts of the model than in others. It is important to undertake convergence testing, that is, to test a model with varying mesh resolutions, under a variety of load conditions, to make sure that the mesh is acceptable. Automatic mesh-generation software makes it feasible to generate meshes of varying coarseness for such testing.

7.2.2.6 Software

The software used for creating 3-D models may be divided into three classes: (1) CAD software intended for design and manufacturing, (2) software for the artistic "design" of naturalistic objects, and (3) software specifically intended for the reconstruction of natural objects. Class 1 is taken here to include the model-generation facilities built into many finite-element packages. Currently, none of these three classes is very well equipped to handle complex natural systems. In class 1, some CAD software can deal with relationships among parts in complex assemblies, but (a) the multiple parts are simply adjacent or in contact, or perhaps occasionally bonded at simple interface surfaces; and (b) such software is not well suited to modeling arbitrary natural shapes. Software in class 2 is generally intended only for visualization; relationships among multiple parts, if handled at all, are generally limited to interactions between geometric control points. Software in class 3 is generally intended only for visualization and for quantification of properties such as length and volume. Some software does exist for deriving finite-element models from imaging data but such software generally does not attempt to model the physical interactions and relationships among multiple component parts with widely varying sizes and properties.

7.3 Material Modeling

To take full advantage of the power of the finite-element method, one should have *a priori* information about the material properties, rather than simply adjust parameters to fit particular experimental results. There are many different materials involved in the outer and middle ear, including air, bone, ligament, tendon, muscle, cartilage, synovial fluid, epithelium, mucosa, fat, and nerve, as well as specialized structures such as the lamina propria of the eardrum and the fibrocartilaginous ring. The connective-tissue components include various forms, both dense and loose, and with both regularly and irregularly organized fibers.

Until recently the only explicit measurements of middle ear material properties were for the eardrum (e.g., von Békésy 1949; Kirikae 1960; Wada et al. 1996) and the interpretation of even those data requires great care (e.g., Fay et al. 2005). In the past few years, material properties have been measured for other middle ear structures (e.g., Cheng and Gan 2007; Soons et al. 2010) and a variety of new techniques have been used for the eardrum (e.g., Luo et al. 2009; Zhang and Gan 2010; Aernouts and Dirckx 2012). With care, values can also be estimated from measurements in supposedly similar tissues elsewhere in the body.

The purpose here is not to review these different measurements but to summarize the ways in which material properties are represented in finite-element models. Issues of tissue nonuniformity and inhomogeneity are not addressed here. They can be handled either by benign neglect, by averaging, or by applying different material properties to different elements in the finite-element mesh.

Also ignored here are many biomechanical issues related to phenomena of living tissue, such as metabolic processes in general; actively maintained chemical

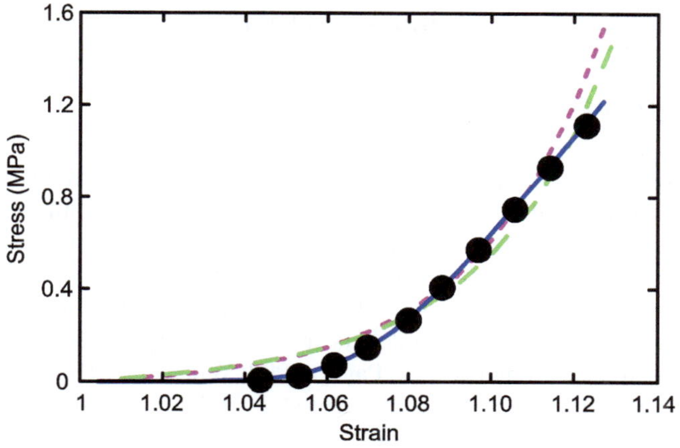

Fig. 7.3 Comparison of experimental data (*black dots*) with three different material models (*solid blue*, *long-dash green*, and *short-dash magenta lines*). See text for discussion

gradients (leading, for example, to gas exchange as discussed by Dirckx, Chap. 5); development, growth, and remodeling; healing and osseo-integration; and many sources of variability related to genetics, environment, and history.

The simplest form of material property is linear elasticity, for which the material properties are specified by Young's modulus and Poisson's ratio. The material may be isotropic or not. For dynamic problems, one also needs parameters for mass density and for damping. Mass density is relatively easy to estimate. Damping is more difficult and is often represented by rather arbitrary parameters, such as the α and β of Rayleigh damping (e.g., Funnell et al. 1987). Zhang and Gan (2011) recently used a simple viscoelastic representation based on experimental measurements.

Nonlinearities may arise for geometric reasons, even if the stress–strain relationship of the material is linear. In this case a St. Venant–Kirchhoff material model can be used, which is formulated like a nonlinear material but has the same Young's modulus and Poisson's ratio as for a linear formulation.

Many different finite-element formulations are available for truly nonlinear hyperelastic materials. The results with different material models are sometimes similar and the available experimental data are not always good enough to justify a preference for one model over another. As an example, Fig. 7.3 shows stress–strain curves for three different material models compared with experimental data. The curves for the Ogden model (magenta, short dashes) and the Veronda–Westmann model (green, long dashes) are very similar. The solid blue curve for the model of Decraemer et al. (1980), based on a simple structural model, is also similar but does appear to fit the data better than the curves for the other two models, which are purely phenomenological.

7.4 Model Verification and Validation

7.4.1 Introduction

ASME (originally known as the American Society of Mechanical Engineers) has formulated general guidance for the iterative verification and validation of computational models (ASME 2006; Schwer 2007). The guidelines are aimed primarily at the modeling of designed and manufactured systems but are also relevant for natural systems such as the middle ear. The building of confidence in the results of computational models is analogous to the building of confidence in experimental measurements (e.g., Parker 2008; Winsberg 2010). In this section, brief overviews of model verification and model validation are presented, followed by discussions of the specific topics of uncertainty analysis and parameter fitting.

7.4.2 Model Verification

Model verification is considered to include both code verification and calculation verification. Code verification involves checking that the mathematics of the model have been correctly implemented in the software. This may be considered to be the responsibility of the software developer, but wise modelers may want to check things for themselves by comparison with analytical solutions comparable to the real model, by the use of "manufactured solutions" (Roache 2002), or by comparison with other software. It is also useful to explore simulation results in depth, looking for odd behavior, although this will tend to be biased by expectations.

The second step, calculation verification, is very much the responsibility of the modeler, and includes running the model with different mesh resolutions and, for time-domain solutions, different time steps, to make sure that the discretization is fine enough that it does not affect the results too much (cf. section "Volume Mesh Generation").

7.4.3 Model Validation

Model validation is an evaluation of how closely the behavior of a model matches the experimentally measured behavior of the system being modeled, with the experimental data *not* having been used in formulating or refining the model. The match is expressed quantitatively in terms of some validation metric that is appropriate to the intended use of the model. Ideally the match is expressed not only as a measure of the difference, but together with a measure of the uncertainty of the results and a confidence level, for example, "the relative error between the experiment and simulations was 18 ± 6 % with [an] 85 % confidence level" (Schwer 2007, p. 251). The estimation of uncertainty is addressed in Sect. 7.4.4. Note that there are two different types of uncertainty here: epistemic (or reducible) uncertainty, due to lack of knowledge about parameters, and aleatory (or stochastic, or irreducible) uncertainty, due to inherent randomness (e.g., Helton et al. 2006). These concepts are not really mutually exclusive and usage of the terminology is rather inconsistent (e.g., Moens and Vandepitte 2005, p. 1529).

Validation metrics (or adequacy criteria) can be expressed in terms of either response measures (raw model outputs) or response features derived from those outputs (e.g., Mayes 2009). For a static middle ear simulation, for example, interesting response measures might be the umbo displacement or the complete spatial displacement pattern of the eardrum. Response features might be the maximum displacement on the eardrum, or the location of that maximum, or the ratio of the maximum displacement to the umbo displacement. For a dynamic linear model, response measures could be the magnitudes and phases of the frequency response at multiple frequencies, and response features might be the average low-frequency magnitude and the frequency of the lowest resonance.

A validation metric quantifies the difference between a response measure or feature as produced by the model and the same measure or feature as measured experimentally. Obviously the available choice of metrics depends on what experimental data are available. There are many ways of formulating validation metrics, including correlations and sums of differences. It is desirable to have a relatively small number of metrics, for ease of interpretation and decision making. If one is interested in matching the shape of a frequency-response function, rather than using the frequency-by-frequency differences between simulated and measured magnitudes, one might use the means of the squared differences over selected frequency bands, or frequency shifts between simulated and measured resonance frequencies.

Whether the model matches the experimental data well enough depends on the purpose of the modeling. In general one wants some confidence that the modeling approach can be trusted to make predictions beyond the specific details of the model and experiment for which the validation is done, but this is a difficult matter of judgment. (Obviously one must always keep in mind the ranges of frequency, load, and displacement for which the underlying assumptions of the model are valid.) If it is decided that the match is not good enough, then either the model or the experiment may be revised and refined. Revision of the model (model updating) may involve not only the computational model itself (parameter fitting) but also the underlying conceptual model (choices of what physical phenomena to include) and mathematical model (how the physical phenomena are formulated).

7.4.4 Uncertainty Analysis

Uncertainty analysis has two related purposes: (1) to characterize the uncertainty of a model's output; and (2) to determine which model parameters are mainly responsible for that uncertainty (sensitivity analysis). The amount of uncertainty in the output is important because it gives insight into how much faith to put in conclusions based on the simulation results and how much weight to give them when making decisions. If a model predicts a 50 % improvement in some clinical outcome, but the model uncertainty is $\pm$ 70 %, clearly the model will not be used to try to influence a clinician's practice. Sensitivity analysis, that is, determining which parameters account for most of the output uncertainty, is important because it gives guidance in deciding how to try to improve the model. The improvement can be made either by adjusting parameter values or, preferably, by obtaining better a priori estimates of the parameter values. If a model is very sensitive to a particular parameter then that parameter is a good candidate for additional experimental efforts to determine its value. On the other hand, if a model is insensitive to a parameter then that parameter can just be fixed and attention can be directed elsewhere.

Sensitivity analysis is discussed in terms of a parameter space. If a model has three parameters, then it has a 3-D parameter space. If it has k parameters then it has a k-dimensional parameter space. For each parameter there will be a best-guess estimate for its value (the baseline value), plus a range of values that it might reasonably have, and perhaps a probability distribution of values within that range or some other characterization of the possible values (e.g., Helton et al. 2006). Some parameters may have much narrower ranges of likely values than others. For a soft biological tissue, for example, which is mostly water, the mass density parameter is known to within a much smaller tolerance than the stiffness parameter.

To estimate the uncertainty of a model, one should ideally run simulations for all possible combinations of many different values of every parameter, to see how the model output changes. If one uses n different values of each of k parameters, one would need n^k simulations. If there are, say, four parameters, and one uses only the minimum, best-guess and maximum value for each, then already $3^4 = 81$ different simulations are needed. This is the full-factorial method of choosing combinations of parameter values, and it quickly becomes impractical for the numbers of parameters often encountered, for a reasonably generous number of values per parameter, and particularly when the model is computationally expensive, which finite-element models often are.

It is therefore desirable to reduce the number of parameter combinations. The most common approach is the one-at-a-time method: first one parameter is varied over its range with all of the other parameters at their baseline values, then that parameter is returned to its baseline value and a second parameter is varied, and so on. This approach certainly reduces the number of simulations, but its great drawback is its failure to provide any information about parameter interactions. For example, suppose that when parameter a is at its baseline value then increasing parameter b from its baseline value increases the model output, but that when a has some other value then increasing *b* actually decreases the model output. This is an interaction between the two parameters. Such interactions are not uncommon in complex systems and obviously they will have a substantial impact on the uncertainty of the model's behavior.

Clearly it is necessary to obtain a more complete sampling of points in the parameter space. A distinction can be made between preliminary screening analyses and more complete, quantitative analyses. A number of strategies have been used for selecting the points, such as random sampling, quasi-random sampling, importance sampling, Latin Hypercube sampling, and the Morris (or elementary-effects) method (e.g., Helton et al. 2006). Campolongo et al. (2011) describe a strategy that involves doing multiple one-at-a-time parameter variations. Thus, a modeler can do the usual one-at-a-time analysis around the baseline parameter values; then do a few more one-at-a-time analyses around other points to obtain screening information about interactions; and then do a larger number of one-at-a-time analyses (perhaps with some parameters omitted) for a full quantitative result.

The actual results of an uncertainty analysis may simply be visualized, for example, as scatter plots, or they may be subjected to sophisticated statistical machinery (e.g., Helton et al. 2006).

7.4.5 Parameter Fitting

7.4.5.1 Introduction

Parameter fitting is part of model updating, and consists of trying to find the set of parameter values that causes a model to best fit some experimental data. It is also known as parameter identification or model calibration; some authors have used the term "model validation" but that term should be reserved for the broader activity described in Sect. 7.4.3. Parameter fitting is usually preceded by a sensitivity analysis to provide insight into which parameters are most important for the fitting. For a small number of parameters it may be feasible to try to find the best fit by manually adjusting parameters, but it is often necessary to use some algorithmic approach. This involves two steps: choice of a cost function (section "Cost Function") and the actual algorithm for minimizing that function (section "Minimization Algorithms").

Before these two issues are addressed, it is important to mention the strategy of trying to reduce the number of parameters by using different pathological or experimental conditions. For example, Zwislocki (1957) exploited the pathological conditions of otosclerosis and of an interrupted incudostapedial joint to simplify his middle ear circuit models. Experimentally, methods that have been used for the middle ear include removing structures (e.g., Wever and Lawrence 1954, pp. 124 ff.); blocking the motion of the stapes (e.g., Margolis et al. 1978) or malleus (e.g., von Unge et al. 1991); and draining the liquid from the cochlea (e.g., Lynch et al. 1982). (The second of these should perhaps be worded as "attempting to block" because it is not always easy to produce the desired easy-to-model effect (e.g., Ladak et al. 2004).) This approach involves the assumption that the effects of the remaining parameters are not affected by the experimental change. This is particularly serious for lumped-parameter models—draining the cochlea, for example, may change the mode of vibration of the stapes, thereby changing its effective inertia and the effective stiffness of the annular ligament. The problem may also be present, to a lesser extent, in finite-element models.

7.4.5.2 Cost Function

For most minimization algorithms, the cost function to be minimized must be a single number, so for a particular model the various validation metrics of interest must be combined together, often as some sort of weighted average. In the best possible scenario, a cost minimum will be found such that all of the validation metrics have very small values. In real life, however, the minimum may correspond to parameter values that make some of the metrics very good and others very bad, so the formulation of a cost function will be a delicate matter.

There is a family of multiobjective minimization algorithms (e.g., Erfani and Utyuzhnikov 2011) that address the existence of multiple, conflicting cost functions. They do so by producing a family of solutions rather than a single solution, so the final decision is left to the user. The delicate decision making is thus done at the end rather than at the beginning of the process.

7.4.5.3 Minimization Algorithms

One can visualize the minimization problem for two parameters as searching for the lowest point on a response surface, with the x and y coordinates corresponding to the parameter values, and the z value (height) corresponding to the value of the cost function. The brute-force method of minimization is just to calculate the cost function at closely spaced points over the whole surface, but this is impractical for more than a few parameters or for any but the simplest cost functions. When evaluation of the cost function depends on running a high-resolution dynamic finite-element simulation, and especially if it is nonlinear, then each exploratory step is expensive.

As a result, various strategies have been devised to try to reduce the number of parameter-value combinations that must be tried. A major problem is the distinction between the global minimum of a function and possible local minima. The minimization strategy may appear to have found a minimum, but it may be just a small valley on a high plateau, with a much lower minimum in some region that has not been explored. The most common approach to this problem is to try the minimization algorithm multiple times from multiple starting points. Another problem is in deciding what step sizes to use when varying parameters. If the step size is too large, some narrow deep valleys may be missed, but excessively small step sizes will be impractically time consuming. One family of strategies attempts to continuously adapt the step size to the shape of the surface in the immediate vicinity.

There are many minimization algorithms available, and they continue to multiply. One major subdivision is between those that require an explicit formulation of the derivative of the cost function with respect to each parameter, and those that do not. For models of any complexity, it is much easier if one does not need explicit derivatives, but the price is generally increased computational time. Another major division is between deterministic algorithms, which use some sort of sequential strategy to patiently seek locations with lower cost, and stochastic algorithms, which use a shotgun approach. Marwala (2010) compares a variety of minimization algorithms for finite-element model updating.

7.4.5.4 Discussion

Parameter fitting in general is difficult, and becomes dramatically more difficult as the number of parameters increases, so it is highly desirable to reduce the number of uncertain parameters, and to understand which parameters have the largest impacts. If a fitting algorithm fails to find an acceptable fit for a given model and set of parameter ranges, it may be that no acceptable fit exists, but it is possible that the algorithm has missed it. A fitting algorithm may also fail to recognize a situation where there are many combinations of parameter values that give equally good results.

Compared with models of artificial systems, models of natural systems are likely to have much more parameter variability and uncertainty, and modelers are likely to rely more on parameter fitting than on the rest of the validation process as defined at the beginning of Sect. 7.4.3. Another point about models of natural systems is that there may be considerable uncertainty about the geometry as well as about material properties. It is important to keep in mind that the predictive power of a model is likely to be much reduced if its parameters have had to be fitted to obtain agreement with the available experimental data.

Extensive parameter fitting tends to negate the whole philosophy of finite-element modeling. This is particularly true if parameter values are set without regard to physiological plausibility. For example, in the absence of a specific rationale to the contrary, all ligaments in a model should have the same material properties. If it is necessary to give them different properties to fit specific experimental data, then something is wrong somewhere.

Ideally, modelers should be blind to the specific experimental data that will be used to validate their model (ASME 2006, p. 7), so as to strengthen confidence in its predictive power. For example, Funnell and Laszlo (1978) claimed that their model structure and material properties were established a priori, but since there was no formal blinding of the modelers to the experimental data, one cannot be sure that there was not some bias, unconscious or otherwise, in the definition of the model. Such blinding is not often practical, but it can at least be acknowledged as an ideal.

7.5 Models of the Outer and Middle Ear

7.5.1 Summary of Modeling Approaches

The interconnections of the various anatomical units of the outer, middle, and inner ear may be represented as in Fig. 7.4. Three of the blocks are in the form of two-port networks. All sound energy must flow through the first block, which stands for whatever portion of the external ear lies between the sound source and the middle ear itself. The energy leaving this block must pass through the eardrum into the middle ear cavities. In the process some of the energy enters the ossicular chain, whence it passes into the cochlea.

The sections that follow will provide summaries of how the five major blocks have been modeled, generally starting with circuit models and ending with finite-element models. Figure 7.5 is a pastiche of most of the circuit models that have been described in the literature, organized in blocks corresponding to those in Fig. 7.4. No single published model has included so many components, but the figure gives an idea of the potential complexity. Many of the groups of components in the figure are common to more than one published model. Two noteworthy features are the three-piston eardrum model of Shaw and Stinson (1986) and the multipart air-cavity model of Onchi (1961); both of these are mentioned again below. Most authors have established the structures of their models based on the mechanical structure of the system, but Wever and Lawrence (1954, pp. 394 ff.) established the simplest model structure that they could find that would fit their data, and then treated the relationships between model elements and anatomical structures as being "of course a matter of conjecture."

There are places in circuit models where it is strictly required to put ideal transformers, both to represent transitions between acoustical and mechanical parts of a system, and to make explicit the various lever mechanisms of the middle ear. In practice either they can be included explicitly, or their effects can be absorbed into the parameter values. In the discussions here, specific circuit

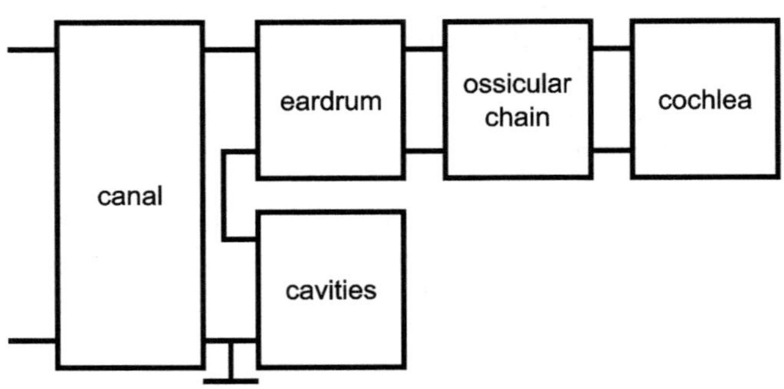

Fig. 7.4 Block diagram of the ear canal, middle ear, and cochlea

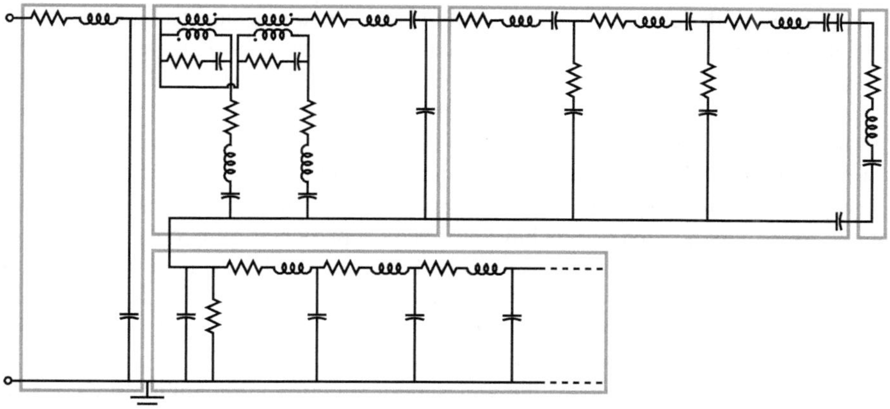

Fig. 7.5 Circuit model of outer, middle, and inner ear, organized into the same major blocks as shown in Fig. 7.4. This figure combines features from a variety of published circuit models

parameter values will not be given in any case, because they vary considerably from species to species and from individual to individual, and according to the structure selected for the model and the data used to estimate the parameters.

Most models are parametric, although Teoh (1996, p. 132) effectively used a nonparametric model of the pars tensa, ossicular chain and cochlea, in combination with parametric models of the other parts of the system.

In discussions of middle ear function in terms of analogous electrical circuits, it is important to know the frequency range over which it is legitimate to use lumped-parameter models. For example, Beranek (1954) says that for a closed tube to be modeled by a capacitor, the length (in meters) must be less than $30/f$ (where f is the frequency in Hz) for an error of 5 %. For a length of, say, 10 mm, which is typical of the middle ear, this constraint corresponds to an upper frequency limit of 3 kHz, which is rather low. Note, however, that an error of 5 % corresponds to only 0.5 dB. If one can accept an error at the highest frequency of 1 dB, the upper frequency limit can be extended to over 10 kHz. An acceptable error of 2 dB means an upper limit of almost 15 kHz.

The middle ear is linear up to sound pressures of at least 120 or 130 dB SPL (e.g., Guinan and Peake 1967). Linearity will be assumed here for responses to sounds but not for responses to large quasi-static pressures.

7.5.2 Ear Canal

At low frequencies the ear canal can be modeled as a simple rigid-walled cavity characterized just by its volume. In a circuit model, this can be corrected for in the experimental data (e.g., Zwislocki 1957) or it can be explicitly represented by just a single capacitance. At higher frequencies the wavelength starts to become comparable to the canal length and standing-wave patterns start to form along the canal. As a first approximation this can easily be modeled as a single mass-spring combination to produce the first natural frequency (e.g., Onchi 1949, 1961) or the canal can be modeled analytically as a uniform transmission line (Wiener and Ross 1946). A one-dimensional modified horn equation can be used to model the effects of the nonuniformity of the transverse canal dimensions (Khanna and Stinson 1985) as well as the effect of the distributed acoustical impedance of the eardrum (Stinson and Khanna 1989). Such effects have also been approached using coupled mechanical and acoustical finite-element models. For example, Koike et al. (2002) compared the effects of a more or less realistic canal shape (curved, but with the drum not tilted in a realistic way) with those of a simple cylindrical canal. At 7 kHz, the variation of the pressure across the drum was less than 2 dB.

At even higher frequencies, the wavelength becomes comparable to the transverse canal dimensions and the pressure starts to be nonuniform across the canal (e.g., Stinson and Daigle 2005). At very high frequencies, the precise orientation and shape of the canal termination at the eardrum also become important. Rabbitt and Holmes (1988) modeled this analytically using asymptotic approximations. Tuck-Lee et al. (2008) used a special adaptive finite-element approach to facilitate the calculations for high frequencies. In a finite-element model of the human ear canal, Hudde and Schmidt (2009) found acoustical modes that raise interesting questions about the notion of a midline axis that is often assumed in canal modeling.

Although Tuck-Lee et al. (2008) did allow for some absorption in the walls of the canal and cavities, most models of the canal have assumed that its walls are effectively rigid. It appears that this is not a reasonable approximation in human newborns. For the newborn human canal, in which the bony canal wall has not yet formed, Qi et al. (2006) found that the response of a hyperelastic finite-element model to large static pressures (as used in tympanometry) is strongly affected by the flexibility of the canal wall. Preliminary results indicate that this is also true of the response to auditory frequencies (Gariepy 2011).

Most ear-canal modeling has treated the common experimental situation in which the sound source is characterized by sound-pressure measurements with a probe microphone in the canal. Small probe tubes can have significant effects in some circumstances (e.g., Zebian et al. 2012). In modeling the response of the ear to free-field sound with the microphone outside the canal, it is necessary to take into account the shape of the pinna (e.g., Hudde and Schmidt 2009) and perhaps even the shape and dimensions of the whole head.

7.5.3 Air Cavities

In some species, the middle ear air cavities include multiple chambers with relatively narrow passages between them (see Rosowski, Chap. 3). Circuit modeling of the cavities for any given species is relatively straightforward: identify the distinct chambers, associate each one with a capacitor, and associate the

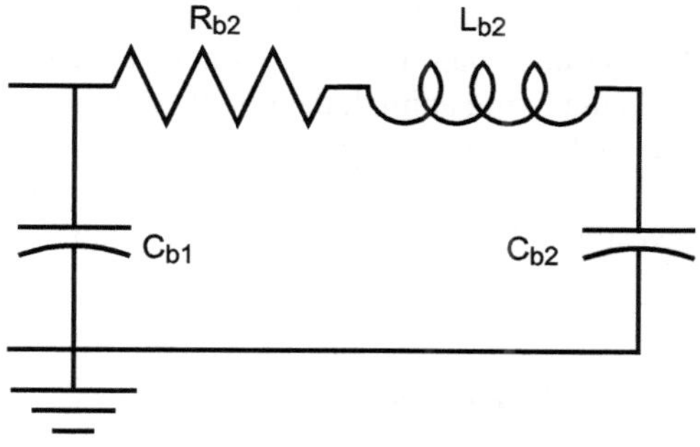

Fig. 7.6 Circuit model for middle ear cavities with one main chamber (C_{b1}), one secondary chamber (C_{b2}), and a narrow passage between them (R_{b2} and L_{b2})

passages between the chambers with series R-L branches. Figure 7.6 shows a model with one main chamber and one secondary chamber. The capacitance values can be calculated from the chamber volumes. The resistance and inductance values will often be determined from measured frequency characteristics because the estimation of R and L parameters for short narrow tubes is an approximate business at best (e.g., Beranek 1954, Chap. 5) and the intercavity passages are more irregularly shaped than simple tubes. The interconnected chambers will introduce resonances. Since the capacitances are known, the inductances can be estimated from the resonance frequencies, and the resistance can then be estimated from the width and sharpness of the resonance.

For the human, C_{b1} represents the tympanum and epitympanum, while C_{b2} represents the mastoid antrum and air cells. Although the air cells have a complex form and could be more accurately modeled (Onchi 1961; Stepp and Voss 2005), it is often sufficient simply to include their volume with that of the antrum because at high frequencies, where their form is more critical, the increasing reactance of L_{b2} will tend to isolate them from the tympanum (Zwislocki 1962). In the cat, C_{b1} and C_{b2} represent the ectotympanic and entotympanic cavities, respectively (Møller 1965; Peake and Guinan 1967). In the guinea pig, they represent the tympanum and epitympanum, respectively (Funnell and Laszlo 1974). In the rabbit there is only one cavity (Møller 1965). Zwislocki (1962) added an extra resistor for energy absorption in the tympanic cavity, Eustachian tube and mastoid air cells but did not find it necessary for the guinea pig (Zwislocki 1963).

The finite-element model of Gan et al. (2006) included an explicit model of the middle ear cavities as well as of the canal, but the sound-pressure differences between locations within the cavities were very small at frequencies up to 10 kHz, the maximum frequency considered. As mentioned previously for the canal, Tuck-Lee et al. (2008) used a special algorithm for the air cavities; they also used a special approach (involving perfectly matched layers) for modeling the common experimental condition of opened air cavities, avoiding the need for explicitly modeling the infinite (or at least very large) surrounding air space. Their analysis of the effects of having two cavities communicating through a small opening supported previous suggestions (Puria 1991; Huang et al. 2000) that this configuration in the cat avoids a notch in the frequency response at around 10 kHz that could interfere with acoustic cues used for sound localization.

In the living animal, the middle ear cavities are criss-crossed by mucosal strands and folds, some of them carrying blood vessels or nerves or even connective-tissue fibers (e.g., Palva et al. 2001). Their possible acoustical effects have never been investigated but are assumed to be small.

7.5.4 Eardrum

7.5.4.1 Helmholtz Versus von Békésy

The first attempted quantitative model of eardrum function was the "curved-membrane" hypothesis of Helmholtz (1868). This was a distributed-parameter analytical model that depended critically on the curvature of the eardrum. It also made use of anisotropy, assuming differences between the radial and circumferential directions. The model was later elaborated by Esser (1947) and by Guelke and Keen (1949), and an error in Helmholtz' calculations was pointed out by Hartman (1971). In the meantime, however, von Békésy (1941) had made capacitive-probe measurements of eardrum vibrations and described the eardrum as vibrating, at frequencies up to about 2.4 kHz, "as a stiff surface along with the manubrium" with a very flexible region around the periphery (as translated von Békésy 1960, p. 101). This view of the eardrum, as being mostly a rigid structure tightly coupled to the malleus, dominated subsequent modeling for many years.

7.5.4.2 Lumped-Parameter Models

In the circuit model of Onchi (1949, 1961), the human eardrum was represented by a single mass attached by springs to the tympanic annulus and to the manubrium. Zwislocki (1957) represented the human eardrum with two parts, one branch in parallel with the ossicular chain, corresponding to "the compliance and the resistance of the eardrum . . . when the ossicular chain is rigidly fixed" (with negligible inertia) and a second branch in series with the ossicular chain, corresponding to "the portion of the eardrum that may be considered rigidly coupled to the malleus" (and incorporating the effect of the middle ear air cavity). The first branch allows sound energy to pass through the eardrum directly into the middle ear air cavities without driving the ossicles. Møller (1961) represented the human eardrum in essentially the same way. Zwislocki (1962) refined his model by adding an inductor to represent eardrum mass. For higher frequencies where "the eardrum vibrates in sections" he suggested that "a transmission line would probably constitute the best analog" but he confined himself to adding an empirically chosen series resistor-capacitor combination in parallel with the inductor. The extra resistor and capacitor were not found necessary for the guinea pig (Zwislocki 1963).

Møller (1965) modeled the cat and rabbit eardrum as a single branch in parallel with the branch representing the ossicular chain and cochlea, with the part of the eardrum tightly coupled to the malleus being implicitly included in the latter branch. The detailed nature of each branch was not specified. Peake and Guinan (1967), in their model for the cat, did not find it necessary to include a parallel branch for the eardrum.

In the early 1970s, eardrum vibration-pattern measurements by laser holography made it clear that the mode of eardrum vibration described by von Békésy was incorrect (Khanna and Tonndorf 1972). Even at low frequencies, no part of the drum acts like a rigid plate. In recognition of this new evidence, and with the goal of extending eardrum circuit models to higher frequencies, Shaw and Stinson (1983) reinterpreted the nature of the two eardrum "piston" components, associating them with (1) the small part rigidly coupled to the manubrium and (2) all of the rest of the eardrum. They also added an explicit coupling element between the two parts. They later refined the model further by adding a third "piston" (Shaw and Stinson 1986). Kringlebotn (1988) did not include multiple branches for the eardrum itself but did include a parallel branch after the series branch, to represent coupling between the eardrum and the manubrium. A branch was also included to represent the suspension of the eardrum at the tympanic annulus; this will be mentioned again in section "Finite-Element Models." In an attempt to deal with higher frequencies, Puria and Allen (1998) presented a delay-line model of the eardrum; it was further explored by O'Connor and Puria (2008) and extended by Parent and Allen (2007, 2010).

Lumped-circuit models of the eardrum continue to be useful in some applications. For example, Teoh et al. (1997) made good use of such a model to elucidate the effects of the large pars flaccida of the gerbil eardrum. Two-port models have also been used productively (e.g., Shera and Zweig 1991; O'Connor and Puria 2008). Such models are particularly appropriate for describing experimental data that describe eardrum behavior by a single number such as umbo displacement or, especially, acoustical impedance. However, it is clear that lumped models cannot model the spatial vibration patterns of the eardrum, nor can they address questions arising from those patterns.

7.5.4.3 Analytical Models

To address spatial patterns of the eardrum, in addition to the work by Esser (1947) and Guelke and Keen (1949) mentioned in section "Helmholtz Versus von Békésy," analytical models were formulated by Frank (1923), Gran (1968, Chap. 2), and Wada and Kobayashi (1990), but all were forced to make many oversimplifications, including the critical one of taking the eardrum as flat rather than conical. Asymptotic analytical models have been more successful (Rabbitt and Holmes 1986; Fay 2001) and are mentioned again later. Goll and Dalhoff (2011) recently presented a 1-D string model of the eardrum that can be viewed as a distributed variant of the lumped delay-line models mentioned in section "Lumped-Parameter Models."

7.5.4.4 Finite-Element Models

In direct response to the new holographic spatial-pattern data of Khanna and Tonndorf (1972), the finite-element method was applied to the middle ear, with particular attention to the eardrum (Funnell and Laszlo 1978; Funnell 1983). These 3-D models were based on a review of the anatomical, histological, and biomechanical nature of the eardrum (Funnell and Laszlo 1982). All of the mechanical parameters except damping were based on a priori estimates, so very little parameter fitting was required. (The possibility of doing this is one of the strengths of the finite-element method.) The displacement patterns and frequency responses calculated with these models were qualitatively similar to those observed experimentally by Khanna and Tonndorf (1972).

These first finite-element models were for the cat middle ear because of the high-quality experimental data that were available. Subsequently, Williams and Lesser (1990) published a finite-element model of the human eardrum and manubrium, but the ossicular chain and cochlea were not modeled, and the model did not produce reasonable natural frequencies with realistic parameters. Wada et al. (1992) published the first model of the human middle ear that included both the eardrum and the ossicles. See Volandri et al. (2011) for a recent survey of finite-element models of the human eardrum. In many of these models, values for many of the parameters were established by fitting the model behavior to specific experimental data, rather than by a priori estimates. In models with many parameters, unless the fitting is done very carefully and with very good data, it can lead to parameter values that have questionable physical significance. It has happened, when published models were revised somewhat or compared with different data, that some of the material-property parameters changed dramatically with no rationale other than data fitting.

The 3-D shapes of the eardrum in these early models were qualitatively based on rather imprecise and coarse shape measurements (e.g., Kojo 1954; Funnell 1981). Representing the shape parametrically (Funnell and Laszlo 1978) made it possible to evaluate the effects of changing both the depth of the cone and the degree of curvature; Rabbitt and Holmes (1986) represented the shape with a function describing the deviation from a conical shape. Much better shape measurements became available later using moiré topography (Dirckx et al. 1988), and a method was developed of incorporating the measurements directly into finite-element models (Funnell and Decraemer 1996). Both these results and the earlier parametric studies showed that 3-D eardrum shape has a significant effect on the behavior of middle ear models, indicating the importance of good shape measurements and of models that reflect those measurements. Beer et al. (1999) used a parameterized shape based on 40 points measured with a scanning laser microscope as described by Drescher et al. (1998). Sun et al. (2002a, 2002b) based their model geometry on serial histological sections, but this is problematic for the shape of a very delicate structure such as the eardrum, which is very vulnerable to distortion during histological processing. Fay et al. (2005, 2006) assumed a conical shape near the manubrium and a toroidal shape near the annulus, fitted to the moiré data of Funnell and Decraemer (1996).

The eardrum thickness in the early models either was assumed to be constant or was made variable to correspond to the rather coarse observations that were available, such as the measurements at ten locations on histological sections by Uebo et al. (1988). The best thickness measurements so far have been done using confocal microscopy for cat and gerbil (Kuypers et al. 2005a, b) and later for human (Kuypers et al. 2006), and they have started to appear in finite-element models (Tuck-Lee et al. 2008 for cat; Maftoon et al. 2011 for gerbil).

Kringlebotn (1988), Wada et al. (1992), and Williams et al. (1996) all included a spring term at the boundary of the eardrum. However, experimental measurements seem to indicate that the drum can be thought of as having zero displacement at the boundary (e.g., Gea et al. 2009) and the thickness of the fibrocartilaginous ring is so much greater than that of the drum that its displacements can be expected to be much smaller. The need for nonzero displacements at the boundary may result from inadequacies in the model, such as inappropriate eardrum curvature or rigid ossicular joints. It is true, however, that the drum thickens gradually toward the boundary, and the fibrocartilaginous ring itself tapers down (albeit rapidly) to the thickness of the drum, so the question of the boundary condition depends on exactly where the boundary is taken to be.

In most species, the eardrum and malleus are tightly connected together along the whole length of the manubrium, but in the human ear this is not the case. The soft connection in the middle region of the manubrium has been explicitly modeled (Koike et al. 2002; Sun et al. 2002b) but does not appear to make much difference.

7.5.4.5 Layers and Fibers

In many models the eardrum has been assumed to be uniform throughout its thickness and to be isotropic. This has been in spite of ultrastructural observations suggesting both nonuniformity, because of the layered structure of the eardrum, and anisotropy, because of highly organized fiber orientations (e.g., Lim 1995). (Note that Schmidt and Hellström (1991) described the fiber layers as being somewhat different in the guinea pig than in rat and human.) It has also often been assumed that there is negligible resting tension in the eardrum. It is still not clear how acceptable these assumptions are; certainly anisotropy, for example, can have significant effects (e.g., Funnell and Laszlo 1978).

The mechanical properties of the eardrum will depend on the mechanical properties of the fibers and of the ground substance in which the fibers are embedded; on the numbers, orientations, and packing of the fibers; on the mechanical coupling between the fibers and the ground substance; and on the variable thicknesses of the different layers. Rabbitt and Holmes (1986) included these features in their asymptotic analytical model. They pointed out that the anisotropic arrangement of the fibers may lead to more anisotropy in the in-plane (membrane) mechanical properties than in the transverse (bending) mechanical properties. It is important to keep this in mind when interpreting the results of different methods for measuring eardrum mechanical properties. Fay et al. (2005, 2006) also took the layers, fibers, and variable thicknesses into account. They emphasized the gradual transition from a very flexible, nearly isotropic region near the annulus to a stiffer, anisotropic region near the manubrium.

Finite-element modeling has been done of the effects of holes in the eardrum (Gan et al. 2006, 2009), and a model of the effects of slits in the drum has been used to address the relative contributions of the radial and circular fibers (Tuck-Lee et al. 2008). They discussed the apparent need in their model for a shear stiffness that is higher than might be expected for a material for which the stiffness is assumed to arise primarily from stiff parallel fibers. However, there is still much to be learned about the biochemistry and mechanics of collagenous materials, and of the eardrum in particular (e.g., Broekaert 1995; Buehler 2008).

As noted previously (Funnell and Laszlo 1982), the outermost layer of the epidermis, the stratum corneum, is thin but very dense and it may have some effect on the mechanical properties of the eardrum. Yuan and Verma (2006) reported Young's moduli for the stratum corneum that are comparable to some estimates for the eardrum fiber layers, and the thicknesses of the two layers in the eardrum may also be comparable.

7.5.5 Ossicular-Chain Models

7.5.5.1 Lumped-Parameter Models

Circuit models of the ossicular chain generally include mass (inductor) elements for the malleus (possibly including the tightly coupled part of the eardrum), incus, and stapes. As a first approximation the malleus and incus are considered to rotate about an axis joining the anterior mallear[1] and posterior incudal processes, so the mass term actually represents the moment of inertia of the bones as they move about that axis. If the stapes is considered to move like a piston, then its mass term actually corresponds to its mass, but if it is considered to rotate about one part of the annular ligament (e.g., Møller 1961), then its mass term corresponds to a moment of inertia.

A circuit model may include spring (capacitor) elements corresponding to the suspensory ligaments (and possibly the tensor tympani and stapedius muscles) and to the malleus-incus (incudomallear) and incus-stapes (incudostapedial) joints. If one of the joints is considered to be rigid, then the capacitor is omitted and the inductors corresponding to the two ossicles can be combined. For example, for the human ear, Møller (1961) and Peake and Guinan (1967) included flexibility of the incudomallear joint, but Zwislocki (1957, 1962) did not. The guinea pig malleus and incus are actually fused together so no joint is required (Zwislocki 1963). None of these early models included flexibility of the incudostapedial joint.

[1] The adjective corresponding to malleus is mallear (cf. Latin adjective *mallearis*), not malleal as sometimes seen, and certainly not malleolar, which is the adjective for malleolus.

Damping elements (resistors) are included in various places to represent energy dissipation. Sometimes they are associated with inductors and sometimes with capacitors; the latter seems more appropriate.

The effects of the tensor tympani and stapedius muscles were represented explicitly by Onchi (1949, 1961), but more often they have been represented by changes in the elasticities and resistances of the branches representing the malleus and the stapes. Although it is possible that the muscles affect the effective masses of the ossicles, for example, by modifying the rotational axis of the malleus and incus, such effects are probably secondary to the changes of elasticity and resistance.

One disadvantage of circuit models is that their parameter values may need to be changed to accommodate changes in modes of vibration because of frequency, muscle contractions, different applied loads, or other effects. This greatly limits their predictive power. As is the case for the eardrum, however, both circuit models (e.g., O'Connor and Puria 2008) and two-port models (e.g., Shera and Zweig 1992a, b) of the ossicular chain continue to have value as concise representations of experimentally observed phenomena.

7.5.5.2 Distributed-Parameter Models

Early finite-element modeling of the middle ear concentrated on the eardrum because its inherently distributed nature represented the weakest part of available lumped-parameter middle ear models. In these models the axis of rotation was taken as fixed (e.g., Funnell and Laszlo 1978; Wada et al. 1992). Distributed models of the ossicles, ligaments, and muscles were needed, however, to cope with changes of vibration mode such as those resulting from muscle contractions (e.g., Pang and Peake 1986), and the need became even more clear as it became more and more obvious that the 3-D motions of the ossicular chain are very complex, with the position of the ossicular axis of rotation varying greatly with frequency and even within one cycle (e.g., Decraemer et al. 1991).

Eiber and Kauf (1994) described a model in which the ossicles were represented as distributed rigid bodies but the ligaments and joints were represented by lumped-parameter springs and dashpots. Hudde and Weistenhöfer (1997) described a model based on a 3-D generalization of circuit modeling, combining some features of the two-port and rigid-body approaches. These kinds of models are intermediate between circuit models and finite-element models. (The finite-element method can also incorporate perfectly rigid bodies by the use of "master" and "slave" degrees of freedom [e.g., Funnell 1983], thus reducing the computational cost when parts of the model are assumed to have negligible deformations.)

A finite-element model of the cat middle ear was developed based on a 3-D reconstruction from serial histological sections (Funnell and Funnell 1988; Funnell et al. 1992), with 20-μm sections and every second one used. Complete models of three different human middle ears were created from histological sections by Gan's group (Gan et al. 2004; Gan and Wang 2007). The usual practice was followed of cutting 20-μm sections and using only every tenth one, which of course limits the resolution for small structures. Much higher resolution was obtained by cutting 1-μm sections and using every one (Funnell et al. 2005), but this was feasible only for a small portion of the middle ear.

Finite-element models of the ossicles have also been based on magnetic-resonance microscopy (e.g., Van Wijhe et al. 2000) and X-ray microCT (e.g., Hagr et al. 2004). Other 3-D middle ear models that could be adapted to finite-element modeling have been created recently, based on histology (Wang et al. 2006; Chien et al. 2009) and OPFOS (Buytaert et al. 2011).

Illustrating the complementary nature of the different types of data, Elkhouri et al. (2006) created a model of the gerbil middle ear based primarily on magnetic-resonance microscopy data, but used microCT data for some small bony features, histological data to clarify some even smaller features, and moiré data for the shape of the eardrum. Although MR imaging distinguishes soft tissues better than X-ray CT does, it is possible to identify and model soft tissues with the latter (e.g., Sim and Puria 2008; Gea et al. 2009).

Even when detailed 3-D models of the ossicles were used, the shapes of the ligaments have often been very approximate. The annular ligament in particular, being so narrow, has often been represented by some number of springs whose stiffnesses were estimated either by using its dimensions and assumed material properties or by fitting to experimental data.

So far this section has focused on the generation of geometry for models of the ossicular chain. In terms of mechanics, a few examples will now be given of the kinds of issues that cannot be addressed with lumped-parameter models. To begin with, it is interesting to note that although in circuit models of the ossicular chain the incudostapedial joint has often been omitted, it must be present in some form in 3-D models because it represents the place where the (more or less) rotational motions of the malleus and incus are converted into the (more or less) translational motion of the stapes. Having a rigid joint would cause a very unphysiological constraint. The incudomallear joint, on the other hand, can be and often is omitted from both circuit and finite-element models as a first approximation. When included, both joints are generally treated rather simplistically. It may be important for some purposes to explicitly model the ligamentous joint capsules, the synovial fluid, the cartilaginous joint surfaces, and perhaps even the transition from bone to calcified cartilage to uncalcified cartilage.

The ossicles are generally considered to be rigid, but detailed modeling has indicated that there may be significant flexibility in the manubrium of the malleus (Funnell et al. 1992), in the stapes (Beer et al. 1999), and in the bony pedicle between the long process of the incus and the lenticular plate (Funnell et al. 2005, 2006).

Finite-element models are able to produce complex ossicular motions, and permit exploration of the conditions required for rotation about the supposed anatomical incudomallear axis, for example, or for piston-like motion of the stapes. Such models can be used, for example, to replicate and to predict the different effects of different ligaments (Dai et al. 2007) and to explore a hypothesized high-frequency twisting mode of the malleus (Puria and Steele 2010).

7.5.6 Cochlea

The load presented by the cochlea was considered by Zwislocki (1962) to have both resistive and reactive components, but to be mainly resistive, and it has often been modeled as a simple dashpot. Puria and Allen (1998) modeled it as a nonparametrically specified frequency-dependent impedance. In the early finite-element models there was a single damping representation for the whole model so the cochlear load was insufficient, but Koike et al. (2002) introduced additional damping for it. Recently middle ear models have been coupled to explicit finite-element models of the cochlea (Gan et al. 2007; Kim et al. 2011).

7.5.7 Nonlinearity

The middle ear behaves linearly for the purposes of normal hearing but becomes nonlinear in response to very loud sounds, to explosions, and to the large quasi-static pressures involved in large changes of altitude and in clinical tympanometry. The finite-element method was used by Stuhmiller (1989) to study the effects of blast, but with a linear model, and by Wada and Kobayashi (1990) to study the effects of tympanometric pressures, but with the eardrum assumed to be flat and circular. Price and Kalb (1991) and Pascal et al. (1998) attempted to model nonlinearity with a circuit model by empirically modifying parameter values and outputs. More recently, models have included both nonlinearities and realistic geometries (Ladak et al. 2006; Qi et al. 2006; Wang et al. 2007). Homma et al. (2010) modeled the effects of large static pressures, but by adjusting the material properties for the different pressures rather than by using nonlinear properties. Nonlinear simulation of the eardrum is complicated by its curvature, which may lead to wrinkles and unstable "snap-through" for large positive ear-canal pressures.

7.6 Summary

Apart from reviewing a few basic concepts involved in modeling, this chapter has summarized some of the work that has been done on modeling normal outer-ear and middle ear function. Modeling has also been done of various special conditions. In particular, finite-element models have been used to simulate such things as middle ear prostheses and implants, eardrum abnormalities, ventilation tubes, reverse transmission and otoacoustic emissions, liquid in the middle ear cavity, and bone conduction. Such studies may be where the true value of middle ear modeling lies, but the conclusions can only be as valid as the models themselves, and the models are far from definitive at this point. Modeling studies of the effects of anatomical variability have been few, but the great variability of human ears (e.g., Todd 2005) makes this an important factor in the interpretation of model results.

Apart from a few models for cat and gerbil, the only finite-element middle ear models have been for human ears even though the best experimental data for validation are from other species. More nonhuman models would probably be valuable in the development of better validated models. In the guidelines for model validation discussed in Sect. 7.4.3, the recommendation is that "validation of a complex system should be pursued in a hierarchical fashion from the component level to the system level" (Schwer 2007, p. 247). This has not been the approach for the middle ear, although recent experimental and modeling work on individual components (cf. Sect. 7.3) may lead in that direction.

Modeling has not yet been done of the effects of the smooth muscle present in the fibrocartilaginous ring surrounding the eardrum (Henson et al. 2005). This would be interesting for its own sake, and is also related to the open question of how much tension there is in the eardrum under normal operating conditions and how important it is.

Another open question is about the high-frequency behavior of the middle ear. Some experimental data (e.g., Olson 1998; Puria and Allen 1998) have shown a high-frequency middle ear response that has a relatively flat magnitude and a phase lag that increases more or less linearly with frequency. These are the characteristics of a delay line, and some modeling has been done that starts from the pure-delay hypothesis (cf. Voss et al., Chap. 4). Naturally, such models can reproduce delay-like behavior, but it may not be entirely clear how they are related to actual physical and mechanical mechanisms. Not all experimental data show the delay-like behavior. Moreover, some data that have been interpreted as having delay-like high-frequency magnitudes can be equally well described (from a different bias) as combinations of roll-offs and peaks. Most experimental data are not clean enough to be unambiguous.

Measurements at very high frequencies are very difficult to do reliably and repeatably. For one thing, sound-pressure fields become very nonuniform over very short distances. For another thing, the vibration modes of the middle ear become extremely complex and frequency-dependent. The very notions of input and output become difficult to define.

Finite-element models will necessarily show delay-like behavior if (1) that is how the middle ear really behaves and (2) the models are adequate representations of the middle ear. If a finite-element model does not satisfactorily replicate some set of experimental data, it does not condemn finite-element modeling as an approach, but rather indicates that that particular model was missing some features. It is possible, of course, that the missing features include unrecognized artefacts and errors of the experiments.

It is sometimes stated that a disadvantage of finite-element models is that they have too many parameters. Two things must be kept in mind, however. First, the number of parameters is directly related to the amount of realism included in the model. Because only as much realism needs to be included as is necessary to match the behavior of interest as closely as desired, the number of parameters might be claimed to be the minimum necessary. The second thing to keep in mind is that the real concern is not the number of parameters but rather the number of parameters that have to be assigned values by fitting to the experimental data being modeled. In principle, all of the parameters of a finite-element model can be assigned values based on experimental data (e.g., measurements of material properties) that are completely distinct from the situation being modeled, so the number of free parameters can be very small. In practice, of course, finite-element models, like others, are sometimes given additional free parameters because of a lack of knowledge or to make up for shortcomings in the models.

Another disadvantage sometimes ascribed to finite-element models is that their complexity can hinder insight into the fundamental principles involved in a system. It is certainly true that some simple lumped models have provided substantial insight into middle ear mechanics. If, however, the relevant fundamental principles include the asymmetrical distributed nature of the system, then complex models are required to elucidate it.

A third disadvantage ascribed to finite-element models is their computational expense, and that can indeed become a serious issue, with some simulations being very time-consuming. Again, however, it depends on the degree of realism required for the purposes of the model. If the realism is not necessary, it can be removed. If it is necessary, then, in the words of an esteemed graduate-studies supervisor, slow work takes time.

Acknowledgments This work has been funded by the Canadian Institutes of Health Research, the Natural Sciences and Engineering Research Council (Canada), and the Research Fund of Flanders (Belgium). We wish to acknowledge the profound contributions of Shyam M. Khanna to middle ear mechanics. W. R. J. Funnell and W. F. Decraemer in particular wish to thank him for his tremendous support as we were beginning our careers and for continuing to be an outstanding role model as a scientist and as a person. We also wish to express our sense of loss at the premature passing of Saumil Merchant. He was an important part of the interface between the clinic and us engineers and physicists.

References

Aernouts, J., & Dirckx, J. (2012). Static versus dynamic gerbil tympanic membrane elasticity: Derivation of the complex modulus. *Biomechanics and Modeling in Mechanobiology*, 11(6), 829–840.

American Society of Mechanical Engineers (ASME). (2006). *Guide for verification and validation in computational solid mechanics*. New York: American Society of Mechanical Engineers.

Beer, H.-J., Bornitz, M., Hardtke, H.-J., Schmidt, R., Hofmann, G., Vogel, U., Zahnert, T., & Hüttenbrink K.-B. (1999). Modelling of components of the human middle ear and simulation of their dynamic behaviour. *Audiology and Neuro-Otology*, 4(3–4), 156–162.

Bell, D. C. (2009). Contrast mechanisms and image formation in helium ion microscopy. *Microscopy and Microanalysis*, 15(2), 147–153.

Beranek, L. L. (1954). *Acoustics*. New York: McGraw-Hill.

Bourke, P. (1997). Polygonising a scalar field using tetrahedrons. *Geometry, Surfaces, Curves, Polyhedra*. Retrieved from: http://paulbourke.net/geometry/polygonise/#tetra. Accessed July 19, 2011.

Broekaert, D. (1995). The tympanic membrane: A biochemical updating of structural components. *Acta Oto-Rhino-Laryngologica Belgica*, 49, 127–137.

Brown, J. A., Torbatian, Z., Adamson, R. B., Van Wijhe, R., Pennings, R. J., Lockwood, G. R., & Bance, M. L. (2009). High-frequency ex vivo ultrasound imaging of the auditory system. *Ultrasound in Medicine & Biology*, 35(11), 1899–1907.

Buehler, M. J. (2008). Nanomechanics of collagen fibrils under varying cross-link densities: Atomistic and continuum studies. *Journal of the Mechanical Behavior of Biomedical Materials*, 1(1), 59–67.

Buytaert, J. A. N., Salih, W. H. M., Dierick, M., Jacobs, P., & Dirckx, J. J. J. (2011). Realistic 3D computer model of the gerbil middle ear, featuring accurate morphology of bone and soft tissue structures. *Journal of the Association for Research in Otolaryngology*, 12(6), 681–696.

Campolongo, F., Saltelli, A., & Cariboni, J. (2011). From screening to quantitative sensitivity analysis. A unified approach. *Computer Physics Communications*, 182(4), 978–988.

Carlton, P. M., Boulanger, J., Kervrann, C., Sibarita, J.-B., Salamero, J., Gordon-Messer, S., Bressan, D., Haber, J.E., Haase, S., Shao, L., Winoto, L., Matsuda, A., Kner, P., Uzawa, S., Gustafsson, M., Kam, Z., Agard, D.A., & Sedat, J.W. (2010). Fast live simultaneous multi-wavelength four-dimensional optical microscopy. *Proceedings of the National Academy of Sciences of the USA*, 107(37), 16016–16022.

Cerveri, P., & Pinciroli, F. (2001). Symbolic representation of anatomical knowledge: Concept classification and development strategies. *Journal of Biomedical Informatics*, 34(5), 321–347.

Cheng, T., & Gan, R. Z. (2007). Mechanical properties of stapedial tendon in human middle ear. *Journal of Biomechanical Engineering*, 129(6), 913–918.

Chien, W., Northrop, C., Levine, S., Pilch, B. Z., Peake, W. T., Rosowski, J. J., & Merchant, S. N. (2009). Anatomy of the distal incus in humans. *Journal of the Association for Research in Otolaryngology*, 10(4), 485–496.

Dai, C., Cheng, T., Wood, M. W., & Gan, R. Z. (2007). Fixation and detachment of superior and anterior malleolar ligaments in human middle ear: Experiment and modeling. *Hearing Research*, 230(1–2), 24–33.

Danzl, R., Helmli, F., & Scherer, S. (2011). Focus variation—A robust technology for high resolution optical 3D surface metrology. *Strojniski Vestnik/Journal of Mechanical Engineering*, 57(3), 245–256.

Decraemer, W. F., & Khanna, S. M. (1999). New insights into vibration of the middle ear. In *The function and mechanics of normal, diseased and reconstructed middle ears* (pp. 23–38). Presented at the 2nd International Symposium on Middle-Ear Mechanics in Research and Otosurgery, Boston, October 21–24.

Decraemer, W. F., Maes, M. A., & Vanhuyse, V. J. (1980). An elastic stress–strain relation for soft biological tissues based on a structural model. *Journal of Biomechanics*, 13(6), 463–468.

Decraemer, W. F., Khanna, S. M., & Funnell, W. R. J. (1991). Malleus vibration mode changes with frequency. *Hearing Research*, 54(2), 305–318.

Decraemer, W. F., Dirckx, J. J. J., & Funnell, W. R. J. (2003). Three-dimensional modelling of the middle-ear ossicular chain using a commercial high-resolution X-ray CT scanner. *Journal of the Association for Research in Otolaryngology*, 4(2), 250–263.

De Mey, J. R., Kessler, P., Dompierre, J., Cordelières, F. P., Dieterlen, A., Vonesch, J.-L., & Sibarita, J.-B. (2008). Fast 4D Microscopy. In *Methods in cell biology*, Vol. 85: *Fluorescent proteins* (pp. 83–112). Philadelphia: Elsevier.

Denk, W., & Horstmann, H. (2004). Serial block-face scanning electron microscopy to reconstruct three-dimensional tissue nanostructure. *PLoS Biology*, 2(11), 1901–1909.

Dirckx, J. J. J., Decraemer, W. F., & Dielis, G. (1988). Phase shift method based on object translation for full field automatic 3-D surface reconstruction from moire topograms. *Applied Optics*, 27(6), 1164–1169.

Drescher, J., Schmidt, R., & Hardtke, H. J. (1998). Finite-Elemente-Modellierung und Simulation des menschlichen Trommelfells. *HNO*, 46(2), 129–134.

Egolf, D. P., Nelson, D. K., Howell, H. C., III, & Larson, V. D. (1993). Quantifying ear-canal geometry with multiple computer-assisted tomographic scans. *The Journal of the Acoustical Society of America*, 93(5), 2809–2819.

Eiber, A., & Kauf, A. (1994). Berechnete Verschiebungen der Mittelohrknochen unter statischer Belastung. *HNO*, 42(12), 754–759.

Elkhouri, N., Liu, H., & Funnell, W. R. J. (2006). Low-frequency finite-element modeling of the gerbil middle ear. *Journal of the Association for Research in Otolaryngology*, 7(4), 399–411.

Erfani, T., & Utyuzhnikov, S. V. (2011). Directed search domain: A method for even generation of the Pareto frontier in multiobjective optimization. *Engineering Optimization*, 43(5), 467–484.

Esser, M. H. M. (1947). The mechanism of the middle ear: II. The drum. *Bulletin of Mathematical Biophysics*, 9, 75–91.

Fay, J. (2001). *Cat eardrum mechanics*. Ph.D. thesis, Stanford University.

Fay, J., Puria, S., Decraemer, W. F., & Steele, C. (2005). Three approaches for estimating the elastic modulus of the tympanic membrane. *Journal of Biomechanics*, 38(9), 1807–1815.

Fay, J. P., Puria, S., & Steele, C. R. (2006). The discordant eardrum. *Proceedings of the National Academy of Sciences of the USA*, 103(52), 19743–19748.

Fercher, A. F. (2010). Optical coherence tomography – development, principles, applications. *Zeitschrift für Medizinische Physik*, 20(4), 251–276.

Frank, O. (1923). Die Leitung des Schalles im Ohr. *Sitzungsberichte der mathematisch-physikalischen Klasse der Bayerischen Akademie der Wissenschaften zu München*, 1923, 11–77.

Funnell, S. M., & Funnell, W. R. J. (1988). An approach to finite-element modelling of the middle ear. In *Proceedings of the 14th Canadian Medical & Biological Engineering Conference* (pp. 101–102). Montréal, June 28–30.

Funnell, W. R. J. (1981). Image processing applied to the interactive analysis of interferometric fringes. *Applied Optics*, 20(18), 3245–3250.

Funnell, W. R. J. (1983). On the undamped natural frequencies and mode shapes of a finite-element model of the cat eardrum. *The Journal of the Acoustical Society of America*, 73(5), 1657–1661.

Funnell, W. R. J. (1984). On the choice of a cost function for the reconstruction of surfaces by triangulation between contours. *Computers and Structures*, 18(1), 23–26.

Funnell, W. R. J., & Decraemer, W. F. (1996). On the incorporation of moiré shape measurements in finite-element models of the cat eardrum. *The Journal of the Acoustical Society of America*, 100(2), 925–932.

Funnell, W. R. J., & Laszlo, C. A. (1974). Dependence of middle-ear parameters on body weight in the guinea pig. *The Journal of the Acoustical Society of America*, 56(5), 1551–1553.

Funnell, W. R. J., & Laszlo, C. A. (1978). Modeling of the cat eardrum as a thin shell using the finite-element method. *The Journal of the Acoustical Society of America*, 63(5), 1461–1467.

Funnell, W. R. J., & Laszlo, C. A. (1982). A critical review of experimental observations on ear-drum structure and function. *ORL: Journal for Oto-Rhino-Laryngology and Its Related Specialties*, 44(4), 181–205.

Funnell, W. R. J., Decraemer, W. F., & Khanna, S. M. (1987). On the damped frequency response of a finite-element model of the cat eardrum. *The Journal of the Acoustical Society of America*, 81(6), 1851–1859.

Funnell, W. R. J., Khanna, S. M., & Decraemer, W. F. (1992). On the degree of rigidity of the manubrium in a finite-element model of the cat eardrum. *The Journal of the Acoustical Society of America*, 91(4), 2082–2090.

Funnell, W. R. J., Heng Siah, T., McKee, M. D., Daniel, S. J., & Decraemer, W. F. (2005). On the coupling between the incus and the stapes in the cat. *Journal of the Association for Research in Otolaryngology*, 6(1), 9–18.

Funnell, W. R. J., Daniel, S. J., Alsabah, B., & Liu, H. (2006). On the coupling between the incus and the stapes. In *Auditory mechanisms: Processes and models, Proceedings of the Ninth International Symposium* (pp. 115–116). Portland, OR, July 23–28, 2005.

Funnell, W. R. J., Maftoon, N., & Decraemer, W. F. (2012). Mechanics and modelling for the middle ear. Retrieved from: http://audilab.bme.mcgill.ca/mammie/. (Accessed November 11, 2012).

Gan, R. Z., & Wang, X. (2007). Multifield coupled finite element analysis for sound transmission in otitis media with effusion. *The Journal of the Acoustical Society of America*, 122(6), 3527–3538.

Gan, R. Z., Feng, B., & Sun, Q. (2004). Three-dimensional finite element modeling of human ear for sound transmission. *Annals of Biomedical Engineering*, 32(6), 847–859.

Gan, R. Z., Sun, Q., Feng, B., & Wood, M. W. (2006). Acoustic-structural coupled finite element analysis for sound transmission in human ear—Pressure distributions. *Medical Engineering & Physics*, 28(5), 395–404.

Gan, R. Z., Reeves, B. P., & Wang, X. (2007). Modeling of sound transmission from ear canal to cochlea. *Annals of Biomedical Engineering*, 35(12), 2180–2195.

Gan, R. Z., Cheng, T., Dai, C., Yang, F., & Wood, M. W. (2009). Finite element modeling of sound transmission with perforations of tympanic membrane. *The Journal of the Acoustical Society of America*, 126(1), 243–253.

Gariepy, B. (2011). *Finite-element modelling of the newborn ear canal and middle ear*. M. Eng. Thesis, McGill University, Montréal.

Gea, S. L. R., Decraemer, W. F., Funnell, W. R. J., Dirckx, J. J. J., & Maier, H. (2009). Tympanic membrane boundary deformations derived from static displacements observed with computerized tomography in human and gerbil. *Journal of the Association for Research in Otolaryngology*, 11(1), 1–17.

Gehrmann, S., Höhne, K. H., Linhart, W., Pflesser, B., Pommert, A., Riemer, M., Tiede, U., Windolf, J., Schumacher, U., & Rueger, J. M. (2006). A novel interactive anatomic atlas of the hand. *Clinical Anatomy*, 19(3), 258–266.

Georgiev, T., Lumsdaine, A., & Chunev, G. (2011). Using focused plenoptic cameras for rich image capture. *IEEE Computer Graphics and Applications*, 31(1), 62–73.

Goll, E., & Dalhoff, E. (2011). Modeling the eardrum as a string with distributed force. *The Journal of the Acoustical Society of America*, 130(3), 1452–1462.

Gran, S. (1968). *Analytische Grundlage der Mittelohrmechanik. Ein Beitrag zur Anwendung der akustischen Impedanz des Ohres*. Ph.D. thesis, Universität Oslo.

Guelke, R., & Keen, J. A. (1949). A study of the vibrations of the tympanic membrane under direct vision, with a new explanation of their physical characteristics. *The Journal of Physiology*, 110, 226–236.

Guinan, J. J., & Peake, W. T. (1967). Middle-ear characteristics of anesthetized cats. *The Journal of the Acoustical Society of America*, 41(5), 1237–1261.

Hagr, A. A., Funnell, W. R. J., Zeitouni, A. G., & Rappaport, J. M. (2004). High-resolution X-ray computed tomographic scanning of the human stapes footplate. *The Journal of Otolaryngology*, 33(4), 217–221.

Hart, D. P., Frigerio, F., & Marini, D. M. (2010). Three-dimensional imaging using an inflatable membrane. Retrieved from: http://www.google.ca/patents/US20100042002. (Accessed November 11, 2012).

Hartman, W. F. (1971). An error in Helmholtz's calculation of the displacement of the tympanic membrane. *The Journal of the Acoustical Society of America*, 49(4B), 1317.

Helmholtz, H. L. F. (1868). Die Mechanik der Gehörknöchelchen und des Trommelfells. *Pflügers Archiv für die gesammte Physiologie (Bonn)*, 1, 1–60.

Helton, J. C., Johnson, J. D., Sallaberry, C. J., & Storlie, C. B. (2006). Survey of sampling-based methods for uncertainty and sensitivity analysis. *Reliability Engineering & System Safety*, 91(10–11), 1175–1209.

Henson, M., Madden, V., Raskandersen, H., & Henson, O. W., Jr. (2005). Smooth muscle in the annulus fibrosus of the tympanic membrane in bats, rodents, insectivores, and humans. *Hearing Research*, 200(1–2), 29–37.

Henson, M. M., Henson, O. W., Jr., Gewalt, S. L., Wilson, J. L., & Johnson, G. A. (1994). Imaging the cochlea by magnetic resonance microscopy. *Hearing Research*, 75(1–2), 75–80.

Homma, K., Shimizu, Y., Kim, N., Du, Y., & Puria, S. (2010). Effects of ear-canal pressurization on middle-ear bone- and air-conduction responses. *Hearing Research*, 263(1–2), 204–215.

Huang, G. T., Rosowski, J. J., & Peake, W. T. (2000). Relating middle-ear acoustic performance to body size in the cat family: measurements and models. *Journal of Comparative Physiology A: Sensory, Neural, and Behavioral Physiology*, 186(5), 447–465.

Hudde, H. (1983). Estimation of the area function of human ear canals by sound pressure measurements. *The Journal of the Acoustical Society of America*, 73(1), 24–31.

Hudde, H., & Schmidt, S. (2009). Sound fields in generally shaped curved ear canals. *The Journal of the Acoustical Society of America*, 125(5), 3146–3157.

Hudde, H., & Weistenhöfer, C. (1997). A three-dimensional circuit model of the middle ear. *Acta Acustica united with Acustica*, 83(3), 535–549.

Jackson, R. P., Chlebicki, C., Krasieva, T. B., & Puria, S. (2008). Multiphoton microscopy imaging of collagen fiber layers and orientation in the tympanic membrane. *Photonic Therapeutics and Diagnostics IV, Proceedings of SPIE - Int. Soc. Opt. Eng.* (Vol. 6842, p. 68421D). San Jose, CA, January 19.

Jang, H. G., Chung, M. S., Shin, D. S., Park, S. K., Cheon, K. S., Park, H. S., & Park, J. S. (2011). Segmentation and surface reconstruction of the detailed ear structures, identified in sectioned images. *The Anatomical Record: Advances in Integrative Anatomy and Evolutionary Biology*, 294(4), 559–564.

Johnson, G. A., Thompson, M. B., Gewalt, S. L., & Hayes, C. E. (1986). Nuclear magnetic resonance imaging at microscopic resolution. *Journal of Magnetic Resonance*, 68(1), 129–137.

Just, T., Lankenau, E., Hüttmann, G., & Pau, H. W. (2009). Optische Kohärenztomographie in der Mittelohrchirurgie. *HNO*, 57(5), 421–427.

Kanzaki, S., Takada, Y., Niida, S., Takeda, Y., Udagawa, N., Ogawa, K., Nango, N., Momose, A., & Matsuo, K. (2011). Impaired vibration of auditory ossicles in osteopetrotic mice. *The American Journal of Pathology*, 178(3), 1270–1278.

Khanna, S. M., & Tonndorf, J. (1972). Tympanic membrane vibrations in cats studied by time-averaged holography. *The Journal of the Acoustical Society of America*, 51(6B), 1904–1920.

Khanna, S. M., & Stinson, M. R. (1985). Specification of the acoustical input to the ear at high frequencies. *The Journal of the Acoustical Society of America*, 77(2), 577–589.

Kim, N., Homma, K., & Puria, S. (2011). Inertial bone conduction: Symmetric and anti-symmetric components. *Journal of the Association for Research in Otolaryngology*, 12(3), 261–279.

Kirikae, I. (1960). *The structure and function of the middle ear*. Tokyo: University Press.

Koester, C. J., Khanna, S. M., Rosskothen, H. D., Tackaberry, R. B., & Ulfendahl, M. (1994). Confocal slit divided-aperture microscope: Applications in ear research. *Applied Optics*, 33(4), 702–708.

Koike, T., Wada, H., & Kobayashi, T. (2002). Modeling of the human middle ear using the finite-element method. *The Journal of the Acoustical Society of America*, 111(3), 1306–1317.

Kojo, Y. (1954). [Morphological studies of the human tympanic membrane]. *Nippon Jibiinkoka Gakkai Kaiho* (*The Journal of the Oto-Rhino-Laryngological Society of Japan*), 57(2), 115–126.

Koning, R. I., & Koster, A. J. (2009). Cryo-electron tomography in biology and medicine. *Annals of Anatomy - Anatomischer Anzeiger*, 191(5), 427–445.

Kringlebotn, M. (1988). Network model for the human middle ear. *Scandinavian Audiology*, 17(2), 75–85.

Kuypers, L. C., Decraemer, W. F., Dirckx, J. J. J., & Timmermans, J.-P. (2005a). Thickness distribution of fresh eardrums of cat obtained with confocal microscopy. *Journal of the Association for Research in Otolaryngology*, 6(3), 223–233.

Kuypers, L. C., Dirckx, J. J. J., Decraemer, W. F., & Timmermans, J.-P. (2005b). Thickness of the gerbil tympanic membrane measured with confocal microscopy. *Hearing Research*, 209(1–2), 42–52.

Kuypers, L. C., Decraemer, W. F., & Dirckx, J. J. J. (2006). Thickness distribution of fresh and preserved human eardrums measured with confocal microscopy. *Otology & Neurotology*, 27(2), 256–264.

Ladak, H. M., Decraemer, W. F., Dirckx, J. J. J., & Funnell, W. R. J. (2004). Response of the cat eardrum to static pressures: Mobile versus immobile malleus. *The Journal of the Acoustical Society of America*, 116(5), 3008–3021.

Ladak, H. M., Funnell, W. R. J., Decraemer, W. F., & Dirckx, J. J. J. (2006). A geometrically nonlinear finite-element model of the cat eardrum. *The Journal of the Acoustical Society of America*, 119(5 Pt 1), 2859–2868.

Leach, R. (2010). *Fundamental principles of engineering nanometrology*. Oxford: Elsevier.

Lee, C.-F., Chen, P.-R., Lee, W.-J., Chen, J.-H., & Liu, T.-C. (2006). Three-dimensional reconstruction and modeling of middle ear biomechanics by high-resolution computed tomography and finite element analysis. *The Laryngoscope*, 116(5), 711–716.

Lee, W.-J., Lee, C.-F., Chen, S.-Y., Chen, Y.-S., & Sun, C.-K. (2010). Virtual biopsy of rat tympanic membrane using higher harmonic generation microscopy. *Journal of Biomedical Optics*, 15(4), 046012.

Liang, J., McInerney, T., & Terzopoulos, D. (2006). United snakes. *Medical Image Analysis*, 10(2), 215–233.

Lim, D. J. (1995). Structure and function of the tympanic membrane: A review. *Acta Oto-Rhino-Laryngologica Belgica*, 49(2), 101–115.

Lowell, P. (1908). *Mars as the abode of life*. New York: Macmillan Co.

Luo, H., Dai, C., Gan, R. Z., & Lu, H. (2009). Measurement of Young's modulus of human tympanic membrane at high strain rates. *Journal of Biomechanical Engineering*, 131(6), 064501.

Lynch, T. J., III, Nedzelnitsky, V., & Peake, W. T. (1982). Input impedance of the cochlea in cat. *The Journal of the Acoustical Society of America*, 72(1), 108–130.

Maftoon, N., Nambiar, S., Funnell, W. R. J., Decraemer, W. F., & Daniel, S. J. (2011). Experimental and modelling study of gerbil tympanic-membrane vibrations. In *34th Midwinter Research Meeting, Association for Research in Otolaryngology* Baltimore, February 19–23.

Margolis, R. H., Osguthorpe, J. D., & Popelka, G. R. (1978). The effects of experimentally-produced middle ear lesions on tympanometry in cats. *Acta Oto-Laryngologica*, 86(5–6), 428–436.

Marwala, T. (2010). *Finite-element-model updating using computional intelligence techniques*. London: Springer.

Mayes, R. L. (2009). Developing adequacy criterion for model validation based on requirements. *Proceedings of the IMAC-XXVII*. Orlando, FL, February 8–12.

Moens, D., & Vandepitte, D. (2005). A survey of non-probabilistic uncertainty treatment in finite element analysis. *Computer Methods in Applied Mechanics and Engineering*, 194(12–16), 1527–1555.

Møller, A. R. (1961). Network model of the middle ear. *The Journal of the Acoustical Society of America*, 33(2), 168–176.

Møller, A. R. (1965). An experimental study of the acoustic impedance of the middle ear and its transmission properties. *Acta Oto-Laryngologica*, 60(1–6), 129–149.

Neudert, M., Beleites, T., Ney, M., Kluge, A., Lasurashvili, N., Bornitz, M., Scharnweber, D., & Zahnert, T. (2010). Osseointegration of titanium prostheses on the stapes footplate. *Journal of the Association for Research in Otolaryngology*, 11(2), 161–171.

Novotny, L. (2011). From near-field optics to optical antennas. *Physics Today*, 64(7), 47–52.

O'Connor, K. N., & Puria, S. (2008). Middle-ear circuit model parameters based on a population of human ears. *The Journal of the Acoustical Society of America*, 123(1), 197–211.

Odgaard, A., Andersen, K., Ullerup, R., Frich, L. H., & Melsen, F. (1994). Three-dimensional reconstruction of entire vertebral bodies. *Bone*, 15(3), 335–342.

Olson, E. S. (1998). Observing middle and inner ear mechanics with novel intracochlear pressure sensors. *The Journal of the Acoustical Society of America*, 103(6), 3445–3463.

Onchi, Y. (1949). A study of the mechanism of the middle ear. *The Journal of the Acoustical Society of America*, 21(4), 404–410.

Onchi, Y. (1961). Mechanism of the middle ear. *The Journal of the Acoustical Society of America*, 33(6), 794–805.

Palva, T., Northrop, C., & Ramsay, H. (2001). Aeration and drainage pathways of Prussak's space. *International Journal of Pediatric Otorhinolaryngology*, 57(1), 55–65.

Pang, X. D., & Peake, W. T. (1986). *How* do contractions of the stapedius muscle alter the acoustic properties of the ear? In *Peripheral Auditory Mechanisms* (pp. 36–43). Berlin: Springer-Verlag.

Parent, P., & Allen, J. B. (2007). Wave model of the cat tympanic membrane. *The Journal of the Acoustical Society of America*, 122(2), 918–931.

Parent, P., & Allen, J. B. (2010). Time-domain "wave" model of the human tympanic membrane. *Hearing Research*, 263(1–2), 152–167.

Parker, W. S. (2008). Franklin, Holmes, and the epistemology of computer simulation. *International Studies in the Philosophy of Science*, 22(2), 165–183.

Pascal, J., Bourgeade, A., Lagier, M., & Legros, C. (1998). Linear and nonlinear model of the human middle ear. *The Journal of the Acoustical Society of America*, 104(3 Pt 1), 1509–1516.

Peake, W. T., & Guinan, J. J. (1967). *Circuit model for the cat's middle ear*. Quarterly Progress Report No. 84 (pp. 320–326). Cambridge, MA: Research Laboratory of Electronics, Massachusetts Institute of Technology.

Polys, N. F., Brutzman, D., Steed, A., & Behr, J. (2008). Future standards for immersive VR. *IEEE Computer Graphics and Applications*, 28(2), 94–99.

Price, G. R., & Kalb, J. T. (1991). Insights into hazard from intense impulses from a mathematical model of the ear. *The Journal of the Acoustical Society of America*, 90(1), 219–227.

Puria, S. (1991). *A theory of cochlear input impedance and middle ear parameter estimation.* Ph.D. thesis, City University New York.

Puria, S., & Allen, J. B. (1998). Measurements and model of the cat middle ear: Evidence of tympanic membrane acoustic delay. *The Journal of the Acoustical Society of America*, 104(6), 3463–3481.

Puria, S., & Steele, C. (2010). Tympanic-membrane and malleus-incus-complex co-adaptations for high-frequency hearing in mammals. *Hearing Research*, 263(1–2), 183–190.

Qi, L., Liu, H., Lutfy, J., Funnell, W. R. J., & Daniel, S. J. (2006). A nonlinear finite-element model of the newborn ear canal. *The Journal of the Acoustical Society of America*, 120(6), 3789–3798.

Rabbitt, R. D., & Holmes, M. H. (1986). A fibrous dynamic continuum model of the tympanic membrane. *The Journal of the Acoustical Society of America*, 80(6), 1716–1728.

Rabbitt, R. D., & Holmes, M. H. (1988). Three-dimensional acoustic waves in the ear canal and their interaction with the tympanic membrane. *The Journal of the Acoustical Society of America*, 83(3), 1064–1080.

Reiser, M. F., Semmler, W., & Hricak, H. (Eds.). (2008). *Magnetic resonance tomography*. Berlin: Springer.

Roache, P. J. (2002). Code verification by the method of manufactured solutions. *Journal of Fluids Engineering, Transactions of the ASME*, 124(1), 4–10.

Salih, W. H. M., Buytaert, J. A. N., Aerts, J. R. M., Vanderniepen, P., Dierick, M., & Dirckx, J. J. J. (2012). Open access high-resolution 3D morphology models of cat, gerbil, rabbit, rat and human ossicular chains. *Hearing Research*, 284(1–2), 1–5.

Sasov, A., & Van Dyck, D. (1998). Desktop X-ray microscopy and microtomography. *Journal of Microscopy*, 191(2), 151–158.

Schmidt, S. H., & Hellström, S. (1991). Tympanic-membrane structure—new views. A comparative study. *ORL: Journal for Oto-Rhino-Laryngology and Its Related Specialties*, 53(1), 32–36.

Schroeder, W., Martin, K. W., & Lorensen, B. (1996). *The Visualization Toolkit: An object-oriented approach to 3-D graphics*. Upper Saddle River, NJ: Prentice Hall PTR.

Schwer, L. E. (2007). An overview of the PTC 60/V&V 10: Guide for verification and validation in computational solid mechanics. *Engineering with Computers*, 23(4), 245–252.

Shaw, E. A. G., & Stinson, M. R. (1983). The human external and middle ear: Models and concepts. *Mechanics of Hearing* (pp. 3–10). Delft: M. Nijhoff (The Hague) & Delft University Press.

Shaw, E. A. G., & Stinson, M. R. (1986). Eardrum dynamics, middle ear transmission and the human hearing threshold curve. In *Proceedings of the 12th International Congress on Acoustics* (pp. B6–6), Toronto, July 24–31.

Shera, C. A., & Zweig, G. (1991). Phenomenological characterization of eardrum transduction. *The Journal of the Acoustical Society of America*, 90(1), 253–262.

Shera, C. A., & Zweig, G. (1992a). Middle-ear phenomenology: The view from the three windows. *The Journal of the Acoustical Society of America*, 92(3), 1356–1370.

Shera, C. A., & Zweig, G. (1992b). Analyzing reverse middle-ear transmission: Noninvasive *Gedankenexperiments*. *The Journal of the Acoustical Society of America*, 92(3), 1371–1381.

Sim, J. H., & Puria, S. (2008). Soft tissue morphometry of the malleus-incus complex from micro-CT imaging. *Journal of the Association for Research in Otolaryngology*, 9, 5–21.

Snavely, N., Seitz, S. M., & Szeliski, R. (2008). Modeling the world from Internet photo collections. *International Journal of Computer Vision*, 80(2), 189–210.

Soons, J. A. M., Aernouts, J., & Dirckx, J. J. J. (2010). Elasticity modulus of rabbit middle ear ossicles determined by a novel micro-indentation technique. *Hearing Research*, 263(1-2), 33–37.

Sørensen, M. S., Dobrzeniecki, A. B., Larsen, P., Frisch, T., Sporring, J., & Darvann, T. A. (2002). The Visible Ear: A digital image library of the temporal bone. *ORL: Journal for Oto-Rhino-Laryngology and Its Related Specialties*, 64(6), 378–381.

Stepp, C. E., & Voss, S. E. (2005). Acoustics of the human middle-ear air space. *The Journal of the Acoustical Society of America*, 118(2), 816–871.

Stinson, M. R., & Khanna, S. M. (1989). Sound propagation in the ear canal and coupling to the eardrum, with measurements on model systems. *The Journal of the Acoustical Society of America*, 85(6), 2481–2491.

Stinson, M. R., & Lawton, B. W. (1989). Specification of the geometry of the human ear canal for the prediction of sound-pressure level distribution. *The Journal of the Acoustical Society of America*, 85(6), 2492–2503.

Stinson, M. R., & Daigle, G. A. (2005). Comparison of an analytic horn equation approach and a boundary element method for the calculation of sound fields in the human ear canal. *The Journal of the Acoustical Society of America*, 118(4), 2405–2411.

Stuhmiller, J. H. (1989). Use of modeling in predicting tympanic membrane rupture. *The Annals of Otology, Rhinology & Laryngology. Supplement*, 140, 53–60.

Subhash, H. M., Nguyen-Huynh, A., Wang, R. K., Jacques, S. L., Choudhury, N., & Nuttall, A. L. (2012). Feasibility of spectral-domain phase-sensitive optical coherence tomography for middle ear vibrometry. *Journal of Biomedical Optics*, 17(6), 060505.

Sun, C.-K. (2005). Higher harmonic generation microscopy. *Advances in Biochemical Engineering/Biotechnology*, 95, 17–56.

Sun, Q., Chang, K.-H., Dormer, K. J., Dyer, R. K., Jr., & Gan, R. Z. (2002a). An advanced computer-aided geometric modeling and fabrication method for human middle ear. *Medical Engineering & Physics*, 24(9), 595–606.

Sun, Q., Gan, R. Z., Chang, K.-H., & Dormer, K. J. (2002b). Computer-integrated finite element modeling of human middle ear. *Biomechanics and Modeling in Mechanobiology*, 1(2), 109–122.

Teoh, S. W. (1996). *The roles of pars flaccida in middle ear acoustic transmission*. Ph.D. thesis, Massachusetts Institute of Technology.
Teoh, S. W., Flandermeyer, D. T., & Rosowski, J. J. (1997). Effects of pars flaccida on sound conduction in ears of Mongolian gerbil: Acoustic and anatomical measurements. *Hearing Research*, 106(1–2), 39–65.
Todd, N. W. (2005). Orientation of the manubrium mallei: inexplicably widely variable. *The Laryngoscope*, 115(9), 1548–1552.
Tuck-Lee, J. P., Pinsky, P. M., Steele, C. R., & Puria, S. (2008). Finite element modeling of acousto-mechanical coupling in the cat middle ear. *The Journal of the Acoustical Society of America*, 124(1), 348–362.
Uebo, K., Kodama, A., Oka, Y., & Ishii, T. (1988). [Thickness of normal human tympanic membrane]. *Ear Research Japan*, 19, 70–73.
Van Wijhe, R., Funnell, W. R. J., Henson, O. W., & Henson, M. M. (2000). Development of a finite-element model of the middle ear of the moustached bat. In *Proceedings of the 26th Annual Conference of the Canadian Medical and Biological Engineering Society* (pp. 20–21). Halifax, October 26–28.
Vogel, U. (1999). New approach for 3D imaging and geometry modeling of the human inner ear. *ORL: Journal for Oto-Rhino-Laryngology and Its Related Specialties*, 61(5), 259–267.
Vogel, U., & Schmitt, T. (1998). 3D visualization of middle ear structures. *Medical Imaging 1998: Image Display, Proceedings SPIE* (Vol. 3335, pp. 141–151). San Diego, February 21.
Voie, A. H., Burns, D. H., & Spelman, F. A. (1993). Orthogonal-plane fluorescence optical sectioning: Three-dimensional imaging of macroscopic biological specimens. *Journal of Microscopy*, 170, 229–236.
Volandri, G., Di Puccio, F., Forte, P., & Carmignani, C. (2011). Biomechanics of the tympanic membrane. *Journal of Biomechanics*, 44(7), 1219–1236.
von Ahn, L., Maurer, B., McMillen, C., Abraham, D., & Blum, M. (2008). reCAPTCHA: Human-based character recognition via Web security measures. *Science*, 321(5895), 1465–1468.
von Békésy, G. (1941). Über die Messung der Schwingungsamplitude der Gehörknöchelchen mittels einer kapazitiven Sonde. *Akustische Zeitschrift*, 6, 1–16.
von Békésy, G. (1949). The structure of the middle ear and the hearing of one's own voice by bone conduction. *The Journal of the Acoustical Society of America*, 21, 217–232.
von Békésy, G. (1960). *Experiments in Hearing*. New York, NY: McGraw-Hill.
von Unge, M., Bagger-Sjöbäck, D., & Borg, E. (1991). Mechanoacoustic properties of the tympanic membrane: A study on isolated Mongolian gerbil temporal bones. *The American Journal of Otology*, 12(6), 407–419.
Wada, H., & Kobayashi, T. (1990). Dynamical behavior of middle ear: Theoretical study corresponding to measurement results obtained by a newly developed measuring apparatus. *The Journal of the Acoustical Society of America*, 87(1), 237–245.
Wada, H., Metoki, T., & Kobayashi, T. (1992). Analysis of dynamic behavior of human middle ear using a finite-element method. *The Journal of the Acoustical Society of America*, 92(6), 3157–3168.
Wada, H., Onda, N., Date, K., & Kobayashi, T. (1996). Assessment of mechanical properties of tympanic membrane by supersonic wave method. *Nippon Kikai Gakkai Ronbunshu, C Hen/ Transactions of the Japan Society of Mechanical Engineers, Part C*, 62(598), 2289–2292.
Wang, H., Northrop, C., Burgess, B., Liberman, M. C., & Merchant, S. N. (2006). Three-dimensional virtual model of the human temporal bone: A stand-alone, downloadable teaching tool. *Otology & Neurotology*, 27(4), 452–457.
Wang, X., Cheng, T., & Gan, R. Z. (2007). Finite-element analysis of middle-ear pressure effects on static and dynamic behavior of human ear. *The Journal of the Acoustical Society of America*, 122(2), 906–917.
Webber, R. L., Webber, S. E., & Moore, J. (2002). Hand-held three-dimensional dental X-ray system: Technical description and preliminary results. *Dentomaxillofacial Radiology*, 31(4), 240–248.

Weistenhöfer, C., & Hudde, H. (1999). Determination of the shape and inertia properties of the human auditory ossicles. *Audiology & Neuro-Otology*, 4(3–4), 192–196.

Wever, E. G., & Lawrence, M. (1954). *Physiological acoustics*. Princeton, NJ: Princeton University Press.

Wiener, F. M., & Ross, D. A. (1946). The pressure distribution in the auditory canal in a progressive sound field. *The Journal of the Acoustical Society of America*, 18(2), 401–408.

Williams, K. R., & Lesser, T. H. (1990). A finite element analysis of the natural frequencies of vibration of the human tympanic membrane. Part I. *British Journal of Audiology*, 24(5), 319–327.

Williams, K. R., Blayney, A. W., & Rice, H. J. (1996). Development of a finite element model of the middle ear. *Revue De Laryngologie - Otologie - Rhinologie*, 117(3), 259–264.

Wilson, J. L., Henson, M. M., Gewalt, S. L., Keating, A. W., & Henson, O. W., Jr. (1996). Reconstructions and cross-sectional area measurements from magnetic resonance microscopic images of the cochlea. *American Journal of Otology*, 17(2), 347–353.

Winsberg, E. (2010). *Science in the age of computer simulation*. Chicago: University of Chicago Press.

Wojtkowski, M. (2010). High-speed optical coherence tomography: Basics and applications. *Applied Optics*, 49(16), D30–61.

Yoo, S. H., Park, K. H., Moon, S. K., Kim, W.-S., & Bae, J. H. (2004). Evaluation of dynamic behavior of the human middle ear with nonhomogeneity by finite element method. *Key Engineering Materials*, 270–273, 2067–2072.

Young, P. G., Beresford-West, T. B. H., Coward, S. R. L., Notarberardino, B., Walker, B., & Abdul-Aziz, A. (2008). An efficient approach to converting three-dimensional image data into highly accurate computational models. *Philosophical Transactions A: Mathematical, Physical, and Engineering Sciences*, 366(1878), 3155–3173.

Yuan, Y., & Verma, R. (2006). Measuring microelastic properties of stratum corneum. *Colloids and Surfaces B: Biointerfaces*, 48(1), 6–12.

Zebian, M., Hensel, J., & Fedtke, T. (2012). How the cross-sectional discontinuity between ear canal and probe affects the ear canal length estimation. *The Journal of the Acoustical Society of America*, 132(1), EL8–EL14.

Zhang, X., & Gan, R. Z. (2010). Dynamic properties of human tympanic membrane – experimental measurement and modelling analysis. *International Journal of Experimental and Computational Biomechanics*, 1(3), 252–270.

Zhang, X., & Gan, R. Z. (2011). A comprehensive model of human ear for analysis of implantable hearing devices. *IEEE Transactions on Biomedical Engineering*, 58(10), 3024–3027.

Zhang, Y.-J. (2006). *Advances in image and video segmentation*. Hershey, PA: IRM Press.

Zwislocki, J. (1957). Some impedance measurements on normal and pathological ears. *The Journal of the Acoustical Society of America*, 29(12), 1312–1317.

Zwislocki, J. (1962). Analysis of the middle-ear function. Part I: Input impedance. *The Journal of the Acoustical Society of America*, 34(9B), 1514–1523.

Zwislocki, J. (1963). Analysis of the middle-ear function. Part II: Guinea-pig ear. *The Journal of the Acoustical Society of America*, 35(7), 1034–1040.

Chapter 8
Diagnostic Measurements and Imaging Technologies for the Middle Ear

Gerald R. Popelka and Lisa L. Hunter

Keywords Acoustic reflex • Admittance • Air conduction • Audiogram • Bone conduction • Computed tomography • Immittance • Impedance • Laser-Doppler vibrometry • Magnetic resonance imaging • Optical coherence tomography • Otoacoustic emissions • Pure tone threshold • Reflectance • Tympanometry

8.1 Introduction

Clinical diagnostic tools and imaging technologies can be used to quantify both the physical and the functional status of the middle ear in humans. Because different pathologies often cause a single structural change and because a single pathology often causes multiple structural changes, few of these diagnostic measures can identify a specific pathology. However, these clinical technologies can provide very precise, repeatable, and quantifiable measures of the structure and function of the entire middle ear and its components in a living human.

Figure 8.1 is an illustration of the human middle ear and its surrounding structures that can be used when discussing current clinical middle ear measurements. For purposes of this chapter, the middle ear consists of the structures encompassing the lateral surface of the tympanic membrane, the middle ear cavity space, the ossicular chain up to the medial surface of the stapes footplate, and all attachments to the ossicles including the tendons of the middle ear muscles. In a

G.R. Popelka (✉)
Department of Otolaryngology–Head and Neck Surgery, Stanford University,
801 Welch Road, Stanford, CA 94305-5739, USA
e-mail: gpopelka@ohns.stanford.edu

L.L. Hunter
Department of Otolaryngology and Division of Audiology, University of Cincinnati,
3333 Burnet Avenue, Cincinnati, OH 45229-3039, USA
e-mail: Lisa.Hunter@cchmc.org

S. Puria et al. (eds.), *The Middle Ear: Science, Otosurgery, and Technology*,
Springer Handbook of Auditory Research 46, DOI 10.1007/978-1-4614-6591-1_8,

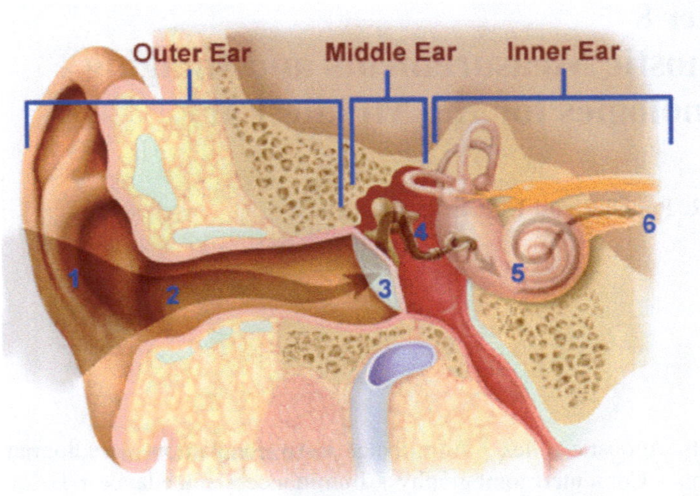

Fig. 8.1 Schematic representation of the right peripheral human ear illustrating the outer ear that consists of the acoustic structures lateral to the middle ear including the pinna (*1*) and the external ear canal (*2*); the middle ear itself, which consists of mechanical structures including the conical shaped tympanic membrane (*3*) and the ossicles (*4*); and the inner ear, which consists of the cochlea with its sensory structures (*5*) and the connected neural structures (*6*)

living human, measurement apparatus can be placed at various positions in the external ear canal, requiring that the characteristics of the remaining ear canal become a component of the measurement. The influence of the external ear canal is reduced the closer the apparatus is positioned to the tympanic membrane but at the expense of a more invasive, uncomfortable, and risky measurement. Generally, it is not practical to place measurement apparatus within the middle ear, except during surgery, or to place measurement apparatus on the medial side of the middle ear. However, it is possible to use the sensory functions of the cochlea as a measurement sensor to obtain responses on the medial side of the middle ear. Medical imaging can add additional structural information and in some cases functional information.

Clinical measurement methods began many years ago and continue to evolve. The oldest methods are the most standardized and the recent methods the least standardized, though both are valid, reliable, and useful. Virtually all of the measures are controlled by governmental regulations or institutional policies, and current audiology textbooks cover the standard clinical measures (Roeser et al. 2007; Katz et al. 2009). The purpose of this chapter is to outline current and emerging clinical middle ear diagnostic measures and imaging technologies, to provide a reasonable understanding of the underlying measurement concepts, and to illustrate the quantitative information that can be obtained for the human middle ear. Quantifiable measures are emphasized, with the full understanding that medical practice also includes many subjective measures that are not elucidated here.

Prior to performing any middle ear measures, one should examine the ear canal with an otoscope or binocular microscope to ensure there are no cerumen blockages, foreign bodies, drainage, tympanic membrane perforations, or an external ear canal that easily collapses. All of these conditions can affect middle ear measurements and should be corrected or noted. In the case of drainage or other obvious pathologic conditions, the measures may need to be deferred pending medical assessment and intervention. A cerumen blockage of less than 50 % of the ear canal volume may or may not affect some of the middle ear measures but should be noted in either case to allow proper interpretation of the results.

Four fundamental measurement approaches provide quantifiable information about the middle ear in humans. A particular approach is selected based on a variety of factors including the specific type of information desired and practical considerations including convenience, availability of measurement devices, minimization of patient time, and minimization of invasiveness. In many cases the results of the different approaches are interpreted in aggregate.

8.1.1 Behavioral Measures

Behavioral measures are the primary, oldest, most common, and least invasive of quantifiable measures of middle ear function. Quantifiable behavioral measures are based on the application of a controlled signal that activates the sense of hearing that originates in the cochlea and a voluntary response from a cooperative subject such as raising the hand or pressing a button. Calibrated acoustic signals are presented by air conduction from one of a variety of transducers positioned at or near the entrance to the ear canal. Calibrated vibratory signals are presented from one of a variety of vibratory transducers applied to the skull, typically on the surface of the skin and the subdermal soft tissue that overlies the mastoid process resulting in a signal reaching the cochlea by bone conduction. The behavioral response, a voluntary indication of whether or not the person heard the stimulus, is elicited based on activation of the same sensory structures in the cochlea whether the stimulus is delivered by air conduction or by bone conduction. Thus the measurement point for both cases is just beyond the middle ear. The response to an air conduction signal includes the effects of the middle ear because the stimulus must pass through the middle ear to reach the cochlea. The response to a bone conduction signal minimizes, though does not eliminate, the middle ear effects because the signal bypasses the middle ear and travels to the cochlea directly by bone conduction. Comparison of air-conduction and bone-conduction behavioral responses provides quantitative information on middle ear function.

Both the air- and bone-conduction transducers are calibrated on a decibel scale in hearing level (dB HL) with well-known standards such that the air- and bone-conduction signals produce equal responses at an equal dB HL in individuals with normal middle ear function. In the presence of abnormal middle ear function, the level of the air conduction stimulus that reaches the cochlea will be attenuated compared to level of the bone conduction stimulus that reaches the same cochlea. The amount of the stimulus elevation in decibels to overcome this attenuation is a quantification of the energy loss through the middle ear in cases of abnormal middle ear function. This diagnostic information can be obtained at multiple frequencies and will accurately quantify middle ear function regardless of the absolute hearing sensitivity of the cochlea as long as the hearing sensitivity of the cochlea itself does not change between the two measures. For all cooperative subjects willing and able to provide behavioral responses, it is very reasonable to assume that cochlear function is stable whether the cochlea is normal or abnormal. The only exception is the rare occurrence of cochlear pathology associated with fluctuating hearing, and even in these cases, the change in cochlear function occurs much slower than the time it takes to complete the two measures.

8.1.2 Physical Measures

Physical measures comprise the second category of quantifiable information of the middle ear. These were developed later than the behavioral measures of middle ear function but are equally common and only slightly more invasive. Physical measures are based on the application of a controlled acoustic signal and controlled static air pressure in the external ear canal while simultaneously measuring the resultant sound pressure level that develops in the ear canal. The sound pressure level that develops in the ear canal in turn is dependent on the acoustic characteristics of the external ear canal itself and the acoustic characteristics at the lateral surface of the tympanic membrane, which in turn are related to the physical characteristics of the tympanic membrane and the middle ear structures.

Several conditions can alter the measured acoustic characteristics at the lateral surface of the tympanic membrane. These include the applied static air pressure in the ear canal, both positive and negative relative to ambient air pressure, many abnormal middle ear conditions related to abnormal development, and many middle ear pathologies. Changes in the physical characteristics of the tympanic membrane and the middle ear structures in turn result in measurable changes in the ear canal sound pressure level. The changes in the resultant measured ear canal sound pressure can be converted to a variety of acoustic units that employ the impedance analogy (impedance, reactance, and resistance), the admittance analogy (admittance, susceptance, and conductance) and acoustic reflectance, all under the generic umbrella term of acoustic immittance measures. Both magnitude and phase of the measured sound pressure level can be determined. The raw measured value represents the effects of the combined acoustic characteristics of the external ear canal and the lateral surface of the tympanic membrane, which can be further separated into values that represent the acoustic characteristics of each separately. Because the measured acoustic immittance values vary as a function of the applied ear canal air pressure, additional analysis can provide a very good estimate of the middle ear air pressure, even though no pressure transducer is placed in the middle ear. These measures are considered physical rather than physiologic because they are based strictly on the physical characteristics of the external ear canal and middle ear structures and can even be obtained in a cadaver.

8.1.3 Physiological Measures

Physiological measures of the middle ear include a variety of involuntary responses that result from calibrated and controlled stimuli that induce specific involuntary physiological activity. The involuntary physiological responses range from reflexive contractions of the middle ear muscles detected from acoustic measures in the external ear canal to acoustic signals generated by the cochlear structures, also detected in the external ear canal, to neural activity from the peripheral auditory neural system (cochlea and cranial nerve VIII) and central auditory neural system (brainstem and higher neural systems) detected from electrical signals measured with surface electrodes on the head. Though changes in these physiological responses in relation to changes in the controlled stimuli can be analyzed to infer functional status of the neural mechanisms themselves, that is not the focus of this chapter. However, assuming that the neural functional status does not change between measures, or as a result of other interventions, these involuntary physiological responses can also be analyzed to obtain quantifiable measures of the functional status of the middle ear structures. In essence, the voluntary behavioral measures mentioned earlier can be replaced by these involuntary physiological measures.

8.1.4 Imaging

Imaging technology in general, and for the middle ear in particular, is currently undergoing many new developments and is evolving. Entirely different imaging technologies for the middle ear either are becoming available or are being adapted specifically for clinical use. The details of some of these technologies and how the findings relate to research are reviewed by Funnell, Maftoon, and Decraemer, Chap. 7). Consequently, clinical middle ear imaging is the least standardized of the measures and the most likely to change in the near future.

The full range of clinical medical imaging techniques can be used for the middle ear depending on the purpose. The middle ear and some of its structures can be observed directly with an endoscope in a limited number of cases. Because the middle ear contains bone, soft tissue, and air-filled space, conventional medical imaging can provide quantifiable images of the structures of the middle ear as well. A variety of measures are based on ionizing radiation (X-ray) that optimizes visualization of bone. Others are based on magnetic resonance imaging (MRI) that optimizes visualization of soft tissues. Optical methods are able to visualize both bone and soft tissues. All imaging methods are more invasive than behavioral measures or physical measures but for different reasons. The ionizing radiation used in X-ray imaging raises concerns about direct damage to tissues, a serious enough risk that this approach is not used for research on normal subjects. The magnetic signals used in magnetic resonance imaging (MRI) do not necessarily have a risk of damage to tissues directly, but also can be considered invasive because the individual must often be positioned for long periods of time in the very small spaces of an imaging scanner that in turn increases the risk of claustrophobia or related psychological reactions. Optical methods may require apparatus positioned deep in the ear canal. In some cases, imaging conducted for unrelated medical problems can also be reformatted to provide structural images of the middle ear. If additional procedures are employed, some functional information can be obtained as well.

8.2 Middle Ear Behavioral Measures

8.2.1 Auditory Threshold

The most common behavioral measure of middle ear function is based on auditory sensitivity as quantified by the threshold of hearing. To be precise, the threshold of hearing is defined as the level of an acoustic signal that is heard 50 % of the time in a series of controlled presentations. The threshold level is directly related to the status of the entire peripheral auditory system, including the sensitivity of the cochlea. A limiting factor is the ambient acoustic noise level making it necessary to conduct the measures in sound attenuating booths. The measurement process involves systematic increases in the level of a calibrated signal to where it is audible 100 % of the time followed by systematic decreases in the level of the signal to where it is audible 0 % of the time. The threshold level is defined as the level of the calibrated stimulus in the range between complete audibility and complete inaudibility, usually defined as audible 50 % of the time. The decibel range between audibility and inaudibility is only a few decibels so that stimulus step sizes of 5 dB are adequate for most measures. More importantly, changes in threshold associated with relevant changes in the middle ear usually exceed this range substantially. It is not uncommon for a middle ear condition to change auditory threshold by 70 dB.

Threshold measures are influenced by a variety of factors that can cause behavioral results to be quite variable. These include factors beyond the peripheral auditory system such as psychological status (e.g., motivation to provide a correct response, fatigue, cognitive ability, developmental age), physiological status (stability of cochlear function related to certain disease processes), effects of certain pharmaceuticals, and exposure to excess acoustic stimulation. The measures also are influenced by the measurement protocol (e.g., instructions on how to respond, the presentation sequence of the stimuli). Though these factors do affect the variability of the threshold measures, when controlled and specified, the threshold values are fairly accurate and reliable for most subjects measured under a specific protocol.

The most common protocol for obtaining a threshold measure for both clinical and research purposes is called a modified Hughson Westlake protocol (Roeser et al. 2007). This protocol includes presentation of a series of calibrated signals, both well above and well below threshold, a 5-dB step size, and enough presentations at threshold levels to determine the 50 % audibility point, usually from three to five presentations and produces acceptable test–retest variability. Most middle ear pathologies and abnormalities result in threshold changes that exceed this variability by a substantial amount such that any criticisms of the subjective nature of the measures are easily mitigated. If finer resolution is required, smaller step sizes and automated algorithms can also be implemented at the expense of greatly increasing measurement time.

8.2.2 Measurement Parameters

The frequency range of hearing for humans is commonly stated as from 20 Hz to 20 kHz. This wide range places significant demands on the stimulus transducer that plays a substantial role in accurately quantifying the thresholds and must be considered when interpreting auditory threshold measures. For air conduction stimuli, the output of a loudspeaker positioned and calibrated in space without the subject present (sound field) is an effective way to specify a sound and a result that accurately includes the effect of the external ear canal in its natural state. This is the optimal choice for obtaining measures that include the complete external ear canal, including the pinna effects that become prominent at higher frequencies and reduce the substantial stimulus variability largely related to the inconsistent acoustic leaks between the transducer and the ear for earphone measures. Smaller insert earphones can be inserted and coupled tightly to the more regularly shaped ear canal with a replaceable soft tip that creates and maintains a consistent acoustic seal between the transducer and the ear canal and improves the reliability of the calibrated stimulus and therefore the reliability of the threshold measures. However, the remaining length of the ear canal is not clearly determined and easily can be altered based simply on how deeply the insert earphone is positioned in the ear canal, resulting in generally small but unknown changes in the level of the applied acoustic signal. Circumaural earphones fit completely around the pinna and are more likely to establish a tight acoustic seal that allows for much more consistent stimulus delivery. These earphones have the added benefit of much better high frequency output than other transducers and allow accurate threshold measures to be obtained up to the maximum frequency for humans (20 kHz) (Ballachanda 1997). However, the output of a loudspeaker calibrated in an undisturbed sound field can be substantially altered by the presence of the subject and by the surrounding acoustic environment that almost always includes reflective surfaces inside the small booths intended to attenuate ambient acoustic noise. The use of sound field signals also makes it more difficult to determine which of the two ears is responding. An earphone more precisely establishes which ear is being stimulated and can have signal calibration improvements compared to the sound field case. Older style earphones that fit on top of the ear (supraaural) do not present consistent stimulus levels because of the unspecified and uncontrolled acoustic coupling between the irregularly shaped pinna and the earphone cushion.

The bone conduction transducers also have limitations related to maximum frequency output and coupling factors. All current conventional bone conduction transducers employ electrodynamic technology that uses a rather large moving rod positioned in a coil to impart the vibratory signal to the skull. Because this rod has significant mass, the output of the bone conduction transducer decreases as frequency increases, resulting in a maximum practical measurement frequency of up to only about 6 kHz (Popelka et al. 2010). The effective output of a bone conduction transducer is also limited because it must be coupled to the skull via the skin surface rather than directly to the bone of the skull. The highly variable soft tissue between the transducer and the skull attenuates the level of the signal to the skull and introduces stimulus variability across subjects. The surface area of the portion of the transducer that is held against the skull and the coupling force of the transducer to the head can be controlled to reduce the stimulus variability associated with these two factors. The surface area of the portion of the transducer held against the head can be defined in shape (1.75 mm^3area with specified radius of the contact area and the radius of the edge) as well as the coupling force (Toll et al. 2011).

In spite of these variability issues, quite accurate air-conduction and bone-conduction thresholds can be determined in most patients and research subjects. At a minimum, all of these measurement factors must be delineated under a specified protocol to produce the least variable threshold measures and to provide sufficient information for the measures to be repeatable.

8.2.3 Inherent Threshold Measurement Limitations

The presence of two hearing organs in a patient or subject (right and left cochleae) raises a concern about the participation of the nontest ear during the measurement of the test ear. The amount of the test signal that reaches the nontest ear from the test ear is called interaural attenuation and is determined by a variety of factors associated with the subject and with the transducer. A measurement signal from a bone conduction oscillator placed at one skull location, typically on the mastoid process of the ear of interest, can reach both cochleae at the same level because both cochleae are rigidly coupled to the same skull, resulting in an interaural attenuation of 0 dB. Conversely, an air conduction signal from an insert earphone placed deep in the ear canal of the ear of interest can reach the cochlea of the opposite ear but only at a greatly reduced level, resulting in an interaural attenuation of greater than 100 dB. Interaural attenuation values range from as little as 0 dB to as high as

100 dB and are frequency and transducer specific. If the interaural attenuation is low enough to implicate participation of the nontest ear, the nontest ear can be prevented from contributing to the measured threshold with the addition of an acoustic signal to the nontest ear that covers up or masks its function and thereby removes its participation in the measurement of the test ear. The application of masking to block the nontest ear must be performed carefully because the masking signal itself can affect the hearing sensitivity in the test ear either directly based on the same interaural attenuation values outlined earlier, or indirectly because the neural pathways of each ear have connections between them within the central nervous system. In most cases, effective masking can be applied to the nontest ear such that the threshold measures represent only the function of the ear being measured.

The normal bony cochlea is an enclosed fluid-filled system with rigid walls except for two openings, the oval window that contains the footplate of the stapes and the adjacent round window that is covered by a flexible membrane. These two openings allow the applied signal to generate fluid movements within the cochlea that in turn deflect the cilia on the sensory cells within the cochlea and initiate the sense of hearing. A breach of the rigid walls can result in a third opening that can alter the normal fluid movements in a way that may differ between air- and bone-conduction stimulation. A condition called superior vestibular semicircular canal dehiscence is an abnormal opening in the superior canal of the vestibular system that can act as a third window that in turn can alter the normal fluid dynamics (Rosowski et al. 2004). The specific effects of a dehiscence are currently under investigation. In general, the presence of a superior canal dehiscence results in an improvement of bone conduction thresholds and a decrement in air-conduction thresholds, especially for the frequencies below 1 kHz (Merchant et al. 2007; Merchant and Rosowski 2008). This situation results in a clear difference between air- and bone-conduction thresholds that would suggest a change in middle ear function when this clearly is not the case. However, this condition can also provide a platform for making quantifiable measures that relate to new understanding of the basic mechanisms of air- and bone-conduction transmission.

8.2.4 Audiograms

Figure 8.2 is a representation of auditory thresholds in a typical format called an audiogram. The standard frequencies along the *X*-axis range in octaves from 250 Hz through 8 kHz. Standard intraoctave frequencies are also indicated. The octave representation of the stimulus directly maps onto the length of the cochlear partition, referred to as a tonotopic organization, with high frequencies near the stapes and low frequencies near the apex. Stimulus level is specified on a reverse decibel scale (dB HL) with the reference for 0 dB HL indicated. Because of the normal differential sensitivity of the ear with respect to frequency, the absolute level in decibels re an acoustic reference for air conduction (sound pressure level [SPL])

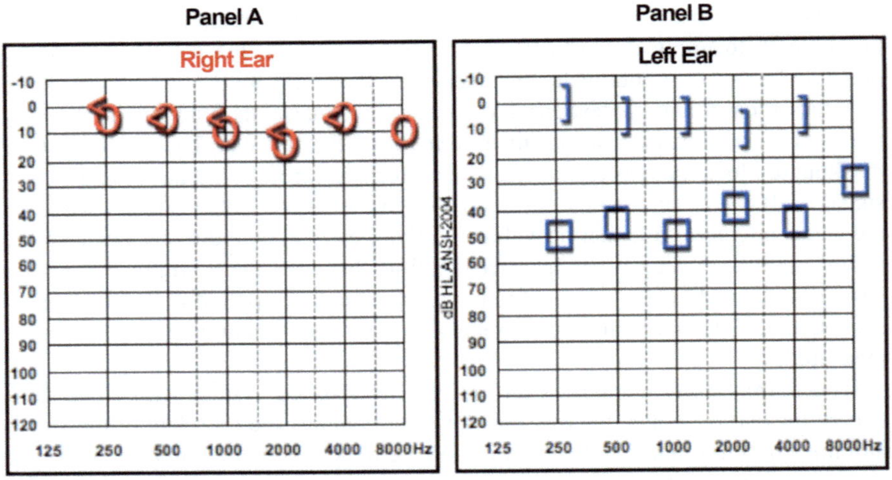

Fig. 8.2 Hearing thresholds in an audiogram format. (*Left*) Right ear thresholds by air conduction (O) and bone conduction (<). (*Right*) Left ear thresholds by air conduction with masking to the opposite ear

and in decibels re a force reference for bone conduction (Newtons) of the signal at each frequency in the audiogram format will differ. However, if the dB HL reference level is specified then the difference between the reference level and HL scales is known and measures on one scale can be converted to the other scale. In terms of auditory function, the dB HL scale is calibrated to be equivalent between air-conduction and bone-conduction behavioral responses in normal ears.

$$\text{Threshold}_{\text{AC}}\,(\text{dB HL}) = \text{Threshold}_{\text{BC}}\,(\text{dB HL}) \tag{8.1}$$

and

$$0\,\text{dB HL} = \text{Threshold}_{\text{Normal Ear}} \tag{8.2}$$

that allows direct comparison between air-conduction and bone-conduction responses. A symbol key on the audiogram identifies the type of conduction (air or bone), the test ear and whether or not the nontest ear was masked. Also indicated are the transducers that were used and the reliability of the behavioral threshold measurements.

The functional status of the middle ear is quantified as the change in transmission through the middle ear described as

$$\begin{aligned}&\text{Threshold}_{\text{AC}}\,(\text{dB HL}) - \text{Threshold}_{\text{BC}}\,(\text{dB HL})\\&\quad = \text{dB loss through the middle ear}\end{aligned} \tag{8.3}$$

Positive numbers represent the attenuation of the signal in decibels through the middle ear for that particular measurement. Occasionally negative decibel values are encountered that normally would suggest an active mechanism. However, in the case of the passive middle ear these negative values are the result of normal variability in the air and bone conduction thresholds and should be ignored.

The results in Fig. 8.2a are from a normal hearing right ear with normal middle ear function. The air conduction thresholds (O) are equivalent to the bone conduction thresholds (>) (in dB HL) such that

$$\text{Threshold}_{\text{AC}}\,(\text{dB HL}) - \text{Threshold}_{\text{BC}}\,(\text{dB HL}) = 0\,\text{dB} \tag{8.4}$$

This threshold difference, commonly referred to as the "air–bone gap" provides direct evidence that the middle ear is functioning normally because there is no loss of the signal through the middle ear compared to the normal case. Note that the interpretation applies only to the middle ear function and not necessarily the actual condition of the middle ear. It is entirely possible that the results would be identical in an ear with a small tympanic membrane perforation, a case in which the middle ear clearly is abnormal but with a particular abnormality that may not affect middle ear sound transmission.

The results in Fig. 8.2b are from the opposite left ear. The symbols indicate that the thresholds were obtained with masking to the opposite nontest ear to prevent it from participating in the measurements that allows an interpretation of the threshold results to accurately reflect only the left middle ear function. The air-conduction thresholds have much larger values than the bone-conduction thresholds, providing direct evidence that the left middle ear is not functioning normally, an interpretation that applies only to middle ear sound transmission. In this case the middle ear abnormality was a perforated tympanic membrane that did affect middle ear sound transmission quantifiably. The relation between a specified middle ear abnormality (location and area of the perforation) and the resultant effect on middle ear function (a transmission loss calculated as the decibel difference between the air- and bone-conduction threshold that can be calculated for each frequency) can be quantified quite precisely.

Though often considered to be subjective because they employ voluntary responses, auditory thresholds can be very reliable and very accurate for quantifying middle ear attenuation as a function of measurement frequency. At a minimum, several measurement variables must be specified for proper interpretation including stimulus units (typically dB HL), a known calibration standard for the transducer that was used and a specified measurement protocol that defines the stimulus sequence and the definition of threshold. The pattern of attenuation through the middle ear across frequency can be useful for understanding middle ear function in relation to observable middle ear conditions. The change in the air–bone gap as the result of an intervention such as surgery can also be used to quantify the efficacy of the intervention.

8.3 Middle Ear and Physical Characteristics

8.3.1 *Acoustic Immittance*

The generic term "acoustic immittance" was coined to include all physical quantities of acoustic impedance and its subcomponents. Acoustic immittance also refers collectively to measures of these physical quantities obtained with a device called an immittance meter that has a probe that delivers a measurement signal to the external ear canal and a microphone that measures residual, unabsorbed sound that indirectly infers middle ear function. A typical clinical immittance measurement system produces a variety of measures including acoustic reflex responses, tympanometry, and multifrequencyimmittance or wideband measures. The clinical acoustic immittance battery provides important diagnostic information about middle ear function and status. Single-frequency tympanometry and acoustic reflex measures have comprised the standard acoustic immittance battery for the past 40 years, while newer procedures using wideband stimuli have been developed recently, including wideband tympanometry.

8.3.2 *Tympanometry*

Tympanometry is a measure of acoustic immittance in the ear canal as a function of a range of applied air pressure in the ear canal that varies from positive to negative, relative to atmospheric pressure. The most common measurement is acoustic admittance (Y_a), or the amount of acoustic energy absorbed by the ear canal, the tympanic membrane and the middle ear, expressed in acoustic millimhos (mmho) with ear canal air pressure expressed in decaPascals (daPa). The reciprocal of acoustic admittance, acoustic impedance, is also used but much less so. The subcomponents of acoustic admittance and acoustic impedance can also be measured. Figure 8.3 shows an acoustic admittance tympanogram from a normal ear. Note that an acoustic impedance tympanogram would be inverted.

To record a tympanogram, a probe assembly attached to an acoustic immittance meter is inserted into the external ear canal. The probe assembly has a pliable tip that seals the probe into the external ear canal to prevent an acoustic leak and to allow air pressure in the ear canal to be varied. The probe assembly has three components: (1) a tube that is attached to an air pressure pump in the immittance meter to vary the air pressure in the ear canal, (2) a miniature loudspeaker that is attached to a signal generator to produce a measurement probe tone, and (3) a miniature microphone to measure the level of the residual or reflected probe tone in the ear canal. A sinusoidal probe tone is delivered from the probe to the ear canal while ear canal air pressure is varied from positive to negative (or in the opposite direction) at a specified rate of change. The meter measures the electrical current needed to maintain a constant sound pressure level (SPL) in the ear canal that is

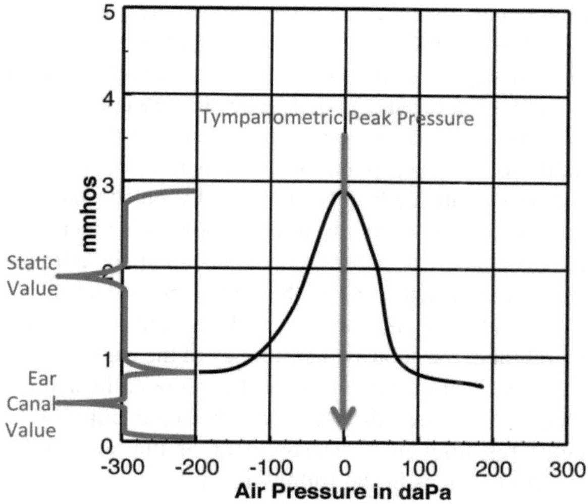

Fig. 8.3 An acoustic admittance tympanogram at 226 Hz from a normal ear showing tympanometric peak pressure, the static value, and the ear canal value

directly proportional to the acoustic admittance magnitude at the probe tip. As the acoustic admittance of the middle ear decreases because of the increased tension in the tympanic membrane caused by the increase (or decrease) in the applied air pressure, the SPL in the ear canal increases (or decreases) proportionally and therefore an increase (or decrease) in electrical current to the probe loudspeaker is required to maintain a constant SPL in the sealed ear canal. This change in the electrical current in response to changes in the SPL measured in the ear canal is directly proportional to the admittance magnitude at the tip of the probe.

8.3.3 Acoustic Impedance and Acoustic Admittance

Acoustic impedance (Z_a in acoustic ohms) of the middle ear system is defined as the total opposition of the system to the flow of the acoustic energy. Acoustic admittance (Y_a in acoustic mhos) is the amount of acoustic energy that flows into the middle ear system. The terms acoustic impedance (Z_a) and acoustic admittance (Y_a) are reciprocal to each other and are described mathematically as:

$$Z_a = \frac{1}{Y_a} \tag{8.5}$$

$$Y_a = \frac{1}{Z_a} \tag{8.6}$$

There are three physical properties that contribute to total acoustic admittance or total acoustic impedance: stiffness, mass, and friction. Stiffness elements in the middle ear system exert effects at low frequencies and are related primarily to the enclosed volume of air in both the external ear canal and the middle ear space. Mass elements in the middle ear system exert high frequency effects and are minimal in normal middle ears. Total acoustic susceptance (or total acoustic reactance in impedance terms) is the vectorial sum of the stiffness and mass elements. If the total acoustic susceptance (B_a) is positive (between 0° and 90° phase), the system is stiffness controlled; if negative (between 0° and −90° phase), the system is mass controlled. The third variable, friction, is independent of frequency and determines the dissipation of acoustic energy, called acoustic conductance denoted by G_a (or acoustic resistance, R_a, in the impedance system).

To calculate the acoustic admittance at the lateral surface of the tympanic membrane, the acoustic admittance of the ear canal (Y_{EC}) is measured and subtracted from the overall measurement. In practice, the acoustic admittance of the ear canal alone is measured directly and subtracted from the total measured acoustic admittance resulting in a measurement of the acoustic admittance of the middle ear system (Y_{ME}), as in the following equation:

$$Y_{ME} = Y - Y_{EC} \tag{8.7}$$

A tympanogram includes the admittance of the external ear canal (Fig. 8.3). The admittance of the external ear canal can be subtracted and is then called an ear canal compensated tympanogram, represented by shifting the tympanogram down to where the admittance value at high negative and high positive air pressures is 0 mmhos. The resulting measurement refers to the acoustic characteristics at the lateral surface of the tympanic membrane (Y_{TM}). The asymmetry in the measured ear canal value at high and low air pressures (Fig. 8.3) is due to slight differences in the volume of the remaining ear canal due to slight displacements of the tympanic membrane at these high positive and high negative air pressures.

8.3.4 Estimate of Middle Ear Air Pressure

As the applied air pressure varies between high negative and high positive values, Y_{TM} will reach its highest value when the air pressures on both sides of the tympanic membrane are equal, resulting in a single peak in the tympanogram. The ear-canal air pressure at which the peak of the tympanogram occurs is the tympanometric peak pressure which is an indirect measure of the air pressure in the middle ear space. Though the tympanometric peak pressure can overestimate the actual middle ear pressure by as much as 100 % (Renvall and Holmquist 1976), it can detect the presence of negative or positive middle ear pressure due to Eustachian tube dysfunction. As the Eustachian tube regains normal function and middle ear effusion resolves, the tympanogram progresses from having no peak (flat tympanogram), to a peak at a high negative pressure (negative tympanometric peak pressure), finally returning to a peak at atmospheric pressure (tympanometric peak pressure near

0 daPa) when Eustachian tube function has returned to normal. Multiple factors produce negative middle ear air pressure including frequent forceful inhaling through the nostrils ("sniffing"), ciliary action in the Eustachian tube, absorption of middle ear gases through increased and excessive diffusion and poor pneumatization of the mastoid. As a result of the multiple mechanisms that contribute to middle ear pressure as well as the inaccuracy of the tympanometric peak pressure measurement relative to actual middle ear pressure, negative tympanometric peak pressure has not been shown to provide reliable diagnostic specificity or sensitivity to diagnosing otitis media but can provide very useful information for quantifying Eustachian tube function and the physical status of the middle ear space.

8.3.5 Ear Canal Volume

A primary purpose of tympanometry is to accurately measure the acoustic admittance at the lateral surface of the tympanic membrane as an indicator of middle ear characteristics. Because the probe tip of the admittance measurement system is lateral to the surface of the tympanic membrane, the acoustic admittance measured at the probe tip jointly reflects the acoustic admittance of the external auditory canal and the acoustic admittance of the middle ear. The accuracy of determining the acoustic admittance at the lateral surface of the tympanic membrane relies upon obtaining an accurate measure of the acoustic admittance of the ear canal between the probe tip and the tympanic membrane. This volume, referred to as V_{EA}, is affected by numerous factors such as the depth of insertion of the probe tip, the dimensions of the ear canal, and the volume occupied by substances in the external ear canal, specifically cerumen. The volume of the ear canal has also been referred to as V_{EC}, or as the acoustic admittance of an equivalent volume of air, V_{EQ}.

The most common method to determine V_{EA} is to use the measured acoustic admittance at a high positive air pressure (200 daPa, 1 daPa $=$ 10 Pa) that drives the acoustic admittance of the middle ear toward zero. At this air pressure point the acoustic admittance measured at the probe tip represents the acoustic admittance of the air in the ear canal, assuming that the ear canal walls are rigid. Under reference conditions using a probe tone of 226 Hz, a 1 cubic centimeter (cc) or 1 milliliter (ml) volume of trapped air has an acoustic admittance of 1 mmho. This measure is called equivalent ear canal volume because the measured acoustic admittance is equivalent to the acoustic admittance of a hard-walled cavity of equivalent volume.

The normal range for V_{EA} is highly dependent on several factors including depth of probe insertion and patient age and gender. Women have smaller ear canal volumes than men at all ages, and ear canal volumes steadily increase with age until the ninth decade, when they start to decrease due to collapsing canals (Wiley et al. 1996). The primary value of the V_{EA} measurement is to ensure that the probe tip is not blocked, and that the tympanic membrane is intact. If the tympanic membrane is not intact there will be no change in the acoustic admittance as a function of air pressure in the canal, resulting in a flat tympanogram and the measure will reflect the volume of both the external ear canal and the middle ear space. In the case of a surgically inserted tympanostomy tube, this information can indicate whether or not the tube is blocked. To interpret a flat tympanogram, it is necessary to know the V_{EA} and compare it to age-appropriate normative values.

8.3.6 *Static Acoustic Admittance*

The acoustic admittance measure after subtracting the acoustic admittance of the ear canal is called peak compensated static acoustic admittance, or simply static admittance, and represents the acoustic admittance at the lateral surface if the tympanic membrane (Y_{TM}). Static acoustic admittance can be determined accurately from the acoustic admittance tympanogram only if the phase of the probe tone is relatively constant when the ear canal is pressurized. At low frequencies (e.g., 226 Hz), phase shifts are negligible. At higher frequencies, phase shifts are more significant and acoustic conductance and acoustic susceptance must be calculated separately.

$$G_{TM} = G_{peak} - G_{tail} \tag{8.8}$$

$$B_{TM} = B_{peak} - B_{tail} \tag{8.9}$$

Peak compensated static acoustic admittance can then be calculated from the compensated acoustic conductance and acoustic susceptance measures:

$$Y_{TM} = \sqrt{G_{TM}^2 + B_{TM}^2} \tag{8.10}$$

Although acoustic admittance (Y_a) tympanometry can provide useful information about middle ear status in adults and children, modern immittance equipment can simultaneously measure B_a and G_a while varying air pressure in the external ear canal, known as multi-component tympanometry (B_a/G_a tympanometry) to obtain additional information. The B_a/G_a tympanograms can be recorded at various probe tone frequencies, but typically include 226, 678, and 1,000 Hz.

8.3.7 *Tympanometric Gradient*

A number of studies have demonstrated that the sharpness of the tympanometric peak is associated with middle ear pathology (Nozza et al. 1992, 1994). Two closely related measures that quantify the sharpness of the tympanometric peak are the tympanometric gradient and tympanometric width. Tympanometric gradient is a measure of the slope of the tympanogram on either side of the tympanometric peak. The most common method for calculating gradient is to calculate the difference in acoustic admittance at the peak and the average of the acoustic admittance values at +50 and −50 daPa relative to the acoustic admittance at peak pressure. The gradient is an index that ranges from 1.0 (flat tympanogram) to very high values depending on the value at the tympanometric peak pressure. The higher the gradient, the sharper and more narrow the tympanogram peak. The presence of middle ear effusion decreases the gradient (and increases the width) of the tympanogram peak. A less common method is to calculate the width of a tympanogram (in daPa) measured at one half the compensated static admittance point. Both measures provide an index of the sharpness of the tympanogram in the vicinity of the peak and quantify the relative sharpness (steepness) or roundness of the peak.

8.3.8 Probe Tone

Vanhuyse et al. (1975) examined tympanometric patterns in adults at various probe tone frequencies and developed a model that predicts the shape of B_a and G_a tympanograms at 678 Hz in normal ears and in ears with various pathologies. The Vanhuyse et al. model categorizes the tympanograms based on the number of peaks or extrema on the B_a and G_a tympanograms and predicts four tympanometric patterns at 678 Hz. The transition between different Vanhuyse patterns can be shifted to higher or lower probe tone frequencies depending on the nature of the middle ear pathology.

Evidence has accumulated that tympanometry using higher probe-tone frequencies (up to and including 1,000 Hz) is more sensitive to changes in middle ear status in infants less than 4 months old compared to 226-Hz tympanometry. Some studies have reported normative data for a variety of young ages, and some have investigated test performance of specific 1,000-Hz admittance criteria in predicting otoacoustic emission screening results (Hunter and Margolis 2011).

The resonant frequency of the middle ear is the frequency at which the total acoustic susceptance is zero and is directly proportional to the square root of stiffness and inversely proportional to the square root of mass. The resonant frequency of the middle ear can be determined using multifrequency tympanometry. The resonant frequency of the middle ear system may be shifted higher or lower compared to healthy ears by various pathologies. Otosclerosis, for example, increases the stiffness of the middle ear system and shifts the resonant frequency of the middle ear system to higher probe tone frequencies.

8.3.9 Wideband Energy Reflectance

Wideband energy reflectance (WBER) is a relatively new middle ear analysis technique, in which complex sounds ranging from 0.2 to 10 kHz or higher are presented into the ear canal and the amount of energy reflected back from the middle ear is calculated. Energy reflectance (ER) has been used in research on human middle ear function for two decades (Keefe et al. 1992; Voss and Allen 1994); however, its application in clinical assessment of the middle ear is still developing. Clinical systems are currently being commercialized. One system is based on the calibration method developed by Voss and Allen (1994; Mimosa Acoustics Corp.) that is FDA approved. A second system is based on the work by Keefe et al. (1992; Interacoustics) that is currently an investigatory research system. Wideband energy reflectance has an advantage over multifrequency tympanometry in that the location of the probe in the ear canal is not as critical as in single-frequency tympanometry, especially at higher frequencies. Further, energy reflectance compared to standard 226-Hz tympanometry may provide a more sensitive measure in evaluating middle ear disorders and conductive hearing loss. Another advantage of reflectance measurements is that the measurement frequency can be up to 10 kHz, with less contamination by standing waves in the ear canal.

Reflectance, *R*(f), refers to the ratio of the incident (forward) and retrograde (backward) pressure waves, while $[R(\mathrm{f})]^2$ is the power reflectance (ER). A value of 0 occurs when all of the sound energy is absorbed by the middle ear and the cochlea while a value of 1.0 occurs when all of the energy is reflected back from the middle ear. The reciprocal of ER is known as power absorption (PA) and when expressed in decibels, is known as transmittance.

Reflectance is mathematically defined as the ratio of 1 minus the product of the admittance (*Y*) and characteristic impedance (Z_0) and 1 plus the product of the admittance and characteristic impedance at different frequencies and static pressures. Normative data on measures of wideband energy reflectance (Rosowski et al. 2012) suggest that the most energy is reflected at the low frequencies, while there are regions of lower reflectance in the mid frequencies, and moderate reflectance at high frequencies. The input admittance (Y_m) is related to reflectance through the following equation:

$$\frac{Y_{\mathrm{m}}}{Y_0} = \frac{1 - R_{\mathrm{m}}}{1 + R_{\mathrm{m}}} \tag{8.11}$$

where $Y_0 = A/\rho c$ (Voss and Allen 1994; Keefe and Simmons 2003), A is the cross-sectional area of the ear canal, ρ is the density of air in the ear canal, and c is the speed of sound. The values ρ and c are constants, while A is estimated based on the size of the probe tip that is selected when the measurement is made. Thus, measurements of wideband energy reflectance rely on several assumptions, including the assumption that the impedance at the lateral surface of the tympanic membrane is similar to that at the microphone, and the cross-sectional area in each subject who uses a specific probe-tip size is approximately the same. At each frequency, Z_s and P_s are calculated from the measurements obtained during a calibration procedure. In this procedure, an ear tip is placed separately into four cavities each with a diameter of 0.74 cm but with different lengths, and two measurements of the pressure response are made within each cavity. For each measurement, the pressure-response is plotted in relation to the noise floor for each frequency. Normative data (Merchant et al. 2010; Rosowski et al. 2012) and data from well characterized disease processes (Voss et al. 2012) are becoming available. Reflectance measures in combination with audiometry may improve the ability to differentiate ossicular fixation from ossicular discontinuity in patients with conductive hearing loss who have an intact tympanic membrane and an aerated middle ear (Nakajima et al. 2012).

8.3.10 Wideband Tympanometry

Wideband tympanometry is being implemented in a research system developed by Douglas Keefe at Boys Town National Research Hospital (distributed by Interacoustics) and is capable of measuring energy reflectance or related parameters at ambient pressure as well as at multiple air pressures. Wideband tympanometry can be used to replace standard tympanometry as well as measure middle ear muscle reflexes, described later. The calibration procedure and the system are similar to the one described and used by Keefe et al. (1992). Figure 8.4 is an illustration of the three-dimensional plot of energy absorbance across a range of frequencies and pressures that is produced, effectively plotting multiple tympanograms. The system also can extract a plot of phase which estimates resonant frequency, and a series of single-frequency tympanograms in Y_a, B_a, and G_a units (Fig. 8.4).

8.3.11 Laser-Doppler Vibrometry

Laser-Doppler vibrometry is based on the concept of measuring the displacement of the tympanic membrane via small wavelength changes in a reflected laser signal. This approach currently is a research tool that is being considered for commercialization that can be useful for differentiation of various ossicular disorders in an ear with an intact tympanic membrane and aerated middle ear. The laser-Doppler vibrometry measures can help differentiate ossicular fixation from ossicular discontinuity in the presence of an air–bone gap (Rosowski et al. 2008). Carefully measured normative values already have been established (Rosowski et al. 2012).

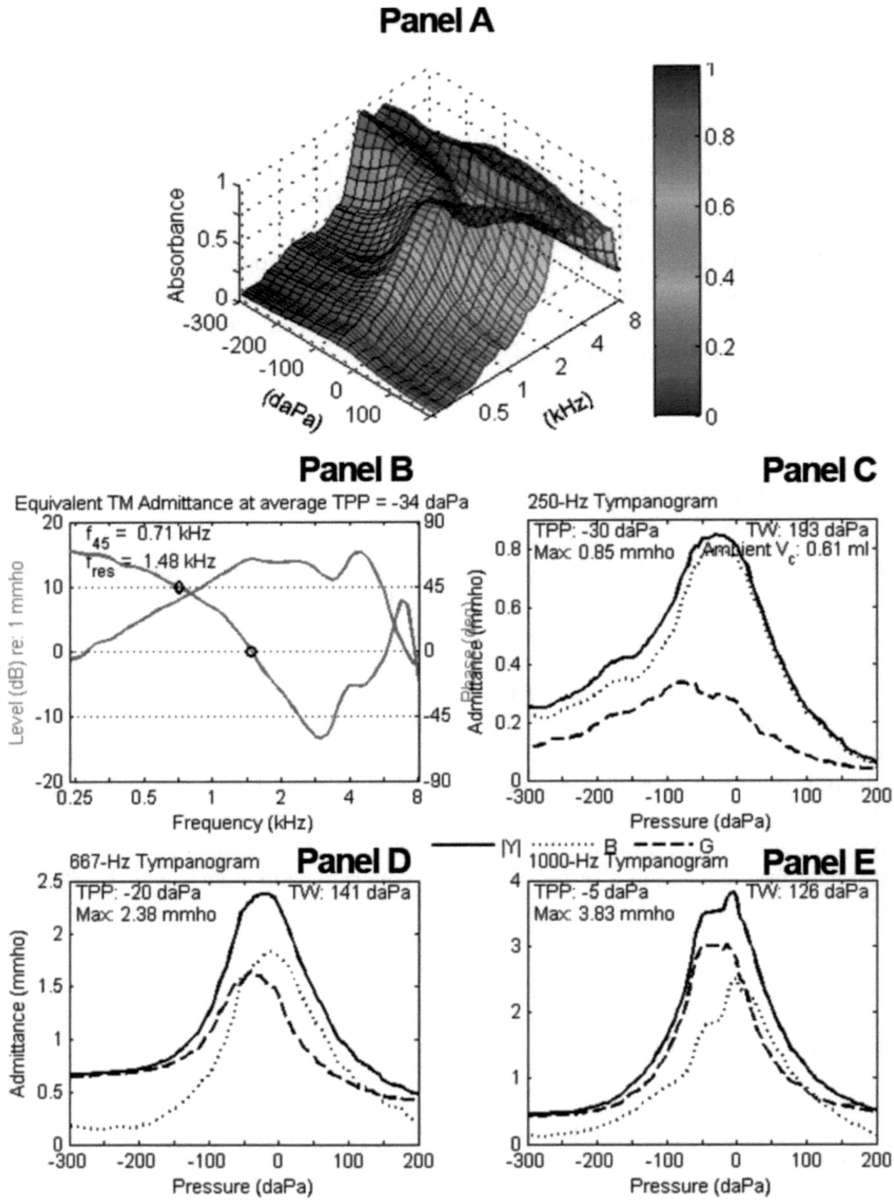

Fig. 8.4 A three-dimensional wideband tympanogram of absorbance as a function of ear canal air pressure across frequency (**a**), sound pressure level (dB) and phase angle (**b**), and single frequency admittance tympanograms (**c–e**). (Interacoustics Wideband Research System)

8.4 Middle Ear and Middle Ear Muscles

8.4.1 Middle Ear Muscle Reflexes

The middle ear contains two muscles, the tensor tympani and the stapedius. A contraction of either of two middle ear muscles can alter middle ear function. Each muscle functions quite differently.

The tensor tympani is located in the bony canal above the osseous portion of the Eustachian tube and originates from the cartilaginous and osseous portions of the Eustachian tube. The muscle terminates in a thin tendon that enters the middle ear space, makes a right angle turn around the cochleariform process, and attaches to the manubrium of the malleus. Neural innervation of the tensor tympani is from the tensor tympani nerve, a motor fiber branch of the mandibular division of the trigeminal nerve (cranial nerve V) and does not receive input from the sensory fibers of the trigeminal ganglion. The muscle contracts reflexively and pulls the malleus medially, in the direction of the normal vibratory motion of the tympani membrane, resulting in an increase in the tension of the tympanic membrane. The increased tension dampens the high-level ossicular vibrations associated with chewing and possibly other high-level internally generated sounds. The tensor tympani muscle also contracts reflexively in response to very high-level external sounds that produce a generalized startle response. Though diagnostic measures can determine if the tensor tympani contracts abnormally, information about the function of this muscle provides only limited information about middle ear function or structure in humans other than that the muscle is present and functions normally, suggesting normal middle ear function (Jones et al. 2008).

The stapedius muscle is smaller than the tensor tympani muscle and is the smallest skeletal muscle in the human body. It originates from a small opening in a cone-shaped prominence on the posterior wall of the middle ear space and terminates in a thin tendon that is attached to the neck of the stapes. Neural innervation of the stapedius muscle is from the first branch of the facial nerve (cranial nerve VII) after it exits the facial nerve canal. The muscle contracts reflexively from auditory input via an ipsilateral or a contralateral pathway. The ipsilateral pathway originates in the cochlea then proceeds to the auditory nerve (cranial nerve VIII) and then to the ipsilateral cochlear nucleus in the brain stem. The pathway then proceeds to the nucleus of the ipsilateral facial nerve (cranial nerve VII) that runs through the internal auditory canal to the ipsilateral stapedius muscle. The contralateral pathway also originates in the cochlea, then to the ipsilateral auditory nerve (cranial nerve VIII), then to the ipsilateral cochlear nucleus in the brain stem. At this point the pathway crosses the brain stem through the trapezoid body to the contralateral superior olivary nucleus, then to the contralateral cochlear nucleus in the brain stem and then to the nucleus of the contralateral facial nerve (cranial nerve VII) that runs through the internal auditory canal to the contralateral stapedius muscle. The reflex is actually consensual with an ipsilateral input resulting in bilateral contractions of the stapedius muscles.

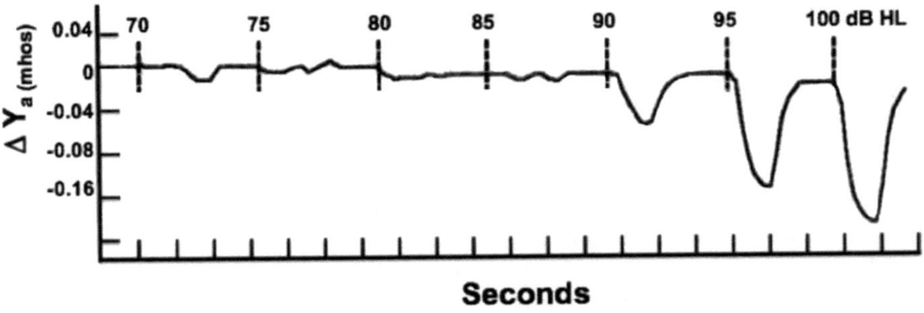

Fig. 8.5 Change in acoustic admittance (Y_a) in the external ear canal as a function of time (seconds). The *vertical dashed lines* indicate the onset of a 1 s duration pure tone at 1,000 Hz at the indicated level

When contracted, the muscle pulls the head of the stapes laterally, orthogonal to the direction of the normal vibratory motion of the tympanic membrane. This muscle contraction also tenses the tympanic membrane and likely controls the amplitude of external sound waves through the middle ear to the cochlea. Because this muscle has a very high ratio of nerve fibers to muscle fibers it is likely that it provides a high degree of controlled tension as opposed to an on-off type of reflex response, suggesting that it may influence functions other than sensitivity such as improving the ability to hear in noise (Pang and Guinan 1997; Arnold et al. 2007). The acoustic stapedius muscle reflex, often called the acoustic reflex arc, is a largely involuntary reflex activated by external sound.

Several important characteristics of the acoustic stapedius muscle reflex response have been determined from studies of electrical potentials measured directly from the muscle. The reflex is inactive for acoustic signals less than about 80 dB HL. The reflex threshold level is defined as the lowest level stimulus that still produces an observable contraction. For higher level stimuli that activate the reflex there is a very short latency between the onset of the stimulus and the beginning of the muscle contraction, approximately 10 ms, reflecting the fact that there are only a few neural synapses in the reflex arc. As the stimulus level is increased the magnitude of the muscle contraction increases until the maximum stimulus input. The magnitude of the contraction is directly proportional to the magnitude of the acoustic stimulus, generally with 1 dB resolution demonstrating clearly that the reflex produces graded responses. These graded responses are likely possible because of the high neural fiber to muscle fiber ratio mentioned earlier.

Figure 8.5 illustrates the relative acoustic admittance (Y_a) in the external ear canal of a subject with normal hearing as a function of time (seconds). Also indicated is a series of 1,000-Hz pure tones delivered ipsilaterally to activate the acoustic stapedius muscle at increasing levels. By time locking the activation signal to the measurement of the ongoing acoustic admittance, the muscle contractions associated with the activation signal easily can be differentiated from acoustic admittance changes not associated with the muscle contraction. For the large response at 90 dB HL, the slight delay between the inset of the stimulus and the rise of the response can be identified. Note that the magnitude of the acoustic admittance change is proportional to the level of the activating signal. Note also that changes in acoustic admittance that initially may appear to be related to the muscle contraction in fact may not be related. The apparent response between 70 dB and 75 dB HL is not due to a muscle contraction because the acoustic admittance change occurs a substantial amount of time after the stimulus. Discounting this aberrant response, a threshold can be identified as occurring between 85 and 90 dB HL.

The stapedius muscle also contracts reflexively to tactile stimuli applied to the skin in the area of the ipsilateral external ear canal. Though a tactile stimulus is much more difficult to control systematically, this mode of stimulation affords a mechanism for contracting the muscle without stimulating the auditory system.

8.4.2 Measurement Parameters

The continuous measurement of the acoustic characteristics in the external ear canal along with an indication of the stimulus levels and durations (Fig. 8.5) allows the middle ear muscle contractions to be detected and quantified effectively, indirectly and noninvasively. As the muscle contracts, the increase in tension of the tympanic membrane results in a proportional decrease in the measured acoustic admittance.

The acoustic admittance at the lateral surface of the tympanic membrane can be measured with a probe sealed to the external ear that contains a miniature loudspeaker for generating a measurement tone. The level of the tone is first calibrated with a known acoustic impedance load typically provided by a hard walled cavity of known dimensions. A 1 cc hard-walled cavity has an acoustic admittance of 1 millimho at 226 Hz. Next, the level of the measurement tone is measured in the canal using the calibrated microphone in the probe. The measured level is determined by the volume of the remaining ear canal between the probe and the tympanic membrane in combination with the acoustic characteristics of the lateral surface of the tympanic membrane. To repeat, the miniature microphone in the probe is used both for calibrating and measuring the level of the measurement tone. The level of the tone in the ear canal is continuously monitored and will change level in direct proportion to the magnitude of the muscle contraction. The same probe system can be used for presenting the activating signals for measurement of the ipsilateral stapedius muscle reflex. A variety of signal processing techniques make sure that the stimulus signal is not directly detected by the admittance measurement system, ensuring that the measurement represents only stapedius muscle activity.

A common measurement protocol uses a low-frequency probe tone (226 Hz) that has a wavelength much longer than the external ear canal insuring that the level of the measurement tone is the same throughout the ear canal and below the frequency range of the signals that are used to activate the ipsilateral reflex. The level of the probe tone is also kept below the threshold of the reflex to prevent the probe tone itself from directly activating the muscle reflex.

A system that records the ongoing acoustic admittance has temporal characteristics faster than the reflex response to allow accurate monitoring of the muscle activity. An indicator of when the activation signal is presented, either with a mark on the recording or with a second channel, will time lock the activation signal presentation and any changes in the measured signal and allow analysis of stapedius muscle reflex activity including its temporal characteristics.

The temporal characteristics of the reflex are characterized by a short latency (about 10 ms), a fairly rapid rise time, and an on time that is not directly related to the on time of the stimulus (Fig. 8.5). The off time, or the time of the response after cessation of the stimulus, represents the natural relaxation of the muscle.

Measurement of the acoustic stapedius muscle reflex behavior can provide useful information on the status of the middle ear. Careful consideration of the results is necessary for correct interpretation because the middle ear status can affect both the level of the activation stimulus reaching the cochlea and the ability to detect the muscle response. Two specific examples can be illustrative.

In the first example, both arches of the stapes are not contiguous because of a disease process. This abnormality will not prevent detection of the muscle contraction from acoustic admittance measures because the muscle tendon is attached lateral to the abnormality. However, this abnormality also causes a large conductive loss through the middle ear such that an ipsilateral stimulus will be greatly attenuated before reaching the ipsilateral cochlea. In this case the ipsilateral acoustic stapedius reflex will be absent. However, if the contralateral ear is normal and used to activate the consensual reflex, the middle ear muscle contraction in the abnormal ear will be detected easily and at normal levels. The combination of the ipsilateral and contralateral stapedius muscle responses provides quantifiable diagnostic information on the functional status of the stapes.

In the second example, a middle ear is filled with an exudate from a disease process that does not result in much of a loss through the middle ear as measured with the auditory thresholds using the air–bone gap. However, the presence of the exudate may decrease the acoustic admittance at the lateral surface of the tympanic membrane preventing detection of the muscle contraction even though the muscle contraction is present. Measurement of behavioral hearing thresholds by both air and bone conduction will provide estimates of the stimulus level reaching the cochlea. Once this is known, ipsilateral and contralateral measures of acoustic stapedius muscle contractions can characterize the status of the middle ear including the status of components within the middle ear, especially when considering the attachment of the stapedius tendon.

Acoustic stapedius muscle reflex thresholds generally are around 85 dB HL for 500-, 1,000-, 2,000-, and 4,000-Hz tonal signals and up to 20 dB better for wideband noise signals. More precise normative values consider the age of the subject, especially for young children, even newborns.

8.5 Middle Ear and Otoacoustic Emissions

The normal cochlea not only detects sounds but also generates sounds. These cochlea-generated sounds are an epiphenomenon associated with active processes within the cochlea. A sound originating from within the cochlea can propagate back through the middle ear and be detected in the external ear canal as an otoacoustic emission. The presence of cochlear active processes was first demonstrated experimentally as otoacoustic emissions in 1978 (Kemp 1978). The otoacoustic emissions arise from a number of different mechanisms within the cochlea but generally are associated with outer hair cell motility. Several lines of evidence suggest that, in mammals, the active processes associated with outer hair cells increase cochlear sensitivity and frequency selectivity.

Otoacoustic emissions will diminish or disappear after damage to the cochlea. Comparison of the measured otoacoustic emission levels with respect to otoacoustic emission levels in normal ears often is used as an indication of the functional status of the outer hair cells. Therefore, otoacoustic emissions have been studied almost exclusively in relation to understanding cochlear mechanisms (Shaffer et al. 2003) or measuring the effects of various cochlear pathologies on cochlear function. However, because the source of the emissions is just medial to the middle ear, and their detection occurs just lateral to the middle ear, measures of otoacoustic emissions also can be used to understand the middle ear properties as well, which is the focus of this chapter.

The levels of the otoacoustic emissions are very low in normal ears, decrease with age (Abdala and Dhar 2012), and become lower still in abnormal ears. Their accurate detection involves consideration of several factors that are independent of the particular type of otoacoustic emission. First, because the responses are very low level acoustic signals, control of the acoustic environment is essential. Usually this is accomplished by making the measures in a sound attenuating booth. Second, the somewhat transient air pressure in the middle ear, normally controlled by the Eustachian tube, can produce small changes in the admittance of the middle ear that in turn can affect both the stimulus level reaching the cochlea and the level of the emission propagating from the cochlea back to the external ear canal. These factors result in variability of the emission levels being measured but can be minimized by ensuring that the otoacoustic emissions measures are made only after the middle ear space is equalized by the Eustachian tube, or that an external air pressure is applied that maintains the air pressure differential across the tympanic membrane at 0 Pa. Detection of the otoacoustic emission can be optimized by this reduction in the variability of the measures.

Otoacoustic emissions have been categorized by various types. One type is associated with no stimulus and the remaining types are defined by the characteristics of an acoustic stimulus that evokes the emission.

Spontaneous otoacoustic emissions are otoacoustic emissions that occur spontaneously with no external stimulus. However, spontaneous otoacoustic emissions are unpredictable regarding their spectral composition, level and occurrence, even in normal ears. Therefore, they do not play much of a role in characterizing middle ear structure or function.

Evoked otoacoustic emissions are otoacoustic emissions that are evoked with an external sound stimulus. Evoked otoacoustic emissions can be measured using three different approaches. In general, the level of the evoked emission is much lower than the level of the evoking signal.

8.5.1 *Transient Evoked Otoacoustic Emissions*

Transient otoacoustic emissions are evoked with a transient stimulus, that is, a very short duration stimulus. The short duration stimulus can be either a brief duration impulse signal (click) or a brief duration tone burst signal. A transient click stimulus will contain spectral energy over the frequency range of the stimulating transducer, generally up to around 4 kHz. A transient tone burst stimulus will contain spectral energy consistent with the frequency of the tone burst. The measurement microphone will pick up the high level stimulus itself and the lower level emission. Because the transient stimulus is much shorter than the time it takes to generate the emission plus the additional time for the emission to travel from the cochlea back to the measurement microphone in the external ear canal, the stimulus portion of the measured signal easily can be separated from the emission portion of the measured signal by starting the measurement just after the stimulus is completed. The rapid onset of a transient stimulus allows time-locked measurement sequences and signal averaging to reduce the unwanted, uncorrelated noise in the measured signal. Thus, controlled stimuli can be used to evoke consistent emissions generated by the cochlea and detected in the external ear canal, affording a measurement of both the level of the stimulus through the middle ear and the level of the emission also through the middle ear but from the opposite direction. Consideration of this measurement paradigm allows direct measures of the middle ear function. The fundamental measurement paradigm for transient evoked otoacoustic emissions is analogous to the measurement paradigm for the ipsilateral acoustic stapedius muscle reflex. A controlled stimulus in the external ear canal travels through the middle ear and generates a constant and repeatable signal as an otoacoustic emission in the cochlea that then travels back through the middle ear where it can be detected in the ear canal providing information about losses through the middle ear from both directions (Puria 2003). Because the signal has a wide spectrum, the transient evoked otoacoustic emission also will have a wide spectrum. Spectral analyses of the response can provide some frequency-specific information though the frequency resolution will not be very high.

8.5.2 Stimulus Frequency Otoacoustic Emissions

Stimulus frequency otoacoustic emissions are evoked with a pure-tone stimulus and are detected by the vectorial difference between the stimulus waveform and the recorded waveform that consists of the sum of the stimulus and stimulus frequency otoacoustic emissions. This category of emission is much more difficult to measure because the emission and the stimulus are at the same frequency. Consequently, much less is known about this otoacoustic emission than the other otoacoustic emissions.

8.5.3 Distortion Product Otoacoustic Emissions

Distortion product otoacoustic emissions are evoked using a pair of primary tones with frequencies f_1 and f_2 with $f_1/f_2 = 1.2$, and levels L_1 and L_2 typically with $L_1 = L_2 = 65$ dB SPL or $L_1 = 65$ dB SPL and $L_2 = L_1 - 10$ dB). The evoked otoacoustic emissions from these stimuli occur at frequencies mathematically related to the primary frequencies, with the highest level distortion product emission $f_{dp} = 2f_1 - f_2$ (the "cubic" distortion product). Because f_{dp} is always different from the frequency of either of the 2 tones that evoke the otoacoustic emission, the otoacoustic emission can be measured in the presence of the evoking tones allowing for constant and simultaneous measurement of the emission. The measurement process can also use signal averaging in either the frequency domain or the time domain to reduce the level of the uncorrelated noise allowing for accurate measures of very low distortion product otoacoustic emissions.

Figure 8.6 illustrates the level of a distortion product otoacoustic emission ($2f_1 - f_2$) as a function of the level of the two tones (f_1 and f_2 with $L_2 = L_1 - 10$ dB) used to evoke the emission, in a normal ear obtained with two different signal averaging algorithms that have different noise floors. An increase in the number of individual spectra during averaging will reduce the variability of the noise floor but maintain the same average level of the noise floor relative to the level of the distortion product. By contrast, an increase in the number of individual waveforms during averaging will maintain the variability of the noise floor but reduce the average noise floor level relative to the level of the distortion product and allow the detection of the emission at lower levels. The choice of signal averaging algorithm can greatly influence the results (Popelka et al. 1993; Nelson and Zhou 1996), making it necessary to specify the signal averaging parameters for proper interpretation of the emission level.

When attempting to define changes in the level of the emission in relation to middle ear conditions, a threshold of the emission can be used. The threshold of the

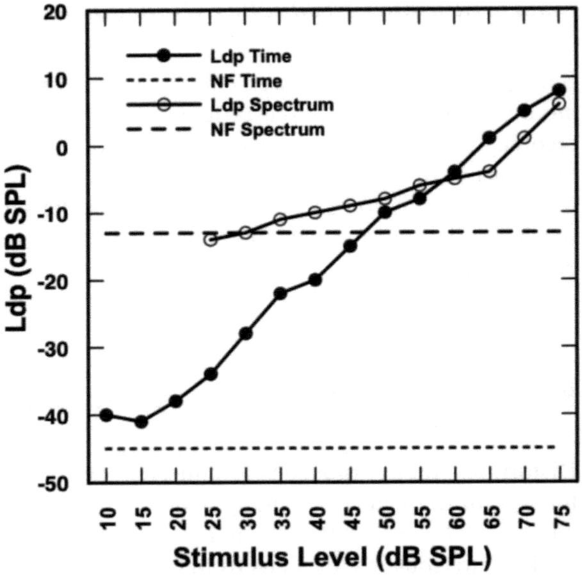

Fig. 8.6 Level of distortion product otoacoustic emissions evoked with a stimulus comprised of a pair of primary tones with frequencies f_1 and f_2 with $f_2/f_1 = 1.2$, and levels L_1 and L_2 with overall level equal to L_1 and $L_2 = L_1 - 10$ dB, illustrating the difference in noise floors between averaging individual waveforms (time domain) and individual spectra (spectrum domain) (Adapted from Popelka et al. 1993)

emission can be defined as the lowest level of the evoking signal that produces an observable emission specified as a predetermined level above the noise floor. However, for repeatability purposes, the exact parameters of the measurement process including the particular signal averaging method and the number of measures per average must be specified.

An alternative and more common analysis is to report changes in the level of the emission counting only those responses that are above a predetermined noise floor level. A related issue for proper determination of the emission level in this manner is that the input–output functions are often not monotonic (Popelka et al. 1993). The nonmonotonic nature of these functions has been studied extensively though they relate more to cochlear function than middle ear function. Once all of these parameters have been specified, changes in the level of an otoacoustic emission reflect the function of the middle ear in the direction of a signal (the emission) on its way from the cochlea through the middle ear into the external ear canal after accounting for the change in the stimulus level reaching the cochlea that evokes the emission. The air–bone gap in dB can be subtracted from the stimulus level (Fig. 8.6) to account for changes in stimulus level reaching the cochlea and the same air–bone gap in decibels can be subtracted from the emission level (Fig. 8.6) to arrive at an expected emission level in cases with a known air–bone gap.

8.6 Middle Ear Imaging

8.6.1 Otoscopy

The basic monocular otoscopy described earlier is necessary to provide information about the external ear canal to ensure that all of the subsequent middle ear measures are in fact measuring the middle ear. Traditional monocular otoscopy can be enhanced with the addition of a pneumatic mechanism for simultaneously inducing air pressure changes manually, usually with a soft bulb that can be squeezed by hand. Pneumatic otoscopy allows for visualization of the movement of the tympanic membrane under dynamic conditions and provides a subjective estimate of its stiffness. Though very useful for medical diagnoses, membrane mobility is observed only for the very low frequency and very large oscillations of the pneumatic system rather than the much smaller and much higher frequency oscillations of auditory signals going through the middle ear system. Otoscopy also can be performed with a binocular microscope that offers three-dimensional viewing. A video camera can be added to otoscopy to provide recorded images that can then be used for more objective measures. With otoscopy, the entire lateral surface of the tympanic membrane may not be visualized because of the tortuous nature of the external ear canal. In cases of a normally transparent tympanic membrane, several middle ear structures can be visualized as well as exudate in the middle ear space. Various rating scales have been devised that can be very useful for reliability across observers and for categorizing middle ear conditions (Casey et al. 2011). Though mandatory for medical diagnosis, otoscopy provides only limited information concerning the actual structure and function of the middle ear anatomy.

8.6.2 Middle Ear Endoscopy

Middle ear endoscopy can be achieved with a thin endoscope inserted into the middle ear. This can be performed surgically with the tympanic membrane raised or through an existing tympanic membrane perforation for a direct view of the middle ear space and its contents. Middle ear endoscopy also can be achieved with a surgical incision in an intact tympanic membrane but is a much more invasive procedure. Bremond et al. (1990) introduced the concept of middle ear endoscopy through the Eustachian tube orifice in the nasopharynx. This method is used less often than transtympanic endoscopy because it can cause tissue irritation and bleeding. The procedure also has had limited use because of the restricted field of view and the presence of disease related exudates in the middle ear that limit observation. In select cases, the method provides useful information including a direct visualization of the ossicular chain. The method may improve in the future as the diameter of the endoscopes decreases and the endoscopes become more flexible with improved optics. Very thin endoscopes currently under development for viewing the cochlea already have narrower diameters and improved optics (Monfared et al. 2006).

8.6.3 X-Ray Imaging

Though a conventional plain-field X-ray image is not useful for middle ear imaging because it is difficult to resolve overlapping structures in a single image, more advanced X-ray imaging can provide useful information. Computed tomography (CT) X-ray imaging uses a narrow collimated X-ray beam from an X-ray tube that rotates around the patient. The tissues of the body differentially attenuate the photons resulting in a gray-scale image that further can be processed with digital processing. The resulting images usually are calibrated with water as a reference and the signal processing can be optimized for viewing specific structures such as bone and soft tissue. Scanning sequentially through the structure of interest produces a series of images that can be viewed separately as very thin "slices" or in a fashion similar to varying the focal plane with a traditional microscope. The resulting series of thin planar images allows visualization of a single structure in the third dimension that improves identifiability of that structure compared to using only a single image. The table on which the patient is positioned can automatically move in concert with the X-ray beam resulting in the ability to image the tissue between the slices that in effect produces very high resolution imaging.

Because CT imaging is ideal for visualizing bone, the erosion of the surrounding bone caused by abnormal soft tissue such as a tumor or cholesteatoma can be observed readily. The extent of bone involvement can be used to quantify the size of the abnormal tissue even though the abnormal soft tissue was not imaged directly. Intravenous contrast agents also can be added that are differentially absorbed by abnormal tissues such as a tumor to allow clearer observation of the abnormal soft tissue itself provided that the soft tissue is sufficiently vascular to deliver the contrast agent. Software tools are provided in the image analysis that can be used for linear or volumetric measures so that the extent of the soft tissue abnormalities can be quantified.

8.6.4 MicroCT Imaging

MicroCT imaging is a form of X-ray tomography that produces cross-sectional planar images simultaneously in multiple orthogonal planes. These planar images are used to recreate a three-dimensional virtual image on a display. The term micro refers to the size of the voxels that are in the micrometer range. The technology has been used for research of the middle ear (Decraemer et al. 2007; Sim and Puria 2008) and is now becoming available for routine clinical use.

A clinical microCT scanner is generally much smaller than a conventional CT scanner. Typically, the patient sits upright with his or her head fixed using a chin rest and the X-ray source and the detector rotate around the patient's head. Note that a typical fan-beam system uses an X-ray source that produces a flat pie-shaped beam collected by a line detector with multiple images obtained by moving the X-ray source axially. By comparison, a cone-beam system uses an X-ray source that produces a cone shaped beam that is collected by charge-coupled device focused on a scintillator material that converts X-ray radiation to visible light, in essence an area detector, to create images that can be used for a three-dimensional reconstruction of the tissue. The cone beam approach limits the amount of exposure to the ionizing radiation and allows image optimization of a specified small volume of tissue. The terms microCT and cone beam imaging tend to be used interchangeably and because the scanner is small, the device often is referred to as an in-office scanner. The largest application is in dentistry but such systems are increasingly being used in otolaryngology for imaging the nasal sinuses and the temporal bone.

Each orthogonal image has a consistent number of image pixels in a regular pattern separated by a specified distance, every 0.125 mm, for example. The resulting scans generally are isotropic so the image planes can be positioned in any direction rather than the traditional axial, coronal and sagittal planes. Post-scan image processing can also stack the individual planar images before rendering a volumetric image. Volume rendering allows a two dimensional projection of a three-dimensional discretely sampled data set. Each volume element, or voxel, is represented by a single value obtained by sampling the immediate area around the voxel.

Different structures with similar threshold density make it difficult to separately visualize them by only adjusting volume rendering parameters. A manual or automated segmentation procedure can remove the unwanted structures from the rendered image. Contrast agents can be used to optimize visualization of bone or in some cases even soft tissue. False coloring can be used to highlight specific structures. Stereo images can be created with colored filters, usually blue and green, to provide the illusion of depth.

Figure 8.7 shows the results of a temporal bone scan of a normal temporal bone centered on the middle ear. The ossicles can be seen clearly at high resolution. The supplied viewer application allows the three orthogonal planar images to be rotated individually for viewing of the rendered image from any direction. Cursor lines can be positioned to select specific regions of interest.

8.6.5 Magnetic Resonance Imaging

Magnetic resonance imaging (MRI) uses a large magnetic field and radio waves to create images. A patient is placed in a uniform magnetic field that aligns the spinning protons of hydrogen atoms, located mostly in the tissues that contain fat and water, in the direction of the magnetic field. Radio frequency pulses are then

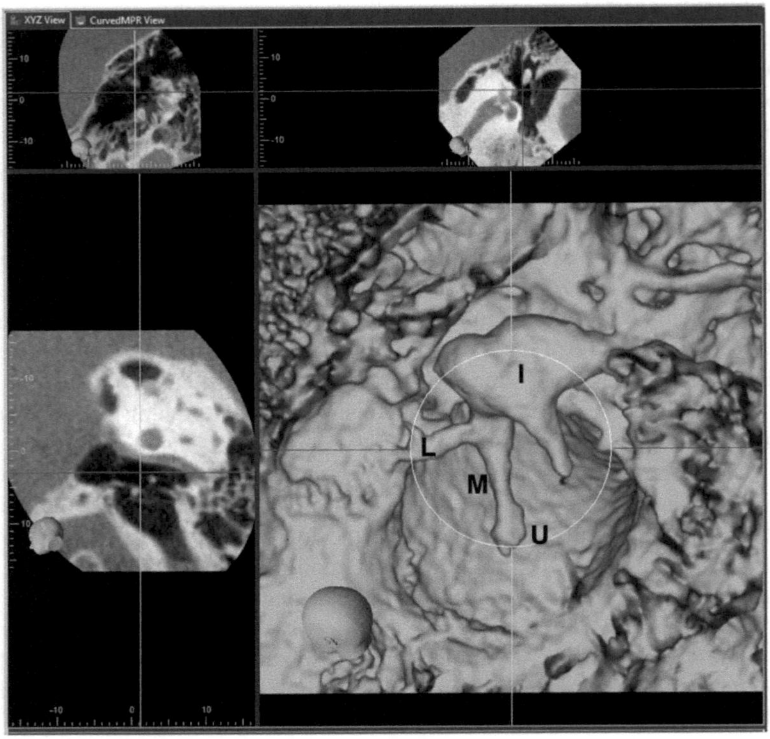

Fig. 8.7 A three-dimensional reconstruction of a normal right middle ear obtained with a cone-beam, in office, clinical microCT scanner (J. Morita Mfg Corp., model 3D Accuitomo) and automatic segmentation (large panel). The imaged volume was a 40 × 30 mm cylinder with a uniform 0.125 mm voxel size. The view is from the inside out. The umbo (*U*) at the end of the manubrium (*M*) of the malleus, the anterior mallear ligament (*L*) and the incus (*I*), are clearly visualized. The three smaller adjacent images show the planes used for the reconstruction and the cursor lines that can be moved for other views. The scan time was approximately 18 s with a radiation exposure about one-seventh of that for a conventional medical CT, similar to the exposure for a conventional dental panoramic image

applied that change the direction of the spinning protons with respect to the direction of the magnetic field that in turn give off energy that can be detected. The quality of the resulting image is dependent on the strength of this radio frequency energy change that is dependent on the strength of the magnetic field and the proximity to the structure of interest of the receiving coil that detects this energy change. Magnetic field strength is specified in Tesla units, with scanners of 0.5–3.0 T common and scanners up to 7 T becoming available. Receiving coils can be placed around the head or on the side of the head for creating images of the temporal bone and its structures.

Because no ionizing radiation is used and no negative effects of magnetism and or radio frequency exposure have been observed, MRI imaging is considered noninvasive with only few considerations. The MRI scanners produce considerable acoustic energy during the pulse sequence that may be a risk to hearing. Psychological reactions in the case of phobias associated with being positioned in the narrow confines of the imaging apparatus also occur. Implanted metallic objects such as heart pacemakers and cochlear implants are also a concern both in terms of patient safety as well as interference with generation of the images.

The detected energy changes are used to generate the image. Small changes in the radio waves and magnetic fields can affect the contrast of the image and the contrast settings can be adjusted to highlight different types of tissue. The imaging plane can be changed in thickness and to any location or direction without moving the patient. As with CT imaging, planar "slices" can be viewed individually or sequentially and used for constructing three-dimensional images.

The images from MRI imaging are very useful for identifying soft tissues, fluid-filled spaces and air-filled spaces. A variety of pulse sequences can be used to enhance or suppress visualization of various structures including soft tissue, air, blood flow, and fluid-filled structures. Contrast agents are also used, often gadolinium, that enhance visualization of abnormal tissues provided that the tissue is sufficiently vascular to absorb the contrast agent. Contrast agents can be nephrotic, so kidney function should be assessed before contrast agent administration. Middle ear soft tissue abnormalities such as cholesteatoma, polyps, granulation tissue, squamous cell tumors, glomus tumors, basal cell tumors, and cholesterol granulomas can be observed by utilizing various MRI sequences (T1, T2, diffusion restricted, post-gadolinium T1) that take advantage of differential imaging characteristics of the various pathologies. However, because MRI images do not identify bony structures very well, they are not useful for visualizing the ossicles or the bone margins that define the middle ear space.

8.6.6 Real-Time Magnetic Resonant Imaging

Real-time MRI imaging can be implemented by a pulse sequence that generates many planar images per second and played back as a movie. This allows imaging of moving structures in real time and is used for a variety of purposes. In otolaryngology, real-time MRI has been used for measuring the movement of structures in the airway including the soft palate, the tongue, and the airway diameter (Barrera et al. 2009, 2010), all important for quantifying functional airway abnormalities associated with sleep apnea or with speech production. With real-time MRI imaging, a tradeoff exists between image resolution and temporal resolution. Real-time imaging at 6 frames/s produces adequate images for larger structures that move relatively slowly such as the soft palate. An increase in the frame rate up to 30 frames/s will allow observation of structures that move faster such as the tongue during speech, but with a penalty of reduction of image resolution. Because the relevant moving structures in the middle ear oscillate in the auditory range (20 Hz–20 kHz) and at extremely small amplitudes, it is unlikely that real-time

MRI will be able to assess the functional status of the tympanic membrane or the ossicles. However, it is potentially possible for Eustachian tube function to be measured in its natural state using real time MRI because the Eustachian tube movements are relatively slow and the tissue displacements are relatively large. The plane of the MRI image can be oriented directly on the oblique angle of the Eustachian tube and the imaging tied to voluntary swallowing. Attempts at this type of middle ear imaging were not successful with a 0.5 T scanner but may prove useful with higher Tesla scanners.

8.6.7 Optical Coherence Tomography

Optical coherence tomography (OCT) produces three-dimensional structural images of tissue with submicrometer resolution, and sound-induced tissue displacement information, from reflections of wide bandwidth light. Clinical optical coherence tomography systems are available in other disciplines such as ophthalmology and cardiology but none are yet available in otology. Adaptations of commercial systems or custom built systems have been used for human auditory research of the middle ear (Djalilian et al. 2008). A patent for an OCT device specifically for the ear was recently issued (USPTO, patent 8115934, February 14, 2012), suggesting that clinical systems will be available for the middle ear in the future.

Optical coherence tomography is a low coherence interferometric method that uses broad-band long wavelength light (near-infrared) that penetrates a few millimeters into soft tissue, and less so in bone, and has critical advantages for clinical measures of the middle ear. It has far greater image resolution than ultrasound or magnetic resonance imaging and greater tissue penetration than confocal microscopy. In contrast to X-ray and MRI imaging, it uses a back-scattered reflected light signal, an echo method similar to ultrasound, but no medium is required so the imaging transducers are not in direct contact with the tissue. Only low signal levels are needed so tissue damage is unlikely. Optical coherence tomography quickly produces sub-surface morphology images at high resolution with no tissue preparation or ionizing radiation. The technology is relatively noninvasive, is able to generate cross-sectional images with micron-scale resolution and has the potential of imaging middle ear structures including the tympanic membrane and its layers, the ossicles, and other middle ear structures such as the middle ear muscle tendons (Pitris et al. 2001) and pathological material in the ear such as cholesteatoma (Djalilian et al. 2010) and possibly biofilms from infectious processes (Xi et al. 2006).

The broad bandwidth light source, a very bright light emitting diode, a femtosecond pulsed laser or even white light, is divided into a direct arm and a mirror reference arm aimed at the tissue surface. A photodetector or a charge coupled device camera measures the combined reflected light from the sample arm and the reference arm. If both arms have the same optical length an interference pattern results based on the characteristics of the tissue. The mirror in the reference arm can be moved to scan the tissue resulting in a reflectivity profile called the time domain method. The sample areas that reflect most of the light create greater interference patterns while the scattered light from other sample areas falls outside the short coherence length and reduces the interference pattern. This reflectivity profile contains information about the spatial dimensions and locations of structures within the tissue. Because the level of reflected light decreases with tissue depth, the image is limited to only a few millimeters below the tissue surface. Wide bandwidth light signals (low coherence interferometry) have interference pattern distances in the micrometer range in contrast to narrow bandwidth signals (high coherence interferometry) with interference pattern distances in the meter range.

A point light on the tissue surface is used to produce an image in the axial dimension (*Z*-axis). The light source can be moved to scan the tissue along a line to produce a two-dimensional cross-sectional image (*X*–*Z* axes) or over an area to produce a three-dimensional volumetric image (*X*–*Y*–*Z* axes scan).

Broad bandwidth interference patterns also can be acquired with simultaneous detection of spectrally separated signals either by time locking the light frequency with a spectral scanning source or by using a dispersive detector such as a grating and a linear detector array, a process called spatially encoded frequency domain optical coherence tomography. This approach immediately generates a depth scan without movement of the reference arm mirror and greatly improves imaging speed. The simultaneous detection of multiple wavelengths determines the scanning range and the full spectral bandwidth determines the axial resolution. A full-depth scan can be acquired with a single exposure but only at the expense of a reduced dynamic range. Optical coherence tomography also has the potential of measuring sound induced tissue displacements at levels and across frequency ranges well within the normal auditory function of humans. Tissue vibrations from a specific structure can be measured by phase locking the signal from the direct arm. In experiments on the much smaller tissues of the inner ear, this approach has been shown to be appropriate for acoustic levels of from 20 to 100 dB SPL and frequencies up to 25 kHz (Chen et al. 2011). The addition of an acoustic transducer should allow accurate functional measurements of middle ear structures (Applegate et al. 2011).

A practical clinical instrument for the middle ear will involve several considerations. The surrounding bone will require all transducers to be placed in the external ear canal to image the middle ear, in particular the tympanic membrane and the structures a few millimeters deeper. The particular methods will be selected based on several considerations. Faster scans reduce movement artifacts but only at the expense of clearer images though other processing such as image stabilization can be considered. The miniaturization of the components also will be a factor because the diameter of the external ear canal limits the space, the tympanic membrane may be only partially visible and the mirror scanning mechanism must be included along with the OCT transducers and the acoustic transducer if functional measures are desired. The scanning parameters also will be a consideration to ensure that the scanning process does not go beyond the edge of the tympanic membrane, suggesting that a simultaneous otoscopic view may be necessary for

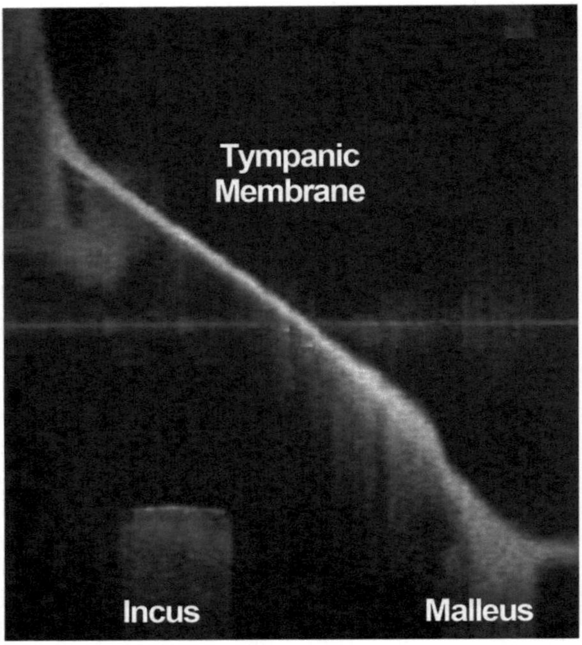

Fig. 8.8 A cross-sectional image of the middle ear of an adult cadaver obtained with optical coherence tomography. The area is orthogonal to the long process of the incus and the handle of malleus. The tympanic membrane, the manubrium of the malleus, and the long process of the incus are identified (3 × 3 mm; resolution, 16 μm) (Courtesy of A. Huynh 2012)

positioning the transducers. Figure 8.8 is a cross-sectional image of the middle ear of an adult cadaver obtained with optical coherence tomography (A. Nguyen-Huynh, personal communication, 2012). The 3 × 3 mm area is orthogonal to the long process of the incus and the handle of malleus and has 16 μm resolution. The tympanic membrane, the manubrium of the malleus, and the long process of the incus can be visualized.

8.6.8 Post-imaging Processing

The raw images (typically in the DICOM format) for CT, microCT, and MRI images can be processed with a variety of software packages (OsiriX, for example) after they have been collected to obtain important additional information. In addition to three-dimensional reconstruction, co-registration of CT and MRI images can be implemented to visualize both bony and soft tissues simultaneously. The structure of interest can be manually segmented from the surrounding tissues in each planar image (Osborn et al. 2011) or computer algorithms can be implemented to segment a particular structure automatically.

An example of post-processing analysis specific to the middle ear can serve as an illustration. Figure 8.9a illustrates a series of planar CT images obtained in a

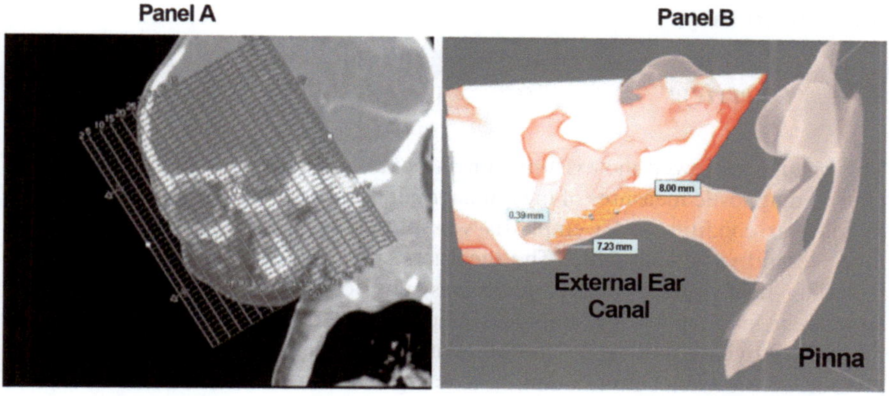

Fig. 8.9 (**a**) A CT image of the head of a newborn showing the individual planar "slices" and the spacing (1 mm) between the "slices."(**b**) A three-dimensional reconstruction (Dextroscope, Inc.) of the left middle ear, external ear canal, and pinna using the CT images in (**a**) and automatic segmentation

newborn. Each planar image covers a thickness of 1 mm of tissue. The CT imaging was performed for medical reasons unrelated to the auditory system, in this case for diagnosing a brain problem so image resolution, location, and other imaging parameters cannot be varied for other purposes. However, because the planar images coincidently covered the peripheral ear, they can be used for post-imaging processing to assess the middle and the external ear. Figure 8.9b shows a three dimensional reconstruction (Dextroscope, Inc.) of the middle ear and the external ear canal of a newborn using the computerized tomographic images in Fig. 8.9a. The three dimensional image was reconstructed using an automated software segmentation algorithm rather than manual segmentation of a single structure such as the middle ear space (Osborn et al. 2011). The regions bounded by bone (middle ear space, the medial portion of the external ear canal) or by cartilage (the lateral portion of the external ear canal and the pinna) are very visible. The regions of soft tissue such as the tympanic membrane are much less visible. Under higher magnification the umbo of the manubrium and the annular ligament can be identified. By making some assumptions about the location of the tympanic membrane using the umbo and annular ligament as landmarks, and using the measurement tools provided by the software, this three-dimensional image allows a fairly accurate measure of the external ear canal volume. Note that this measure of the external ear canal volume was obtained with the structure in its completely natural state. Before this approach, measures of the external ear canal volumes were obtained by injecting substances into the ear canal, a procedure that can be much more invasive and that can actually distort the ear canal dimensions compared to this imaging technique.

8.7 Summary

8.7.1 Overview

This chapter reviewed the existing conventional and current or soon to be current clinical diagnostic measures of the human middle ear and how they relate to specific middle ear conditions. An emphasis was placed on measures that are the least invasive and yet quantifiable. The measures span a very wide range of methods including behavioral voluntary responses, physical attributes of the middle ear structures, involuntary physiological responses, and imaging. Each middle ear measure provides unique or complimentary information and in aggregate can provide considerable information concerning the functional status of the middle ear, the structural status of several middle ear components including the air space, and the effect of contracting either of the two middle ear muscles. A recurring theme was to define a stimulus on the lateral side of the middle ear as close to the tympanic membrane as possible and use various attributes of the cochlea as a sensor immediately adjacent to the medial portion of the middle ear. This approach generally limits the measures to the middle ear and its components alone and attempts to minimize the influence of structures and systems both lateral and medial to the middle ear. A secondary recurring theme was to quantify all of the measures as much as possible, even those prone to variability associated with voluntary responses, noise, and other factors.

8.7.2 Future Directions

The limitations of the current middle ear measures are being addressed with innovative technology. New calibration procedures will allow more accurate specification of high-frequency acoustic stimuli in the external ear canal (Richmond et al. 2011). New bone conduction transducers will allow careful characterization of the middle ear over the entire frequency range (Popelka et al. 2010) and allow testing of new hypotheses of high frequency middle ear function (Puria and Steele 2010). New imaging systems along with new and improved image signal processing will produce increasingly clearer static images of the middle ear. These structural images evaluated in concert with improved functional measures such as those provided by laser Doppler vibrometry, optical coherence tomography, and real-time MRI will provide a more complete structural and functional minimally invasive analysis of the middle ear in humans.

Acknowledgments This work was supported in part by grant R01DC005960 from the National Institute on Deafness and Communication Disorders of the National Institutes of Health. The authors thank Anh Nguyen-Huynh for providing the image in Fig. 8.8 and Gerald Diaz for his help in producing the image in Fig. 8.9.

References

Abdala, C., & Dhar, S. (2012). Maturation and aging of the human cochlea: A view through the DPOAE looking glass. *Journal of the Association for Research in Otolaryngology*, 13, 403–421.

Applegate, B. E., Shelton, R. L., Gao, S. S., & Oghalai, J. S. (2011). Imaging high-frequency periodic motion in the mouse ear with coherently interleaved optical coherence tomography. *Optics Letters*, 36(23), 4716–4718.

Arnold, A., Blaser, B., & Häusler, R. (2007). Audiological long-term results following stapedotomy with stapedial tendon preservation. *Advances in Oto-Rhino-Laryngology*, *65*, 210–214.

Ballachanda, B. B. (1997). Theoretical and applied external ear acoustics. *Journal of the American Academy of Audiology*, *8*(6), 411–420.

Barrera, J. E., Holbrook, A. B., Santos, J., & Popelka, G. R. (2009). Sleep MRI: Novel technique to identify airway obstruction in obstructive sleep apnea. Otolaryngology—Head and Neck Surgery, *140*(3), 423–425.

Barrera, J. E., Chang, R. C., Popelka, G. R., & Holbrook, A. B. (2010). Reliability of airway obstruction analyses from sleep MRI sequences. *Otolaryngology–Head and Neck Surgery: Official Journal of American Academy of Otolaryngology-Head and Neck Surgery*, *142*(4), 526–530.

Bremond, G. A., Magnan, J., Chays, A., & de Gasquet, R. (1990). [Endoscopy of the eustachian tube. 1St evaluation]. *Annales D'oto-Laryngologieetde Chirurgie Cervico Faciale: Bulletin de la SociétéD'Oto-Laryngologiedes Hôpitauxde Paris*, *107*(1), 15–19.

Casey, J. R., Block, S., Puthoor, P., Hedrick, J., Almudevar, A., & Pichichero, M. E. (2011). A simple scoring system to improve clinical assessment of acute otitis media. *Clinical Pediatrics*, *50*(7), 623–629.

Chen, F., Zha, D., Fridberger, A., Zheng, J., Choudhury, N., Jacques, S. L., Wang, R. K., Xiaorui, S., & Nuttall, A. L. (2011). A differentially amplified motion in the ear for near-threshold sound detection. *Nature Neuroscience*, 14(6), 770–774.

Decraemer, W. F., de La Rochefoucauld, O., Dong, W., Khanna, S. M., Dirckx, J. J., & Olson, E. S. (2007). Scalavestibuli pressure and three-dimensional stapes velocity measured in direct succession in gerbil. *The Journal of the Acoustical Society of America*, 121(5 Pt1), 2774–2791.

Djalilian, H. R., Ridgway, J., Tam, M., Sepehr, A., Chen, Z., & Wong, B. J. (2008). Imaging the human tympanic membrane using optical coherence tomography in vivo. *Otology & Neurotology*, 29(8), 1091–1094.

Djalilian, H. R., Rubinstein, M., Wu, E. C., Naemi, K., Zardouz, S., Karimi, K., & Wong, B. J. (2010). Optical coherence tomography of cholesteatoma. *Otology & Neurotology*, 31(6), 932–935.

Hunter, L., & Margolis, R. (2011). Middle ear measurement.In R. Seewald, & M. Tharpe (Eds.), *Comprehensive handbook of pediatric audiology* (pp. 365–388.). San Diego: Plural Publishing.

Jones, S. E., Mason, M. J., Sunkaraneni, V. S., & Baguley, D. M. (2008). The effect of auditory stimulation on the tensor tympani in patients following stapedectomy. *Acta Oto-Laryngologica*, *128*(3), 250–254.

Katz, J., Medwetsky, L, Burkard, R., Hood, L., Eds. (2009). *Handbook of clinical audiology*. Philadelphia: Lippincott Williams & Wilkins.

Keefe, D. H., Ling, R., & Bulen, J. C. (1992). Method to measure acoustic impedance and reflection coefficient. *The Journal of the Acoustical Society of America*, *91*(1), 470–485.

Keefe, D. H., & Simmons, J. L. (2003). Energy transmittance predicts conductive hearing loss in older children and adults. *The Journal of the Acoustical Society of America*, *114*(6 Pt 1), 3217–3238.

Kemp, D. T. (1978). Stimulated acoustic emissions from within the human auditory system. *The Journal of the Acoustical Society of America*, *64*(5), 1386–1391.

Merchant, G. R., Horton, N. J., & Voss, S. E. (2010). Normative reflectance and transmittance measurements on healthy newborn and 1–month-old infants. *Ear and Hearing*, *31*(6), 746–754.

Merchant, S. N., & Rosowski, J. J. (2008). Conductive hearing loss caused by third-window lesions of the inner ear. *Otology & Neurotology*, *29*(3), 282–289.

Merchant, S. N., Rosowski, J. J., & McKenna, M. J. (2007). Superior semicircular canal dehiscence mimicking otosclerotic hearing loss. *Advances in Oto-Rhino-Laryngology*, *65*, 137–145.

Monfared, A., Blevins, N. H., Cheung, E. L., Jung, J. C., Popelka, G., & Schnitzer, M. J. (2006). In vivo imaging of mammalian cochlear blood flow using fluorescence microendoscopy. *Otology & Neurotology*, *27*(2), 144–152.

Nakajima, H. H., Pisano, D. V., Roosli, C., Hamade, M. A., Merchant, G. R., Mahfoud, L., Halpin, C. F., Rosowski, J. J., & Merchant, S. N. (2012). Comparison of ear-canal reflectance and umbo velocity in patients with conductive hearing loss: A preliminary study. *Ear and Hearing*, *33*(1), 35–43.

Nelson, D. A., & Zhou, J. Z. (1996). Slopes of distortion-product otoacoustic emission growth curves corrected for noise-floor levels. *The Journal of the Acoustical Society of America*, *99*(1), 468–474.

Nozza, R. J., Bluestone, C. D., Kardatzke, D., & Bachman, R. (1992). Towards the validation of aural acoustic immittance measures for diagnosis of middle ear effusion in children. *Ear and Hearing*, *13*(6), 442–453.

Nozza, R. J., Bluestone, C. D., Kardatzke, D., & Bachman, R. (1994). Identification of middle ear effusion by aural acoustic admittance and otoscopy. *Ear and Hearing*, *15*(4), 310–323.

Osborn, A. J., Oghalai, J. S., & Vrabec, J. T. (2011). Middle ear volume as an adjunct measure in congenital aural atresia. *International Journal of Pediatric Otorhinolaryngology*, *75*(7), 910–914.

Pang, X. D., & Guinan, J. J. (1997). Effects of stapedius-muscle contractions on the masking of auditory-nerve responses. *The Journal of the Acoustical Society of America*, 102(6), 3576–3578.

Pitris, C., Saunders, K. T., Fujimoto, J. G., & Brezinski, M. E. (2001). High-Resolution imaging of the middle ear with optical coherence tomography: A feasibility study. *Archives of Otolaryngology–Head & Neck Surgery*, *127*(6), 637–642.

Popelka, G. R., Osterhammel, P. A., Nielsen, L. H., & Rasmussen, A. N. (1993). Growth of distortion product otoacoustic emissions with primary-tone level in humans. *Hearing Research*, *71*(1–2), 12–22.

Popelka, G. R., Telukuntla, G., & Puria, S. (2010). Middle-ear function at high frequencies quantified with advanced bone-conduction measures. *Hearing Research*, 263 (1–2), 85–92.

Puria, S. (2003). Measurements of human middle ear forward and reverse acoustics: Implications forotoacoustic emissions. *Journal of the Acoustical Society of America*,113, 2773–2789.

Puria, S., & Steele, C. (2010). Tympanic-Membrane and malleus-incus-complex co-adaptations for high-frequency hearing in mammals. *Hearing Research*, 263(1–2), 183–190.

Renvall, U., & Holmquist, J. (1976). Tympanometry revealing middle ear pathology. *The Annals of Otology, Rhinology, and Laryngology*, *85*(2 Supplement 25 Pt 2), 209–215.

Richmond, S. A., Kopun, J. G., Neely, S. T., Tan, H., & Gorga, M. P. (2011). Distribution of standing-wave errors in real-ear sound-level measurements. *The Journal of the Acoustical Society of America*, 129(5), 3134–3140.

Roeser, R. R., Valente, M. V., & Hosford-Dunn, H. H. D. (2007). *Audiology: Diagnosis* (2nd ed.). New York: Thieme.

Rosowski, J. J., Songer, J. E., Nakajima, H. H., Brinsko, K. M., & Merchant, S. N. (2004). Clinical, experimental, and theoretical investigations of the effect of superior semicircular canal dehiscence on hearing mechanisms. *Otology & Neurotology*, *25*(3), 323–332.

Rosowski, J. J., Nakajima, H. H., & Merchant, S. N. (2008). Clinical utility of laser-dopplervibrometer measurements in live normal and pathologic human ears. *Ear and Hearing*, *29*(1), 3–19.

Rosowski, J. J., Nakajima, H. H., Hamade, M. A., Mahfoud, L., Merchant, G. R., Halpin, C. F., & Merchant, S. N. (2012). Ear-canal reflectance, umbo velocity, and tympanometry in normal-hearing adults. *Ear and Hearing*, *33*(1), 19–34.

Shaffer, L. A., Withnell, R. H., Dhar, S., Lilly, D. J., Goodman, S. S., & Harmon, K. M. (2003). Sources and mechanisms of DPOAE generation: Implications for the prediction of auditory sensitivity. *Ear and Hearing*, *24*(5), 367–379.

Sim, J. H., & Puria, S. (2008). Soft tissue morphometry of the malleus-incus complex from micro-CT imaging. *Journal of the Association for Research in Otolaryngology*, 9(1), 5–21.

Specification for Audiometers.(n.d.). Specification for audiometers. Washington, DC: American National Standards Institute.

Toll, L. E., Emanuel, D. C., & Letowski, T. (2011). Effect of static force on bone conduction hearing thresholds and comfort. *International Journal of Audiology*, *50*(9), 632–635.

Vanhuyse, V., Creten, W., & Van Camp, K. (1975). On the w-notching of tymopanograms. *Scandinavian Audiology*, *4*, 45–50.

Voss, S. E., & Allen, J. B. (1994). Measurement of acoustic impedance and reflectance in the human ear canal. *The Journal of the Acoustical Society of America*, *95*(1), 372–384.

Voss, S. E., Merchant, G. R., & Horton, N. J. (2012). Effects of middle-ear disorders on power reflectance measured in cadaveric ear canals. *Ear & Hearing*,33, 195–208.

Wiley, T. L., Cruickshanks, K. J., Nondahl, D. M., Tweed, T. S., Klein, R., & Klein, B. E. (1996). Tympanometric measures in older adults. *Journal of the American Academy of Audiology*, *7*(4), 260–268.

Xi, C., Marks, D., Schlachter, S., Luo, W., & Boppart, S. A. (2006). High-resolution three-dimensional imaging of biofilm development using optical coherence tomography. *Journal of Biomedical Optics*, 11(3), 34001.

Chapter 9
Surgical Reconstruction and Passive Prostheses

Saumil N. Merchant[†] **and John J. Rosowski**

Keywords Cholesteatoma • Chronic otitis media • Conductive hearing loss • Mastoidectomy • Middle ear aeration • Middle ear reconstructive surgery • Ossicular reconstruction • Ossicular replacement prostheses • Otosclerosis • Stapedectomy • Tympanoplasty

9.1 Introduction

A surgical procedure performed to repair or reconstruct the tympanic membrane (TM) and/or one of more of the ossicles is called *tympanoplasty*. As described in Sect. 9.3.1 there are many subtypes of tympanoplasty. Related terms used by otologic surgeons include myringoplasty (a simple repair of just the TM) and ossiculoplasty (repair or reconstruction of the ossicular chain only). The majority of tympanoplasty procedures are performed for defects of the TM or ossicles as a result of chronic otitis media. Tympanoplasty surgery is often performed in conjunction with *mastoidectomy,* which refers to surgical opening and exenteration of mastoid air cells, typically performed for eradication of infection within the middle ear and mastoid air spaces. The term *stapedectomy* refers to a surgical procedure consisting of removal of the stapes and its replacement by an artificial prosthesis. The main indication of a stapedectomy is to restore hearing in patients with conductive hearing loss due to fixation of the stapes bone caused by otosclerosis.

[†] Deceased

S.N. Merchant
Massachusetts Eye and Ear Infirmary, 243 Charles Street, Boston, MA 02114, USA

J.J. Rosowski (✉)
Eaton-Peabody Laboratories, Massachusetts Eye and Ear Infirmary, 243 Charles Street, Boston, MA 02114, USA
e-mail: John_Rosowski@meei.harvard.edu

S. Puria et al. (eds.), *The Middle Ear: Science, Otosurgery, and Technology*, Springer Handbook of Auditory Research 46, DOI 10.1007/978-1-4614-6591-1_9,

A major objective of tympanoplasty and stapedectomy procedures is restoration of hearing loss caused by middle ear diseases such as chronic otitis media and otosclerosis. These diseases often result in a conductive hearing loss, the severity of which can be quantified by the difference between air- and bone-conduction thresholds on audiometry (so-called air–bone gap). Readers are referred to Chap. 4 by Voss et al. and Chap. 8 by Popelka and Hunter for a discussion of audiometry, air–bone gap, and conductive hearing loss.

The present chapter provides an overview of chronic otitis media and otosclerosis, surgical terminology used, the acoustics and mechanics of these operative procedures, and future directions. Readers should consult clinical and surgical texts for in-depth information and details which are beyond the scope of this chapter (Nadol and McKenna 2005; Brackmann et al. 2010; Merchant and Nadol 2010).

Historically, surgery of the middle ear and mastoid evolved out of the desire of otologic surgeons to combat middle ear infections and to correct the conductive hearing loss caused by disorders such as otitis media and otosclerosis. In many instances, these procedures were developed and improved upon by empirical observations, and by trial and error, as the underlying basic science knowledge of the mechanics of the normal and diseased middle ear was not available. In the pre-antibiotic era, the vast majority of procedures consisted of different types of mastoidectomy, with the main goal being the eradication of infection to prevent meningitis and other feared intracranial complications of otitis media. There was little regard for restoration of hearing at the time. Clinical advances made possible by the introduction of antibiotics and the operating microscope, as well as availability of safe techniques of anesthesia, set the stage for the modern era of middle ear reconstruction in the 1950s when tympanoplasty and stapedectomy were described. The German otologic surgeons Horst Wullstein and Fritz Zollner were at the forefront in the development of techniques of tympanoplasty (Sismanis 2010), while American otologists John Shea and Harold Schuknecht pioneered the modern stapedectomy operation (Handzel and McKenna 2010). Subsequently, many other clinicians and surgeons modified these techniques, and tympanoplasty and stapedectomy became firmly established and adopted on a worldwide basis by the early 1960s. It is pertinent to point out that attempts at similar middle ear procedures had occurred decades earlier in the pre-antibiotic era. For instance, Berthold in Germany described a technique for TM reconstruction in 1878, and Jack and Blake in Boston described a series of patients undergoing stapedectomy in the 1890s (Sismanis 2010).

9.2 Brief Review of Common Disorders Requiring Middle Ear Reconstructive Surgery

Chronic otitis media (COM) is a chronic inflammatory disease of the middle ear and mastoid that may result in partial or total loss of the TM and/or ossicles leading to conductive hearing loss that can be as large as 60–70 dB. COM comprises a

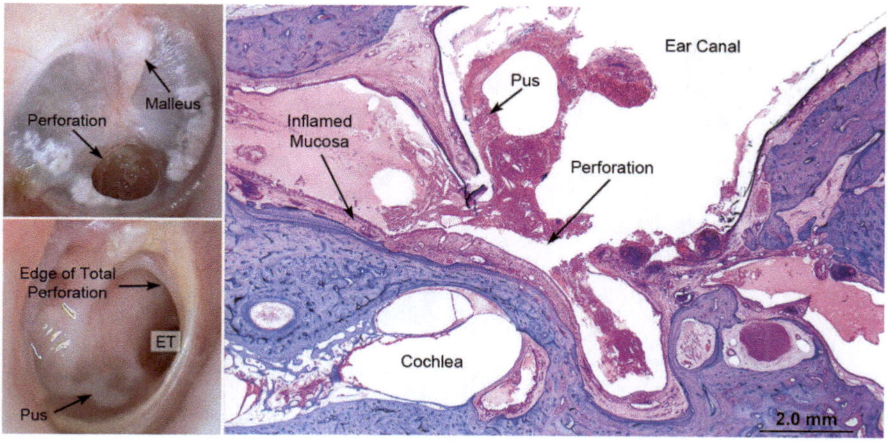

Fig. 9.1 Chronic otitis media (*COM*) without cholesteatoma. (*Left*) Otoscopic images of small and large perforations of the pars tensa of the tympanic membrane (*TM*). (*Right*) shows photomicrograph of a human temporal bone showing multiple pathologic changes in the tympanic cavity including mucosal hypertrophy and inflammation, and formation of pus. Similar changes occur throughout the mastoid air spaces. These changes result in defects in the TM and erosion of the ossicles. *ET* Eustachian tube

spectrum of disorders, which may be broadly classified into two categories: COM without cholesteatoma and COM with cholesteatoma ("cholesteatoma" is a term used to describe a disorder characterized by presence of skin in the middle ear).

The main clinical feature of COM without cholesteatoma is a perforation of the pars tensa of the TM (Fig. 9.1). The hallmark of cholesteatoma is a retraction pocket or a perforation of TM with retention of squamous debris within the middle ear (Fig. 9.2).

Both categories of COM may be accompanied by bacterial infection within the middle ear and mastoid, resulting in purulent otorrhea. Both categories are also often characterized by pathologic changes within the middle ear and mastoid air spaces (Figs. 9.1 and 9.2), including mucosal inflammation, formation of granulation tissue, resorptive osteitis, and erosion of bone, as well as healing responses characterized by deposition of fibrous tissue, formation of new bone, and deposition of hyaline plaques (called tympanosclerosis). In addition, ears with COM often demonstrate abnormalities of middle ear static pressure and dysfunction of the Eustachian tube.

The widespread tissue changes and abnormalities in ears with COM have important implications for tympanoplasty. For example, postoperative mucosal fibrosis, formation of new bone, and development of negative static pressure in the middle ear, which can occur over the course of months or years, can have a detrimental effect on the outcome of tympanoplasty. These factors are responsible for the overall modest nature of tympanoplasty results (Merchant et al. 1998a), especially in comparison to outcomes after stapedectomy for otosclerosis.

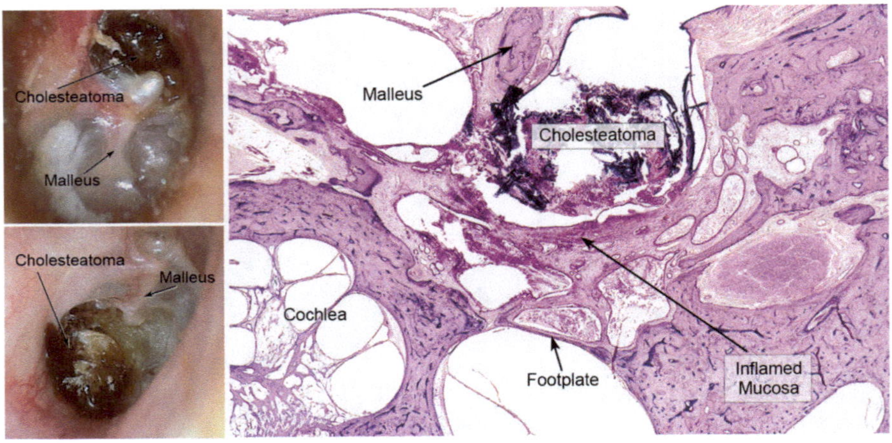

Fig. 9.2 COM with cholesteatoma. (*Left*) Otoscopic images of cholesteatoma of the pars flaccida (*top*) and pars tensa (*bottom*). In both cases, the *dark areas* consist of squamous debris within a retraction pocket of the TM. (*Right*) A photomicrograph of a human temporal bone showing a cholesteatoma of the middle ear. Note the reactive mucosal inflammation and thickening. The cholesteatoma and the reactive mucosal changes lead to erosion and resorption of the ossicles and defects of the TM, all of which result in conductive hearing loss

Otosclerosis is a localized genetic disorder affecting bone of the otic capsule that is characterized by disordered resorption and deposition of bone. Otosclerosis occurs at certain sites of predilection within the otic capsule, one of which is the area anterior to the oval window (Fig. 9.3). An expanding otosclerotic lesion in this location often results in fixation of the stapes, which produces a conductive hearing loss. The physiology of the middle ear and mastoid remains unaffected in patients with otosclerosis. In other words, the TM, malleus and incus, static pressure in the middle ear and mastoid, and mucosa of the middle ear all remain healthy, unlike in cases of COM. As a result, the main surgical challenge is to overcome the mechanical fixation of the stapes and once that is accomplished, long-term results are generally favorable.

9.3 Terminology for Middle Ear Surgical Procedures

9.3.1 Tympanoplasty

Otologic surgeons have developed a large repertoire of tympanoplasty procedures, each designed to correct specific types of anatomical defects for a given disease state of the middle ear. It is common to have more than one method that has been described to correct a given anatomical defect. Wullstein classified tympanoplasty operations as types I through V, based on the concepts of sound transformation at

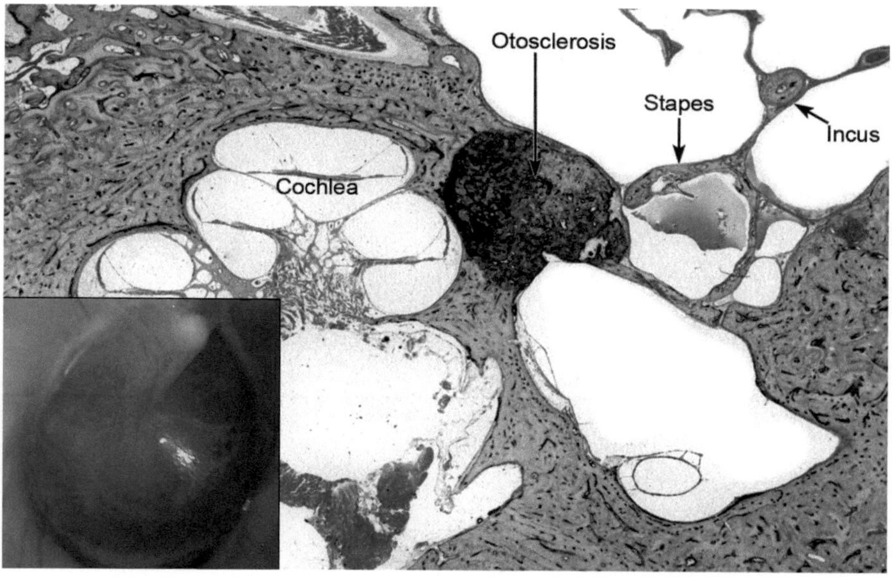

Fig. 9.3 Otosclerosis. Photomicrograph of temporal bone shows an otosclerotic lesion anterior to the oval window that has fixed the stapes. Middle ear structures such as the TM, malleus, incus, and the air spaces are unaffected in otosclerosis. Image in *lower left corner* shows otoscopic appearance of TM, which is typically normal in patients with otosclerosis

the oval window and sound protection of the round window (Wullstein 1956). However, this classification was developed before the advent of ossicular implants; hence, other classification schemes have been developed, for example, the Schuknecht-Nadol modification of Wullstein's classification (Merchant 2005) and the Austin classification (Sismanis and Poe 2010). For the sake of simplicity and ease of understanding, the present chapter relies on describing the reconstruction done in a tympanoplasty rather than adhering to a particular classification scheme.

TM reconstruction is performed when there is a perforation of the TM but the ossicular chain is intact (top left panel in Fig. 9.4). Many different tissue grafts have been described for repair of the TM including temporalis fascia, perichondrium, cartilage, periosteum, and adipose tissue. A large number of surgical techniques have been described for repair of perforations, depending on the size and location of the perforation, as well as presence or absence of additional middle ear pathology. Repair of just the TM is the most commonly performed tympanoplasty in otologic practice.

The ossicular chain is diseased in many cases and has to be reconstructed in a tympanoplasty. A common problem is resorption of the distal part of the long process of the incus. If the remainder of the incus is healthy, it can be removed, reshaped by a surgical drill, and placed back as an incus strut to restore the continuity of the ossicular chain; such a procedure is termed an *incus interposition* (top right panel in Fig. 9.4). If the ossicles are too diseased to be reused, or they are

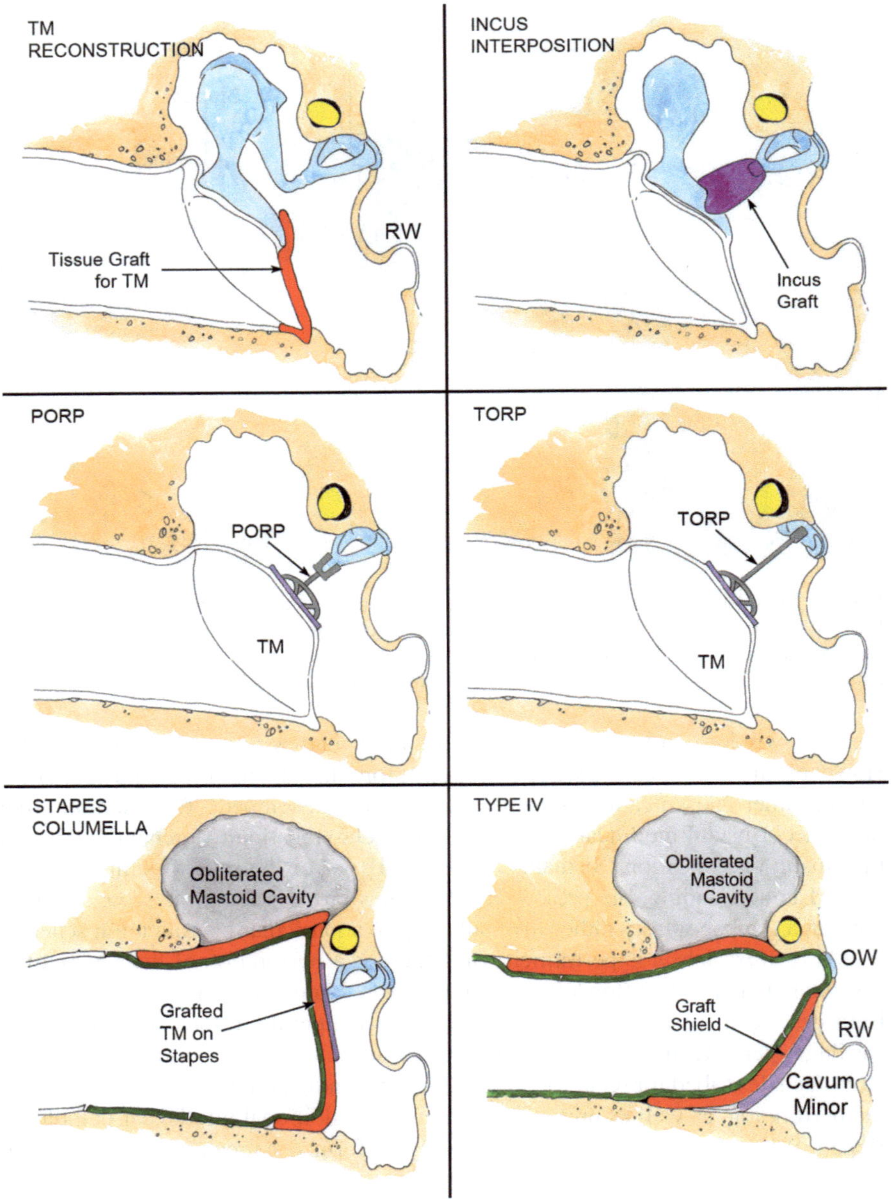

Fig. 9.4 Schematic diagram showing different types of tympanoplasty procedures. See text for explanation. *TM* tympanic membrane, *PORP* partial ossicular replacement prosthesis, *TORP* total ossicular replacement prosthesis, *OW* oval window, *RW* round window

missing (resorbed) because of COM, then synthetic implants (prostheses) can be used to reconstruct the ossicular chain. If the stapes is present, then a prosthesis can be placed from the stapes capitulum to the TM or manubrium; this is termed a partial ossicular replacement prosthesis (PORP; middle left panel in Fig. 9.4). If the stapes is missing, then a prosthesis may be placed between the stapes footplate and the TM or manubrium; this is termed a total ossicular replacement prosthesis (TORP; middle right panel in Fig. 9.4). PORPs and TORPs made of hydroxyapatite or titanium are popular in contemporary otologic practice. A buffer of cartilage is often interposed between a PORP/TORP and the TM to decrease the potential for extrusion of the prosthesis.

When the ossicles are diseased, it is also possible to perform a tympanoplasty without the use of an interposed incus, PORP or TORP. In a *stapes columella tympanoplasty*, the reconstructed TM is advanced in a medial direction so that it is in direct contact with the capitulum of the stapes (bottom left panel in Fig. 9.4). In a *type IV tympanoplasty*, the stapes footplate is allowed to remain directly exposed to incoming sound from the ear canal, and a tissue graft is placed to acoustically shield the round window membrane from sound (bottom right panel in Fig. 9.4). The air space enclosed between the acoustic shield and the round window is called the cavum minor. The cavum minor is aerated via the Eustachian tube.

The choice of tympanoplasty procedure is dictated by the extent of deficiency of the TM and ossicles caused by disease, as well as by the nature of ancillary mastoid surgery, preference of the surgeon, and availability of synthetic materials.

9.3.2 Mastoidectomy

Mastoidectomy refers to a surgical procedure to open the mastoid and drill away the mastoid air cells with the objective of removing infection, or accessing sequestered anatomical areas such as the epitympanum. A common indication for mastoidectomy is COM. There are two broad types of mastoidectomy procedures (Fig. 9.5). A *canal wall-up mastoidectomy* consists of removal of mastoid air cells while preserving the posterior wall of the bony external auditory canal. On the other hand, a *canal wall-down mastoidectomy* consists of removal of the posterior bony canal wall so that the external auditory canal, mastoid, and epitympanum become one common cavity. The bony canal wall is surgically removed to the level of the facial nerve. The size of the resulting mastoid cavity (called "mastoid bowl") is often reduced by obliterating it with tissue such as bone dust, muscle, and fat. Mastoidectomy is often combined with tympanoplasty and the two procedures may be performed at the same time or sequentially.

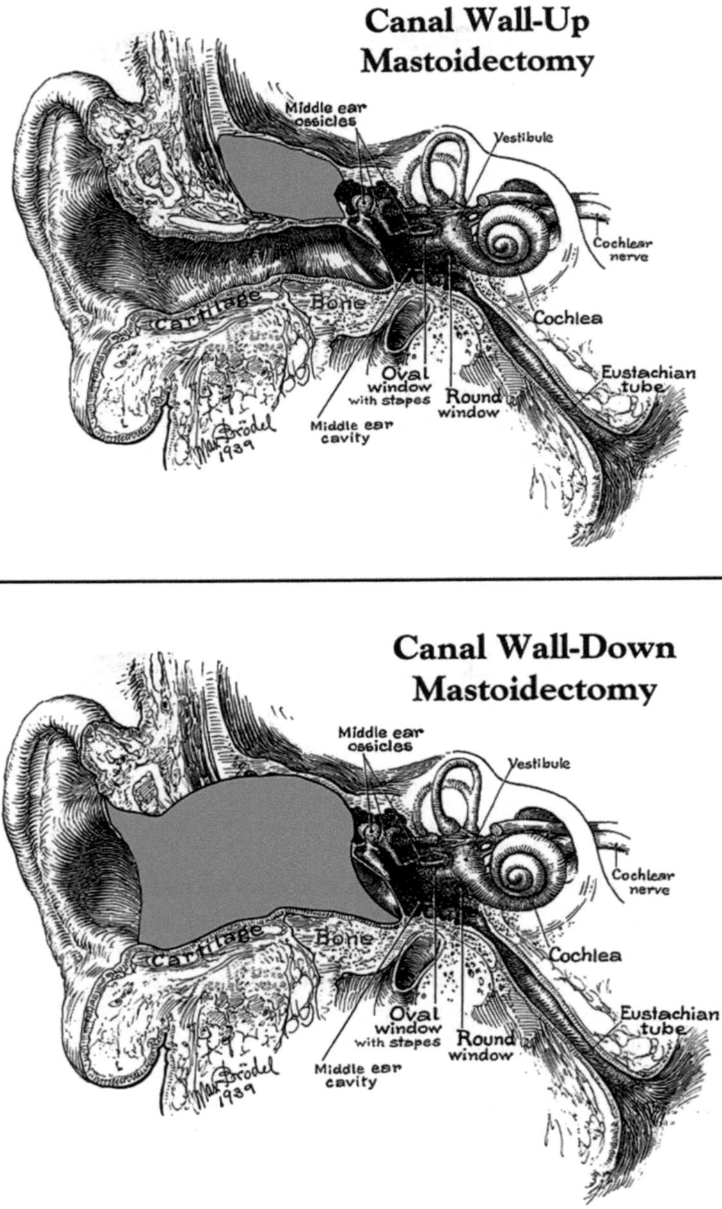

Fig. 9.5 Mastoidectomy. Diagrammatic representation of canal wall-up and canal wall-down mastoidectomy. The uniform *gray areas* show the regions in which the surgeon has opened the mastoid to remove disease. In the canal-wall down procedure, this area is contiguous with the widened ear canal. See also descriptions in text

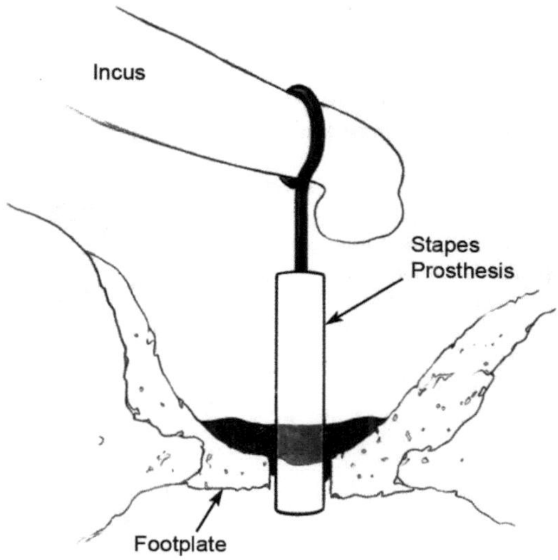

Fig. 9.6 Schematic diagram of stapedotomy, where the superstructure has been removed, an opening made in the footplate, and sound transmission restored by placement of a stapes prosthesis from the incus to the oval window. The *black area* at the *lower end* of the prosthesis represents tissue (e.g., fat) used to seal the opening in the footplate around the piston

9.3.3 Stapedectomy

The term *stapedectomy* is used broadly to include all procedures where the stapes is removed and replaced by a prosthesis. The most common reason to perform a stapedectomy is when the stapes bone is ankylosed due to otosclerosis. Other indications include a stapes fixed by tympanosclerosis (caused by COM), or a stapes that has been compromised due to trauma. When the entire footplate is removed, the procedure is termed a total stapedectomy. When a small opening is made in a portion of the footplate only, the procedure is more accurately called a stapedotomy (Fig. 9.6). Note: In this chapter, the term stapedectomy is used to include both total stapedectomy and stapedotomy procedures. A large variety of stapes prostheses have been described; implants made of Teflon and titanium are in common use.

9.4 Acoustics and Mechanics of Reconstructed Middle Ears

9.4.1 Role of Aeration

Aeration of the middle ear (including the round window) is *critical* to the success of any tympanoplasty procedure. Aeration allows the TM, ossicles, and round window to move. Clinical experience has shown that nonaerated ears often demonstrate 40–60 dB air–bone gaps (Merchant et al. 1998a) because (1) ossicular coupling is greatly reduced (see Voss et al., Chap. 4 for a description of ossicular coupling), and

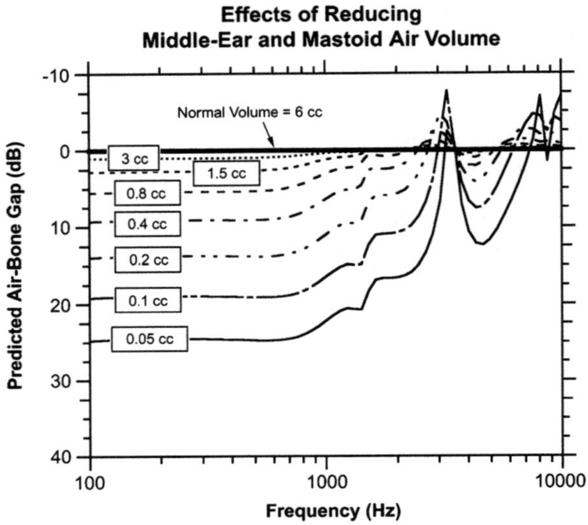

Fig. 9.7 Model predictions of the effects of reducing the volume of the middle ear and mastoid. The normal baseline volume is taken to be 6 cc. Note that reduction of the volume to 0.4 cc is predicted to result in an air bone gap less than 10 dB. Volumes smaller than 0.4 cc are predicted to lead to progressively larger gaps (Modified from Rosowski and Merchant 1995.)

(2) stapes motion is reduced because the round window membrane (which is coupled to the stapes by incompressible cochlear fluids) cannot move freely.

How much air is necessary behind the TM (i.e., within the middle ear and mastoid)? Model analyses of the effects of varying the volume of the middle ear and mastoid predict an increasing low frequency hearing loss as air volume is reduced (Rosowski and Merchant 1995) (Fig. 9.7). The normal, average volume of the middle ear and mastoid is 6 cc; a combined middle ear and mastoid volume of 0.5 cc is predicted to result in a 10 dB conductive hearing loss. Volumes smaller than 0.5 cc should lead to progressively larger gaps, whereas increases in volume above about 1.0 cc are predicted to provide little additional acoustic benefit. Experimental studies using a human temporal bone preparation in which the middle ear and mastoid volume was reduced progressively show results consistent with the model prediction (Gyo et al. 1986; Whittemore et al. 1998).

Static air pressure within the middle ear space is another parameter that can influence middle ear mechanics. Animal studies and measurements of ossicular motion in human temporal bones have demonstrated that middle ear static pressure can have different effects on sound transmission at different frequencies (Murakami et al. 1997). Generally, trans-TM static pressure differences produce decreases in sound transmission through the middle ear for frequencies less than 1,000 Hz, and have less effect at higher frequencies. Also, the effects of such static pressure differences are asymmetric, with larger decreases observed when the middle ear pressure is negative relative to that in the ear canal. The mechanisms by which pressure changes reduce middle ear sound transmission are not well defined, and possible sites of pressure sensitivity include the TM, annular ligament, incudo–malleal joint, and suspensory ligaments of the ossicles. Although some of these structures are drastically altered by tympanoplasty, the acoustic effects of negative and positive middle ear static pressure in reconstructed ears have not been characterized.

9.4.2 TM Reconstruction

Clinical observations indicate that the surgical techniques used to repair the TM can lead to good hearing results with resolution of the conductive hearing loss (Merchant et al. 2003a). However, in up to 30 % of patients there is a residual air bone gap that may vary from 5 to 35 dB even in the presence of an aerated middle ear (Merchant et al. 2000). Although clinical observations suggest that restoration or preservation of the normal TM anatomy can lead to good hearing results, research is needed to define the optimum acoustic and mechanical properties of reconstructed TM. For example: (1) although O'Connor et al. (2008) found that radial fibers are important for sound transmission greater than 4 kHz in the normal TM, little is known of the mechanical significance of the arrangements of structural fibers in reconstructed TM. (2) Although it has been argued that the conical shape of the normal TM plays an important role in middle ear function (Tonndorf and Khanna 1972; Fay et al. 2006), the possible effects of changes in TM shape on postoperative hearing results are not understood. (3) Although many existing models of TM function have been shown to fit some of the available data (Funnell and Decraemer 1996), there are wide differences in the structure of these models, and little effort has been made to compare their significant differences and similarities. With a few exceptions (Tuck-Lee et al. 2008), these models generally have not been applied to the reconstructed TM. Better understanding of the features of TM structure that are critical to its function should lead to improved methods for TM reconstruction.

9.4.3 Ossicular Reconstruction with Incus Interposition, PORPs, and TORPs

The hearing results after ossicular reconstructions vary widely with air–bone gaps ranging from 0 to 60 dB. A large number of studies have evaluated the influence of acoustical and mechanical properties of an ossicular prosthesis including its stiffness, mass, and position; the tension imposed by the prosthesis on the TM and annular ligament; and mechanical features associated with coupling of the prosthesis to the TM and stapes (Goode and Nishihara 1994; Merchant et al. 1998a).

In general, the stiffness of a prosthesis will not be a significant factor as long as the stiffness is much greater than that of the stapes footplate-cochlear impedance. For clinical purposes, prostheses made of ossicles (such as an interposed incus) and many synthetic materials generally meet this requirement.

Model analysis (Rosowski and Merchant 1995) and experimental data (Gan et al. 2001; Bance et al. 2007a) suggest that an increase in ossicular mass does not cause significant detriment in middle ear sound transmission. Increases up to 16 times the ossicle mass are predicted to cause less than 10 dB conductive loss and only at frequencies greater than 1,000 Hz.

The positioning of a prosthesis appears to be important to its function (Murugasu et al. 2005). Measurements in human temporal bone preparations suggest that the angle between the stapes and a prosthesis should be less than 45° for optimal sound transmission (Vlaming and Feenstra 1986; Nishihara and Goode 1994), and that the neck of the malleus is a good contact point for TORPs (Puria et al. 2005). There is also evidence that some variations in positioning produce only small changes. For example, while it is ideal to attach a prosthesis to the manubrium, experimental data show that acceptable results can occur with a prosthesis placed against the posterior–superior quadrant of the tympanic membrane as long as approximately 3–4 mm of the prosthesis' diameter contacts the TM (Goode and Nishihara 1994; Bance et al. 2007b).

The tension the prosthesis creates in the middle ear, which is generally a function of prosthesis length, appears critical in determining the hearing result (Morris et al. 2004). The mechanical impedance of biological structures is inherently nonlinear, and the TM and annular ligament act as linear elements only over the range of small motions (less than 10 μm) associated with physiological sound levels. Larger displacements of the ligament and TM (by a prosthesis that is too long) would stiffen these structures, resulting in a reduction in tympano-ossicular motion and an air–bone gap. Currently, tension cannot be assessed intraoperatively in an objective fashion; a reliable objective test of the tension would be useful to the otologic surgeon.

Coupling refers to how well a prosthesis adheres to the footplate or TM, and the degree of coupling will determine whether or not there is slippage in sound transmission at the ends of a prosthesis. Thus, a prosthesis transmits sound effectively only if there is good coupling at both ends. Clinical observations indicate that it is rare to obtain a firm union between a prosthesis and the stapes footplate. Hence, inadequate coupling at the prosthesis–footplate joint may be an important cause of a persistent postoperative air–bone gap. The physical factors that control coupling have not been determined in a quantitative manner, and further study of this parameter is warranted.

9.4.4 Stapes Columella Tympanoplasty

Large air–bone gaps (40–60 dB) occur as a result of stapes fixation, nonaeration of the middle ear, or both (Mehta et al. 2003; Merchant et al. 2003b). When the stapes is mobile and the middle ear is aerated, the average postoperative air–bone gap is on the order of 20–25 dB, suggesting that there is little middle ear sound pressure gain occurring through the reconstruction. Experimental and clinical studies of the stapes columellar reconstruction have shown that interposing a thin disk of cartilage between the graft and the stapes head improves hearing in the lower frequencies by 5–10 dB. It has been hypothesized that the cartilage acts to increase the "effective" area of the graft that is coupled to the stapes, which leads to an increase in the middle ear gain of the reconstructed ear (Merchant et al. 2003b).

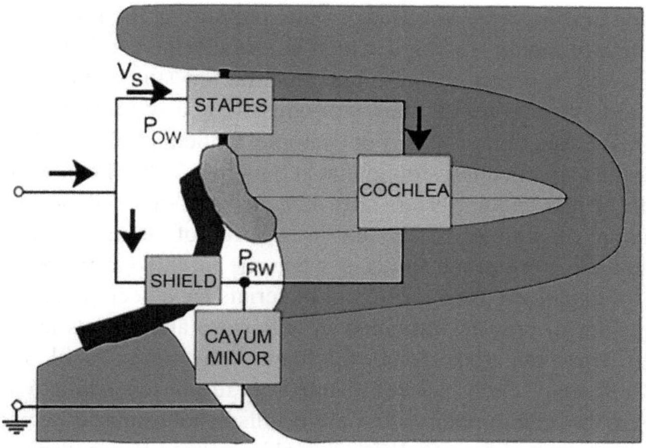

Fig. 9.8 Model of type IV tympanoplasty described by Peake et al. (1992). Stapes velocity, VS, depends on the sound-pressure difference, PWD, between the oval window sound pressure, POW, and the round window sound pressure, PRW. PWD is determined by the acoustic impedance of four structures represented by the *gray boxes*: impedance of the stapes footplate and annular ligament, impedance of the cochlea, impedance of the acoustic graft-shield, and impedance of the cavum minor. See also details in the text

9.4.5 Type IV Tympanoplasty

Peake et al. (1992) described a lumped element model of the type IV reconstruction as shown in Fig. 9.8. Predictions of the model under different conditions were tested against experimental data from a cadaveric temporal bone preparation and against clinical data (Merchant et al. 1995, 1997). A good match was seen between the predictions and the data. The model suggested that an "optimum" type IV reconstruction, as defined by normal footplate mobility, a sufficiently stiff acoustic graft-shield, and adequate aeration of the round window would result in maximum acoustic coupling with a predicted residual air–bone gap of only 20–25 dB. Such an optimum result is indeed consistent with the best type IV hearing results seen clinically. The analyses also predicted that decreased footplate mobility, inadequate acoustic shielding or inadequate round window aeration can lead to hearing losses as large as 60 dB.

9.4.6 Biologic and Pathologic Considerations in Tympanoplasty

In the case of tympanoplasty, an important determinant of the hearing outcome is the mechanical and acoustical adequacy of the reconstruction. In addition, the biology and pathology of COM can have significant effects on hearing results. For example, postoperative mucosal fibrosis, neo-osteogenesis, formation of adhesions and development of negative static pressure in the middle ear can occur over the course of months or years, which in turn can have a detrimental effect on the hearing result. It is instructive to note that the few studies in the literature that assess long-term hearing results show a progressive and systematic decline in initial hearing gain as a function of time. For example, Colletti et al. (1987), in a study of 832 ossiculoplasty procedures, found that at 6 months 77% of ears had an air–bone gap of 20 dB or less, but at 5 years only 42% had such a small air–bone gap.

Proliferation of fibrous tissue and formation of adhesions are significant problems that are more prone to occur when the middle ear mucosa is diseased, removed, or traumatized. Many different materials have been placed in the middle ear in an attempt to prevent formation of adhesions and fibrous tissue (Merchant et al. 1998b). These materials include Gelfoam™, hyaluronic acid, Silastic™, and Teflon™. Gelfoam™ elicits a host inflammatory response leading to its resorption. In some cases, this inflammatory response results in adhesions, especially when the middle ear mucosa is deficient. Further, Gelfoam™ is resorbed within 2 weeks, which is probably insufficient time for mucosal regeneration to occur. Hyaluronic acid is somewhat more difficult to handle than Gelfoam™ and is also absorbed before mucosal regeneration is likely to be completed. Silastic™ and Teflon™ sheeting are relatively inert but they are not resorbed and can sometimes extrude. On occasion, they become engulfed by fibrous tissue leading to a nonaerated ear. Hence, none of the currently available spacer materials is ideal. What is needed is a material that will remain in place for several weeks to allow sufficient time for mucosal regeneration and will then undergo degradation and resorption so the ear can become aerated without fibrosis.

While rates of successful closure of TM perforations are uniformly high, in excess of 90%, a small number of grafted TMs show undesirable pathological changes including proliferation of fibrous tissue and thickening, resorption and excessive thinning, and lack of epithelialization with resulting discharge. The factor or factors controlling such responses are not well understood at present. Similarly, histopathologic responses of the ear to various ossicular grafts and prostheses play an important role in determining outcome of tympanoplasty; these include a foreign body giant cell response to synthetic materials that may lead to breakdown and resorption of parts of an implant (Bahmad and Merchant 2007).

Two significant causes for long-term failure of tympanoplasty are total or partial nonaeration of the middle ear and development of negative static pressure (as previously described). Nonaeration of the middle ear is usually due to Eustachian tube dysfunction and results in TM graft atelectasis, middle ear effusion, fibrocystic sclerosis of the middle ear, or a combination of these changes. Some postoperative ears that are aerated have a tendency to develop negative static pressure in the middle ear. Over the long term, this negative pressure leads to retraction and atelectasis of the reconstructed TM and functional compromise, as well as a predisposition to displacement or extrusion of ossicular prostheses. The negative pressure can also lead to recurrent cholesteatoma. The latter problem is a disadvantage for canal-wall up procedures relative to canal-wall down mastoidectomy.

9.4.7 *Mastoidectomy*

In a canal-wall-down mastoidectomy, the bony tympanic annulus and much of the ear canal is removed, and the TM graft is typically placed onto the stapes head, as well as onto the facial ridge and medial attic wall. This results in a significant reduction in the size of the residual middle ear air space. However, as long as this air space is greater than or equal to 0.5 cc, the resultant loss of sound transmission should be less than 10 dB (see Sect. 9.4.1). Since the average volume of the tympanic cavity is 0.5–1.0 cc (Molvaer et al. 1978), a canal-wall down procedure should create no significant acoustic detriment (in comparison to canal wall-up procedures), so long as the middle ear is aerated. Indeed, clinical studies comparing the acoustic results of canal wall-down verses canal wall-up mastoidectomy have shown no significant differences in hearing between the two conditions (Colletti et al. 1987; Merchant et al. 2003a).

A canal wall-down procedure also results in the creation of a large air space lateral to the eardrum, that is, the air space within the mastoid bowl including the external auditory canal. This mastoid bowl and ear canal air space generates resonances that can influence middle ear sound transmission favorably or unfavorably (Goode et al. 1977). The structure–function relationships between the size and shape of the mastoid cavity, and cavity resonances have not been well defined. An improved understanding of this issue may help otosurgeons to configure mastoid cavities in ways that are acoustically beneficial.

9.4.8 *Stapedectomy*

The output of the middle ear can be quantified by the "volume velocity" of the stapes (Rosowski and Merchant 1995), where volume velocity is the product of stapes linear velocity and the area of the stapes footplate. After a stapedotomy, the effective area of the footplate is reduced to the area of the prosthesis, thereby reducing the volume velocity produced by a given stapes linear velocity. The reduction in effective footplate area also reduces the area of the cochlear fluid over which the force generated by the stapes is applied. Whereas the reduced footplate area leads to a local increase in pressure over the surface of the prosthesis, the average pressure at the cochlear entrance is reduced. The reduction in stapes volume velocity and cochlear sound pressure lead to a decrease in ossicular coupling and the development of an air–bone gap. The smaller the area of the stapes prosthesis is, the greater the air–bone gap. Model predictions of the relationship between piston diameter and residual air–bone gap after stapedotomy were made using a simple lumped element model of the middle ear (Rosowski and Merchant 1995). This analysis predicted the 0.8 mm piston diameter will produce 5 dB better hearing results than the 0.6 mm piston and 10 dB better results than the 0.4 mm piston. These predictions are in general agreement with (1) experimental temporal bone data (Honda and Goode 2004), (2) results of finite element modeling data (Bohnke and Arnold 2007), and (3) clinical observations (Teig and Lindeman 1999; Laske et al. 2011). The predictions made in the simple lumped element model assumed that the effective vibrating footplate surface area after a stapedotomy is no more than the area of the lower end of the prosthesis. In cases of partial or total stapedectomy with placement of a tissue graft and a stapes prosthesis, the effective vibrating surface may be greater than the area of the prosthesis alone, and the model predictions may overestimate the air–bone gap.

9.5 Results After Middle Ear Surgery

Tympanoplasty and mastoidectomy surgery for COM: Such surgery is quite successful in controlling infection and preventing recurrent disease, with success rates in excess of 80–90%. Postoperative hearing results vary widely, depending on extent of TM and ossicular lesions, mucosal disease, cholesteatoma, and Eustachian tube function. When only TM reconstruction is needed, 80–90% of patients will have an air–bone gap of 20 dB or less. When ossicular reconstruction is necessary, long-term closure of the air–bone gap to less than 20 dB occurs in 40–70% of cases when the stapes is intact, and in only 20–55% in which the stapes superstructure is missing (Merchant et al. 1998a).

Stapedectomy surgery for otosclerosis: Results are uniformly good, with long-term closure of the air–bone gap to less than or equal to 20 dB reported in more than 95% of patients.

9.6 Future Directions

There are a number of areas where future research could optimize or improve results after middle ear reconstruction. Some of these areas were mentioned in the earlier sections. These include better understanding of structure–function correlations for reconstructed TMs, better ways to deal with effects of tension in ossicular reconstruction, and improved ways of coupling TORPs to the stapes footplate.

Research efforts have focused on utilizing growth factors and similar drugs to stimulate the closure of TM perforations without the need for surgical repair (Ma et al. 2002); the goal is to be able to apply the treatment topically and induce the edges of the perforation to heal over the opening.

The recent development of real time opto-electronic laser holographic techniques to study motion of the TM offers the prospect of improving our understanding of TM reconstructions and the issue of tension (Rosowski et al. 2009; Cheng et al. 2010). Zahnert and colleagues are developing a novel technique for better coupling of TORPs to the footplate using prostheses that are impregnated

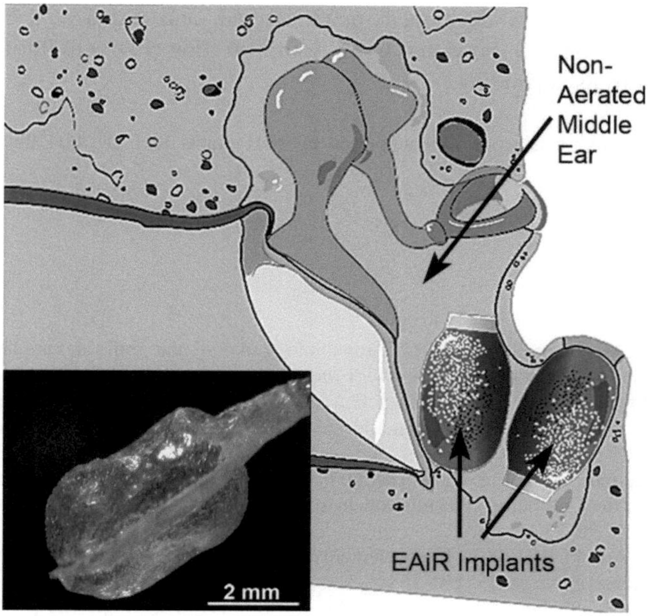

Fig. 9.9 Boston EAiR implant as possible treatment for nonaerated middle ears. See text for explanation

with compounds such as bone morphogenetic protein that induce formation of new bone (Neudert et al. 2010). Another area for future research involves development of ossicular prostheses that would "self-adjust" to changes in position of the TM and changes in static pressure within the middle ear over time, as described by Goode and colleagues (Zhao et al. 2005; Yamada and Goode 2010).

Nonaeration of the middle ear after tympanoplasty surgery is another type of problem for which optimal solutions do not exist in current otologic practice. Such nonaeration is often due to a combination of deposition of fibrous tissue and fluid related to mucosal disease and/or tubal dysfunction. Our research group has been involved in the development of a middle ear implant termed "the Boston EAiR implant" that would restore hearing in such nonaerated ears by providing a semi-permanent compressible air pocket (Fig. 9.9). The EAiR implant is a small air-filled balloon with thin polymer walls. Each implant is an ovoid cylinder 3 × 2 mm in width and 5 mm in length, with a compressibility equivalent to that of an air volume of approximately 20 μL. The implant is designed for placement in the middle ear and is expected to restore hearing by introducing a compressible pocket that allows the TM, ossicles, and round window to vibrate in response to sound. A successful implant must be *biocompatible* (nontoxic to the tissues), *bioinert* (not biodegradable by the host response of the middle ear), *compressible* by sound vibrations, and provide a good *barrier* (it must be impermeable to water and body fluids and prevent loss of air from within the implant). EAiR implants meeting these criteria have been successfully developed, manufactured, and tested at a bench level. Approval from the U.S. Food and Drug Administration (FDA) to initiate prospective clinical trials is pending.

Acknowledgments This work was supported by NIH grants R01 DC 004798 and U24 DC 011943.

References

Bahmad, F., & Merchant, S. N. (2007). Histopathology of ossicular grafts and implants in chronic otitis media. *Annals of Otology Rhinology and Laryngology,* 116, 181–191.

Bance, M., Morris, D. P., & Van Wijhe, R. (2007a). Effects of ossicular prosthesis mass and section of the stapes tendon on middle ear transmission. *Journal of Otolaryngology,* 36, 113–119.

Bance, M., Campos, A., Wong, L., Morris, D. P., & van Wijhe, R. (2007b). How does prosthesis head size affect vibration transmission in ossiculoplasty? *Otolaryngology Head and Neck Surgery*, 137, 70–73.

Bohnke, F., & Arnold, W. (2007). Finite element model of the stapes-inner ear interface. *Advances in Otorhinolaryngology*, 65, 150–154.

Brackmann, D.E., Shelton, C., & Arriaga, M. (Eds.). (2010). *Otologic Surgery*, 3rd edition. Philadelphia: Saunders Elsevier.

Cheng, J. T., Aarnisalo, A. A., Harrington, E., Hernandez-Montes, M. D. S., Furlong, C., Merchant, S. N., & Rosowski, J. J. (2010). The motion of the human tympanic membrane measured with stroboscopic holography. *Hearing Research*, 263, 66–77.

Colletti, V., Fiorino, F. G., & Sittoni, V. (1987). Minisculptured ossicle grafts versus implants: Long term results. *American Journal of Otology,* 8, 553–559.

Fay, J. P., Puria, S., & Steele, C. R. (2006). The discordant eardrum. *Proceedings of the National Academy of Sciences of the U S A,* 103, 19743–19748.

Funnell, W. R. J., & Decraemer, W. M. (1996). On the incorporation of moiré shape measurements in finite-element models of the cat eardrum. *Journal of the Acoustical Society of America,* 100, 925–932.

Gan, R. Z., Dyer, R. K., Wood, M. W., & Dormer, K. J. (2001). Mass loading on the ossicles and middle ear function. *Annals of Otology Rhinology and Laryngology*, 110, 478–485.

Goode, R. L., & Nishihara, S. (1994). Experimental models of ossiculoplasty. In E. Monsell (Ed.), *Ossiculoplasty*: *Otolaryngology Clinics of North America* (pp. 663–675). Philadelphia: Saunders.

Goode, R. L., Friedrichs, R., & Falk, S. (1977). Effect on hearing thresholds of surgical modification of the external ear. *Annals of Otology Rhinology and Laryngology*, 86, 441–451.

Gyo, K., Goode, R. L., & Miller, C. (1986). Effect of middle-ear modification on umbo vibration-human temporal bone experiments with a new vibration measuring system. *Archives of Otolaryngology Head and Neck Surgery*, 112, 1262–1268.

Handzel, O., & McKenna, M. J. (2010). Surgery for otosclerosis. In A. J. Gulya, L. B. Minor, & D. S. Poe (Eds.), *Glasscock-Shambaugh surgery of the ear*, 6th edition (pp. 529–546). Shelton, CT: People's Medical Publishing House-USA.

Honda, N., & Goode, R. L. (2004). The acoustic evaluation of stapedotomy using a temporal bone otosclerosis model. In K. Gyo, & H. Wada (Eds.), *Middle ear mechanics in research and otology* (pp. 203–208). Singapore: World Scientific.

Laske, R. D., Röösli, C., Chatzimichalis, M. V., Sim, J. H., & Huber, A. M. (2011). The influence of prosthesis diameter in stapes surgery: A meta-analysis and systematic review of the literature. *Otology and Neurotology*, 32, 520–528.

Ma, Y., Zhao, H., & Zhou X. (2002). Topical treatment with growth factors for tympanic membrane perforations: Progress towards clinical application. *Acta Oto-Laryngologica*, 122, 586–599.

Mehta, R. P., Ravicz, M. E., Rosowski, J. J., & Merchant, S. N. (2003). Middle-ear mechanics of type III tympanoplasty (stapes columella): I. Experimental studies. *Otology and Neurotology*, 24, 176–185.

Merchant, S. N. (2005). Ossiculoplasty and tympanoplasty in chronic otitis media. In J. B. Nadol & M. J. McKenna (Eds.), *Surgery of the ear and temporal bone*, 2nd edition (pp. 305–324). Philadelphia: Lippincott, Williams & Wilkins.

Merchant, S. N. & Nadol, J. B., Jr. (Eds.). (2010). *Schuknecht's pathology of the ear*, 3rd edition. Shelton, CT: People's Medical Publishing House-USA.

Merchant, S. N., Rosowski, J. J., & Ravicz, M. E. (1995). Middle-ear mechanics of type IV and type V tympanoplasty. II. Clinical analysis and surgical implications. *American Journal of Otology*, 16, 565–575.

Merchant, S. N., Rosowski, J. J, & Ravicz, M. E. (1997). Experimental investigation of the mechanics of type IV tympanoplasty. *Annals of Otology Rhinology and Laryngology*, 106, 49–60.

Merchant, S. N., Ravicz, M. E., Voss, S. E., Peake, W. T., & Rosowski, J. J. (1998a). Middle-ear mechanics in normal, diseased and reconstructed ears. *Journal of Laryngology and Otology*, 112, 715–731.

Merchant, S. N., McKenna, M. J., & Rosowski, J. J. (1998b). Current status and future challenges of tympanoplasty. *European Archives of Oto-Rhino-Laryngology*, 255, 221–228.

Merchant, S. N., Whittemore, K., Poon, B., Lee, C. Y., & Rosowski, J. J. (2000). Clinical measurements of tympanic membrane velocity using laser Doppler vibrometry: Preliminary results, methodological issues and potential applications. In J. J. Rosowski & S. N. Merchant (Eds.), *The function and mechanics of normal, diseased and reconstructed middle ears* (pp. 367–381). Amsterdam: Kugler Publications.

Merchant, S. N., Rosowski, J. J., & McKenna, M.J. (2003a). Tympanoplasty. *Operative Techniques in Otolaryngology–Head and Neck Surgery*, 14, 224–236.

Merchant, S. N., McKenna, M. J., Mehta, R. P., Ravicz, M. E., & Rosowski, J. J. (2003b). Middle-ear mechanics of type III tympanoplasty (stapes columella): II. Clinical studies. *Otology and Neurotology*, 24, 186–194.

Molvaer, O., Vallersnes, F., & Kringelbotn, M. (1978). The size of the middle ear and the mastoid air cell. *Acta Oto-Larygologica*, 85, 24–32.

Morris, D. P., Bance, M., van Wijhe, R. G., Kiefte, M., & Smith, R. (2004). Optimum tension for partial ossicular replacement prosthesis reconstruction in the human middle ear. *Laryngoscope*, 114, 305–308.

Murakami, S., Gyo, K., & Goode, R. L. (1997). Effect of middle ear pressure change on middle ear mechanics. *Acta Oto-Larygologica*, 117, 390–395.

Murugasu, E., Puria, S., & Roberson, J. B. (2005). Malleus-to-footplate versus malleus-to-stapes-head ossicular reconstruction prostheses: Temporal bone pressure gain measurements and clinical audiological data. *Otology and Neurotology*, 26, 572–582.

Nadol, J. B., & McKenna, M. J. (Eds.). (2005). *Surgery of the ear and temporal bone*, 2nd edition. Philadelphia: Lippincott, Williams and Wilkins.

Neudert, M., Beleites, T., Ney, M., Kluge, A., Lasurashvili, N., Bornitz, M., Scharnweber, D., & Zahnert, T. (2010). Osseointegration of titanium prostheses on the stapes footplate. *Journal of the Association for Research in Otolaryngology*, 11, 161–171.

Nishihara, S., & Goode, R. L. (1994). Experimental study of the acoustic properties of incus replacement prostheses in a human temporal bone model. *American Journal of Otology* 15, 485–494.

O'Connor, K. N., Tam, M., Blevins, N. H., & Puria, S. (2008). Tympanic membrane collagen fibers: A key to high-frequency sound conduction. *Laryngoscope*, 118, 483–490.

Peake, W. T., Rosowski, J. J, & Lynch, T. J. III. (1992). Middle ear transmission: Acoustic versus ossicular coupling in cat and human. *Hearing Research*, 57, 245–268.

Puria, S., Kunda, L. D., Roberson, J. B., & Perkins, R. C. (2005). Malleus-to-footplate ossicular reconstruction prosthesis positioning: Cochleovestibular pressure optimization. *Otology and Neurotology*, 26, 368–379.

Rosowski, J. J., & Merchant, S. N. (1995). A mechanical and acoustical analysis of middle-ear reconstruction. *American Journal of Otology,* 16, 486–497.

Rosowski, J. J, Cheng, J. T., & Ravicz, M. E., Hulli, N., Harrington, E. J., Hernandez-Montes, M. S., & Furlong, C. (2009). Computer-assisted time-averaged holography of the motion of the surface of the tympanic membrane with sound stimuli of 0.4 to 25 kHz. *Hearing Research,* 253, 83–96.

Sismanis, A. (2010). Tympanoplasty—tympanic membrane repair. In A. J. Gulya, L. B. Minor, & D. S. Poe (Eds.), *Glasscock-Shambaugh surgery of the ear*, 6th edition (pp. 465–488). Shelton, CT: People's Medical Publishing House-USA.

Sismanis, A., & Poe, D. S. (2010). Ossicular chain reconstruction. In A. J. Gulya, L. B. Minor, & D. S. Poe (Eds.), *Glasscock-Shambaugh surgery of the ear,* 6th edition (pp. 489–500). Shelton, CT: People's Medical Publishing House-USA.

Teig, E., & Lindeman, H. H. (1999). Stapedectomy piston diameter: Is bigger better? *Otorhino-laryngologica Nova*, 9, 252–256.

Tonndorf, J., & Khanna, S. M. (1972). Tympanic membrane vibrations in human cadaver ears studied by time-averaged holography. *Journal of the Acoustical Society of America,* 52, 1221–1233.

Tuck-Lee, J. P., Pinsky, P. M., Steele, C. R., & Puria, S. (2008). Finite element modeling of acousto-mechanical coupling in the cat middle ear. *Journal of the Acoustical Society of America,* 124, 348–362.

Vlaming, M. S. M. G., & Feenstra, L. (1986). Studies on the mechanics of the reconstructed human middle ear. *Clinical Otolaryngology,* 11, 411–422.

Whittemore, K. R., Merchant, S. N., & Rosowski, J. J. (1998). Acoustic mechanisms: Canal wall-up versus canal wall-down mastoidectomy. *Otolaryngology Head and Neck Surgery*, 118, 751–761.

Wullstein, H. (1956). The restoration of the function of the middle ear in chronic otitis media. *Annals of Otology Rhinology and Laryngology,* 65, 1020–1041.

Yamada, H., & Goode, R. L. (2010). A self-adjusting ossicular prosthesis containing polyurethane sponge. *Otology and Neurotology*, 31, 1404–1408.

Zhao, S., Hato, N., & Goode, R. L. (2005). Experimental study of an adjustable-length prosthesis in a temporal bone model. *Acta Oto-Laryngologica*, 125, 33–37.

Chapter 10
Middle Ear Hearing Devices

Sunil Puria

Keywords Battery • Electromagnetic • Hearing aids • Implantable hearing device • Microphone • Multiband compressor • Nonimplantable hearing device • Photonic • Piezoelectric • Sensorineural hearing impairment • Transducer

10.1 Introduction

People like being able to hear—hearing impairment can lead to significant communication problems, especially in noisy places, with spouses and colleagues typically being the first ones to notice. Often a hearing test reveals a sensorineural hearing impairment as opposed to a problem with middle ear conduction; and if the sensorineural hearing impairment is not profound, traditional air-conduction acoustic hearing aids are typically prescribed. Current top-of-the-line acoustic hearing aids are technological wonders with respect to their size, computing power, and number of features, and as a result they are effective for many situations.

However, multiple surveys have shown that the two biggest downsides of acoustic hearing aids are their reduced effectiveness for understanding speech in noisy situations and their less-than-ideal perceived sound quality (Kochkin 2002, 2009). This is true regardless of cost, although the more expensive hearing aids tend to perform better. As a result of these shortcomings, a percentage of people reject and return their hearing aids after first trying them. This, along with people never trying hearing aids to begin with, results in a low adoption rate of less than 25%.

S. Puria (✉)
Departments of Mechanical Engineering and Otolaryngology–Head and Neck Surgery, Stanford University, 496 Lomita Mall, Stanford, CA 94305, USA

EarLens Corporation, 200 Chesapeake Drive, Redwood City, CA 94063, USA
e-mail: puria@stanford.edu

S. Puria et al. (eds.), *The Middle Ear: Science, Otosurgery, and Technology*, Springer Handbook of Auditory Research 46, DOI 10.1007/978-1-4614-6591-1_10,

Hearing impairment thus remains one of the most widespread untreated chronic conditions (Donahue et al. 2010).

In the field of ophthalmology there are a variety of options for correcting imperfect vision. These include simple but effective low-cost eyeglasses, more expensive progressive and transition eyeglasses, contact lenses, and in recent times a host of laser surgical procedures that have proven to be highly successful. The idea of having a comparable range of options to restore hearing, however, is not yet a broadly accepted notion.

Nonetheless, a significant amount of work has been going into the development of devices that restore hearing by directly vibrating the structures of the middle ear or the cochlea, rather than by simply presenting amplified sound in the ear canal. These devices are collectively referred to here as middle ear hearing devices (MEHDs). Most of these devices are implantable hearing devices (IHDs) that require surgery. For the purposes of this chapter, the IHD designation will refer exclusively to the MEHD varieties that require full or partial surgical implantation, although nonimplantable MEHD devices, whose mechanism of action is through nonacoustic direct vibration of the middle ear, are also described.

The basic configurations of acoustic hearing aids and MEHDs are all very similar; they feature an input transducer (e.g., a microphone), a sound processor (typically a multiband compressor) for converting the input signal into an amplified signal suited to the patient's hearing loss, a battery, and an output transducer. One of the primary differences between acoustic hearing aids and MEHDs is in the output transducer: acoustic hearing aids use a tiny loudspeaker, while MEHDs typically use a tiny mechanical actuator. The latter approach has the potential to offer wider bandwidth, an increased gain margin due to reduced feedback, and better sound quality due to reduced distortion. Other surgical hearing devices not covered in this chapter include bone-conduction devices, cochlear implants, and brain-stem implants.

The rest of this chapter is organized as follows. After discussing the pros and cons of MEHDs in Sect. 10.2 and describing the different anatomical sites to which output transducers can be coupled (Sect. 10.3), a new classification system for current and future transducers is introduced based on the number of required anatomical connection points and how those points are linked to one another (Sect. 10.4). Section 10.5 outlines the different methods for transferring power and signal to the output transducer, while Sect. 10.6 describes the various methods and coupling sites for input transduction. The subsequent three Sects. (10.7, 10.8, and 10.9) describe commercial and academic efforts toward the development of different types of partially implantable (PI) and totally implantable (TI) hearing devices, while Sect. 10.10 describes a class of nonimplantable (NI) contact transducers that mechanically actuate the tympanic-membrane surface from the ear-canal side and thus do not require surgery. The functional gains reported for many of these different device classes are compared in Sect. 10.11, followed by concluding remarks in Sect. 10.12.

10.2 Pros and Cons of MEHDs

Why would someone choose an MEHD as opposed to a conventional acoustic hearing aid? For one, there is now evidence that MEHDs can provide better hearing outcomes relative to conventional acoustic hearing aids under some conditions, with a significant number of papers reporting that IHDs provide better speech discrimination relative to acoustic hearing aids under quiet conditions (reviewed by Tysome et al. 2010).

Although it has been shown that the perceived sound quality for both speech and music is improved if one extends the frequency range of a hearing device to 0.1–10 kHz (Moore and Tan 2003), open-canal acoustic hearing aids typically only provide amplification in the 1–4 kHz range, and closed-canal devices provide amplification only in the 0.1–5 kHz range (Moore et al. 2001; Valente 2002). Thus, it is not surprising that acoustic hearing aids are perceived to have low sound quality.

Because MEHDs are not limited by the transducer technology of acoustic hearing aids, however, it is possible in theory to design their output transducers to deliver amplified sound over a wider frequency range than is possible with conventional acoustic hearing aids. In actual MEHDs, this has been achieved to varying degrees. Indeed, recent reports consistently indicate superior sound quality with IHDs relative to acoustic hearing aids (Tysome et al. 2010). There is also evidence that increasing the upper frequency limit of hearing devices can improve the intelligibility of target speech in the presence of competing noise or speech maskers (Carlile and Schonstein 2006; Puria et al. 2008; Moore et al. 2010).

For many people, cosmetic concerns and convenience would provide the most compelling reason for choosing an IHD, much in the way that these concerns lead many to prefer contact lenses or eye surgery over wearing glasses. A number of attempts are being made to develop totally implantable (TI) MEHDs that are completely invisible from the outside, such that others could not become aware of a person's impairment simply by seeing their hearing aid. TI MEHDs can also allow a person to bathe or swim with the device while still receiving amplification. MEHDs can allow the ear canal to be left open or widely vented such that they do not produce occlusion effects (the louder-than-usual perception of one's own voice) and have the potential advantage of providing gain at low frequencies, which is not possible with open-canal acoustic hearing aids.

For subjects with mixed conductive and sensorineural hearing loss, a surgeon can repair the conductive hearing impairment component with a passive prosthesis. However, there is also a class of IHDs called vibroplasty couplers that repair the conductive hearing impairment while at the same time providing amplification to mitigate the subject's sensorineural hearing loss.

Despite the above benefits of IHDs, there are also significant reasons why one might not want to have an IHD. First of all, IHDs require that one undergo surgery, which always carries some inherent risk and in rare cases can lead to complications. Depending on the specific surgical alterations to the body that a given device requires, it may be difficult or impossible for a person to later have the device switched out for a new one with an improved design. In addition, the indications for use dictate that only patients who have moderate-to-severe or profound hearing impairment are candidates, and thus they cannot currently be prescribed for the large number of subjects with mild hearing impairment. Designs featuring an implantable battery may prove inconvenient to recharge and could require additional surgery down the road to replace the battery, and some designs containing implanted magnets may disallow the use of magnetic resonance imaging (MRI) scans and might be prone to electromagnetic interference if the components are not properly shielded. There is also a significant additional financial cost associated with the surgical procedure, on top of the typically high cost of the hearing device itself. With the exception of some countries in Europe, such procedures and devices are typically not covered by medical insurance, although an exception could be made for the class of IHDs intended to treat mixed hearing loss, since the surgical intervention might be required to restore a conductive impairment. An additional drawback of IHDs is that one cannot "play before paying"; that is, one has to make the commitment to undergo surgery and pay for the device before one can even try it out. Because of this mix of potential benefits and drawbacks, IHDs have achieved only partial acceptance in the marketplace. As the technology improves, they will likely become more widely available and more acceptable. NI MEHDs that vibrate the eardrum directly and do not require surgery may be able to overcome some of the disadvantages of IHDs, but this remains to be seen. Perhaps what is most important is that a variety of new hearing-device options are becoming available for otologists and audiologists to offer to their patients.

10.3 Output-Transducer Coupling Sites

A key distinguishing characteristic of the different MEHD designs is the specific structural feature of the middle ear or cochlear anatomy to which the output transducer is coupled. The first reports of directly driving the middle ear to produce the sensation of hearing involved the use of electromagnetic fields (due to a coil in the ear canal, for example) to generate forces on either a single magnet attached to the umbo or on an array of small magnetic particles affixed to the lateral surface of the eardrum (Rutschmann 1959; Goode 1970; Perkins 1996). Another approach, requiring surgery, would be to attach the magnet(s) to the medial side of the eardrum or various parts of the ossicular chain. Owing to the $1/R^2$ decrease in the magnetic field strength with distance (R) from the coil, the energy efficiency of these magnet-and-coil systems may not be adequate for the maximum-output and functional-gain requirements of a hearing device (Perkins et al. 2010). A more energy-efficient approach involves the use of direct mechanical vibration of the head of the malleus, the body of the incus, or, as was done in the case of one of the first successful IHDs (Yanagihara et al. 2001), the head of the stapes.

A more invasive method requires performing a cochleostomy, which involves drilling into the bony capsule of the cochlea so as to directly couple a transducer to the perilymphatic space within the cochlea (Lesinski and Neukermans 1998; Puria and Perkins 2003a). This approach has the potential advantages of higher efficiency, due to there being minimal loss of power in the middle ear structures, and reduced microphone feedback, due to there being less sound traveling back out into the ear canal. Another category of transducers produces direct motion of the cochlear fluids through contact with the round-window membrane. This idea has gained prominence recently because it is less invasive than having to perform a cochleostomy, yet it still retains some of the advantages of directly vibrating the perilymphatic space.

Yet another approach for an MEHD device does not involve direct mechanical coupling to the middle ear or cochlear structures at all, but rather involves producing sound within the middle ear cavity such that the eardrum is vibrated from the medial side (Goode 1970; O'Connor and Puria 2006), or within the round-window (RW) niche so that the cochlear fluid is acoustically stimulated (Puria et al. 2010a).

10.4 Output-Transducer Classification

Each of the different output-transducer coupling sites described above dictates the type and complexity of the surgery required, and to a certain extent the physics involved. For a transducer to be able to transfer mechanical energy to a structure of the middle ear or cochlea, it must have a means of applying a force to that structure relative to a particular inertial reference frame. To make it simpler to understand the physical principles underlying the various MEHD output-transducer types, a classification scheme has been devised based on how the transducer is interposed between its reference frame and the structure being vibrated.

This classification scheme consists of the following four categories, defined according to the number of required connection points between the transducer and structures of the ear, and according to the manner in which those points are linked: (1) zero-connection-point (ZCP) transducers, (2) one-connection-point (OCP) transducers, (3) two-mechanically linked-connection-point (TMLCP) transducers, and (4) two-mechanically unlinked-connection-point (TMUCP) transducers. These categories are described in the text that follows, with illustrated examples shown in Fig. 10.1.

10.4.1 ZCP Transducers

In traditional acoustic hearing aids, sound is generated in the ear canal and the resulting pressure difference between the canal and the middle ear cavity (MEC) then causes the tympanic membrane (TM) to vibrate. Another way of producing

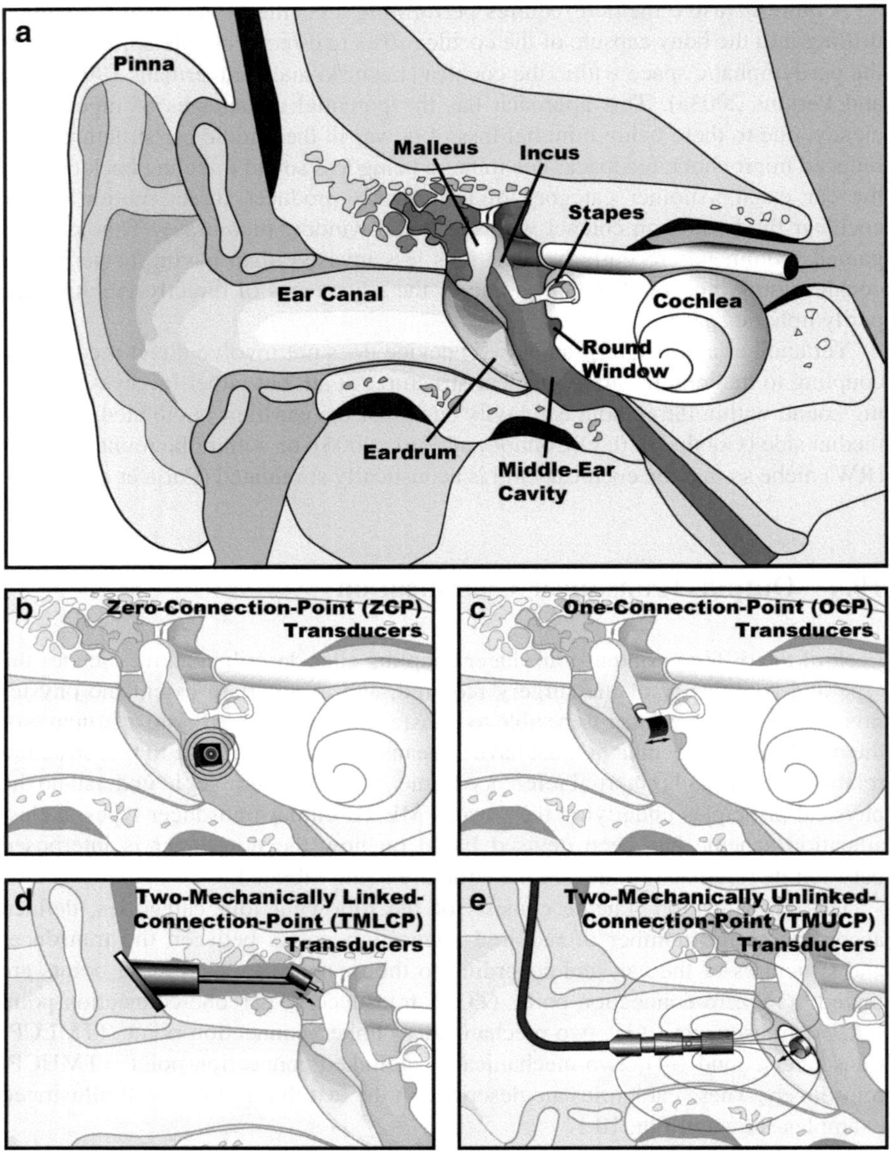

Fig. 10.1 (**a**) Anatomy of the middle ear. (**b–e**) Illustrations of the different types of middle ear transducers, classified according to the number of connection points to the middle ear anatomy necessary for their operation, and whether or not a mechanical linkage is required between two connection points (Reproduced with permission from S. Puria)

this pressure difference is to inject sound directly into the MEC (Fig. 10.1b). This alternate method of vibrating the TM has been demonstrated to produce results comparable to the presentation of sound in the ear canal, and thus could offer a viable means of providing amplification (O'Connor and Puria 2006).

Sound can be produced in the MEC via a hermetically sealed acoustic transducer placed within the MEC space. Such a transducer is classified under the ZCP category because it can, at least in principle, operate without requiring any specific mechanical connections to the structures of the ear: the enclosure of the transducer itself acts as the reference frame for the moving diaphragm responsible for sound production. However, from a practical standpoint it might be preferred to at least partially attach the transducer case to a surface of the MEC so that it does not rattle around due to sudden head motions. While the ZCP category represents the simplest of the IHD types that have been proposed, there are at present no commercial systems of this type being developed.

10.4.2 OCP Transducers

One of the most successful and celebrated MEHD output-transducer designs to date is the floating-mass transducer (FMT), which typically attaches to the incus via a clip (Ball 1996; Ball et al. 1997). Because the FMT mechanism only requires a single point of mechanical attachment to the middle ear, it can be classified as an OCP transducer (Fig. 10.1c). An ingenious and somewhat unanticipated aspect of the FMT is that it pushes against its own mass-inertial reference frame, represented by the body of the device, which itself moves relative to the walls of the MEC. In one embodiment of this concept, the motion is due to forces acting upon a single floating magnet enclosed within a coil assembly. Alternating currents in the coil produce an alternating magnetic field that causes the floating magnet to move back and forth within the coil assembly, but the forces acting upon the mass inertia of the moving magnet in turn cause the coil assembly, and thus the FMT body, to be propelled in the opposite direction. This resulting movement of the FMT body is then transferred efficiently to the attached incus due to an impedance similarity between the FMT and the ossicular chain.

An alternative to the single-magnet FMT approach is a dual-magnet design, in which two repelling magnets are glued together at matching poles (N–N or S–S) to form a single unit with matching poles at the two ends. This produces a strong static magnetic field near the glued surface. The alternating electromagnetic field due to the coil assembly, whose turns are limited to the vicinity of the glued surface, then acts upon the static magnetic field of the dual-magnet unit, causing it to move relative to the coil assembly (Cho 2004; Kim et al. 2007). Like in the single-magnet case, the force exerted on the mass inertia of the dual-magnet unit then similarly causes the FMT body to move in the opposite direction, thus transferring vibration to the incus. One claimed advantage of the dual-magnet design over the single-magnet design is that it does not torque in response to strong uniform magnetic fields, so it is safe for use in an MRI scanner (Song et al. 2000). However, the single-magnet FMT design addresses some of these concerns by designing the coil and incorporating shielding so as to minimize electromagnetic interference.

A very different OCP transducer design is one based on the piezoelectric effect instead of on electromagnetic effects as described previously. In this design, a piezoelectric multilayered stacked actuator is encapsulated in a metal case that is attached to the incus via a clamp. The actuator vibrates due to an applied electric field, and the vibrations are then transferred to the incus through the clamp. This system is also MRI compatible, and has the added benefit of performing better than the single and dual-magnet FMT systems above 5 kHz, at least in human-cadaver temporal bones (Park et al. 2011).

10.4.3 TMLCP Transducers

Another class of devices involves anchoring one end of the output transducer to an immobile structure such as the MEC wall, and mechanically linking the mobile end of the transducer to a structure of the middle ear or cochlea. These TMLCP transducers (Fig. 10.1d) utilize electromagnetic or piezoelectric principles to produce linear motions at their mobile end, such that they could be imagined as a very small high-tech jackhammer.

Although simple in concept, the surgical requirements for this class of transducer are significantly greater than those for the ZCP and OCP systems. This is due to the fact that one end of the device must be anchored to the bone in such a way that the other end can still be precisely aligned and connected to the structure being stimulated. This alignment has to be performed in situ by the surgeon during the surgical procedure, and must be done well enough that the device will stay in place with some degree of precision for many years to come. If the alignment is not done correctly, the device will fail to function and revision surgery will be required. With careful engineering and surgical skill, however, these issues can be well managed.

10.4.4 TMUCP Transducers

A closely related two-point system is one in which the two ends of the transducer are mechanically unlinked subunits, and these are referred to as TMUCP transducers (Fig. 10.1e). One subunit forms the reference frame while the other subunit is attached to a structure of the middle ear or the cochlea. The first subunit then transmits vibration to the second subunit by, for example, establishing a time-varying electromagnetic field that forces the second subunit to move in response.

An example of this transducer type is the direct vibration of the middle ear via a magnet on the eardrum and a coil in the ear canal, as mentioned in Sect. 10.3. The coil typically forms the reference frame, which may be part of a larger assembly placed in the ear canal. An advantage of such a system is that the surgeon does not have to precisely align the two subunits in situ as might be the case for a two-point mechanically linked system. Also, the subunits do not have to be implanted: one of them can be located outside in the ear canal and the other can be placed in contact with the outer eardrum surface without crossing the skin barrier (Perkins et al. 2007, 2010).

An alternative strategy pursed by SOUNDTEC was to implant the magnet near the stapes (Silverstein et al. 2005). The incudostapedial joint (ISJ) was separated, a hermetically sealed magnet with an attached ring was slipped onto the lenticular process of the incus, and the ISJ was then reconnected. This technology is currently marketed as the Maxum Hearing Implant System (Ototronix, Houston, TX). In both systems, changes in the relative positions of the two subunits, for example due to jaw movements, can have the undesirable result of momentarily changing the effective output of the transducers.

10.5 Power and Signal

In the classification scheme described previously, several methods for transferring an amplified signal and power to the output transducer have been used, and new methods are being developed. The most direct and simplest method is via a wired electrical connection from the amplifier to the output transducer.

For implanted transducers, however, the signal has to cross the skin barrier without wires passing through it. The approach used for many decades (for example in cochlear implants) has been to use two aligned coils, with one coil surgically implanted in the temporal bone on the medial side of the skin and the other coil placed outside on the lateral side of the skin. A radio-frequency (RF) carrier signal is transmitted through the skin by magnetic induction from the lateral coil to the medial coil, after which the received signal is demodulated and sent to the implanted output transducer.

For TMUCP transducers, one end of the device wirelessly transmits the power and signal directly to the output end of the device using energy in the form of an electromagnetic field, as was done with the electromagnetic EarLens and SOUNDTEC systems, or possibly instead using energy in the form of ultrasound.

A new and promising concept is to use modulated light to transmit both signal and power. In this case, the photonic signal is received by a photodiode in the ear canal or is transmitted through the eardrum and received by a photodiode inside the middle ear cavity. The photodiode converts and demodulates the light into an electrical signal, which is then used to drive the output transducer (Perkins et al. 2010). While analog acoustic signals and mechanical vibrations alternate in polarity as they compress and decompress the medium in which they are traveling, light, on the other hand, can only be either on or off and has no corresponding state of negative polarity. This, combined with the nonlinearity of photonic components such as laser diodes and photodiodes, results in distortions that must be compensated for using clever and novel analog or digital encoding and decoding methods (Fay et al. 2010; Puria et al. 2010b).

10.6 Input-Transducer Categories and Sites

10.6.1 Acoustic Microphones

Acoustic signals are pressure waves that must first be converted into electrical signals by an input transducer for further processing by the sound processor. The most common input transducer is a microphone, which is a term first coined by Sir Charles Wheatstone (of the Wheatstone bridge). Present-day microphones are electret-type input transducers that are quite small, often with a sub-millimeter input-port diameter. As in acoustic hearing aids, either an omnidirectional or directional microphone is used, the latter having the potential to improve the signal-to-noise ratio (SNR) in certain types of environments. Microphones are typically placed in a behind-the-ear (BTE), ear-level-unit (ELU), or in-the-ear (ITE) component for NI or PI MEHDs. One of the great advantages of using acoustic microphones is that they already have a wide bandwidth, and that the design and manufacturing methods developed during the last few decades, and thus the resulting low costs, can be leveraged.

10.6.2 Subcutaneous Input Transducers

For TI MEHDs, the input transducer has to be placed inside the body. Due to reflections, not all of the sound impinging on the skin ends up vibrating all the way through the tissue, and one of the biggest challenges for TI-MEHD designs has been to build input transducers that can pick up these tissue vibrations effectively. Such transducers are often called subcutaneous microphones, and, as is the case for many transducer designs, they entail tradeoffs between the noise floor, sensitivity, bandwidth, linearity, dynamic range, and size.

One of the key factors that dictates the sensitivity and bandwidth of a subcutaneous microphone is the location site of the input transducer. This is because the location site often defines the type of skin tissue that the sound must pass through, and thus affects the impedance-matching requirements between the air, skin tissue, and subcutaneous microphone. A popular approach has been to take a standard acoustic microphone, couple it to a cavity with a diaphragm, and hermetically seal the entire assembly (Lesinski and Neukermans 1998; Ball et al. 1999). The transducer assembly is then implanted such that the skin is overlaid on top of the diaphragm. A larger diaphragm allows more sound to be collected, resulting in greater sensitivity. A variant of this design involves the incorporation of an acoustic tube between the microphone and the diaphragm contacting the skin, with the acoustic tube coiled inside the cavity of the microphone to save space. The tube parameters are chosen cleverly such that there is a high-frequency boost due to the tube resonance that compensates for the loss across the skin and tissue, thus producing a nearly flat response from 0.1 to about 9 kHz (Jung et al. 2011).

One proposed implantation location is directly above the pinna. This has the advantage of offering a large available area for the transducer diaphragm, and would also allow multiple transducers to be configured as an array for directionality (Puria and Perkins 2003b). One of the disadvantages of this location is that the highly sensitive transducer is prone to responding to loud sounds from direct contact such as by one's own fingers or due to hair movement.

An alternative location for the subcutaneous microphone is underneath the skin of the ear-canal wall. This has the advantage of not producing unintended sounds due to one's fingers or wind noise, especially if the microphone is implanted closer to the eardrum. In addition, the natural canal resonance does not have to be introduced electronically for this location, whereas it does for microphones placed outside the canal entrance. However, as the eardrum vibrates due to the output transducer, the eardrum may produce sound in the ear canal, and this in turn can cause feedback with this microphone location (Puria 2003). This propensity for feedback results in reduced gain margins at the canal resonance frequencies, which limits the usefulness of the aid. The TICA system was an early example of a device incorporating this type of design (Zenner et al. 1998, 2004).

Considering that the human eardrum embodies the refinements of millions of years of design iterations, why not choose a site for the input transducer where it can be put to good use? This would require that a vibration sensor be attached to the malleus or the incus in order to sense the eardrum vibrations, which is what the TI Esteem system (Envoy Medical, St. Paul, MN) does for its input transducer. As implemented, the input transducer consists of a piezoelectric transducer attached to the head of the incus on one end and anchored to the MEC wall on the other end. Thus, it is analogous to the TMLCP output transducer discussed earlier, but is configured as an input transducer. And like the output transducer, the surgical requirements for this class of input transducer are high. Another approach is the development of an accelerometer-based sensor attached to the malleus in an OCP configuration. One of the smallest fabricated accelerometers was shown to work in temporal-bone studies (Park et al. 2007). However, the power requirements of this piezoresistive silicon accelerometer were too high for further commercial development.

10.7 Partially Implantable MEHDs

In subsequent sections of this chapter, systems that use many of the components introduced in the preceding text are described in roughly historical order.

Discussed first are some of the important industry and academic endeavors toward the development of PI MEHDs. The basic configurations of these systems are all very similar; they feature an ELU that includes an acoustic microphone, a multiband-compressor sound processor for converting the microphone signal into an amplified signal suited to the patient's hearing impairment, and a replaceable or rechargeable battery for powering the electronics. The amplified signal is encoded as an RF signal and transmitted using a coil. Because it is external to the body, the ELU can be upgraded whenever the manufacturer makes improvements to the signal-processing algorithms. For example, common changes have been to increase the number of multiband-compressor channels and make improvements to the fitting algorithm. The implanted component contains a second coil that receives the RF signal, demodulates it, and sends it to the output transducer.

10.7.1 The Rion Device E-Type

The Rion Device E-type (RDE), developed in 1983, was likely the first attempt at commercializing an IHD. The output transducer, in the TMLCP class, utilized a piezoelectric ceramic bimorph. It was implanted in Japanese patients who had both conductive and sensorineural hearing loss. Between 1984 and 1989 there were 29 patients implanted with the first version of the RDE. There were complications in some of the patients, and to address these complications an improved second version was produced. The surgical procedures were also improved, and between 1990 and 1997 the second version of the device was implanted in ten additional patients. The Rion Company ceased its middle ear implant operations in 2005 (Komori et al. 2010).

As of 2001, 30 patients were still using the device: 20 patients with the first version of the device and 10 with the second version of the device. Of these patients, 11 continued to have the device in place as of 2010 and were still using it on a daily basis (Komori et al. 2010, 2012). The average usage period of the device (as of 2010) was 16.6 years, and the longest usage time was 21 years for one patient who had no trouble with the initial version of the device. This is remarkable and indicates that there are patients who have had a hearing device implanted in them for a large portion of their adult lives. This might even be an undocumented world record for any implanted electromechanical device inside a human body. Most important is that subjects continued to report that sound produced by the RDE implant was superior to that produced by conventional hearing aids (Komori et al. 2010).

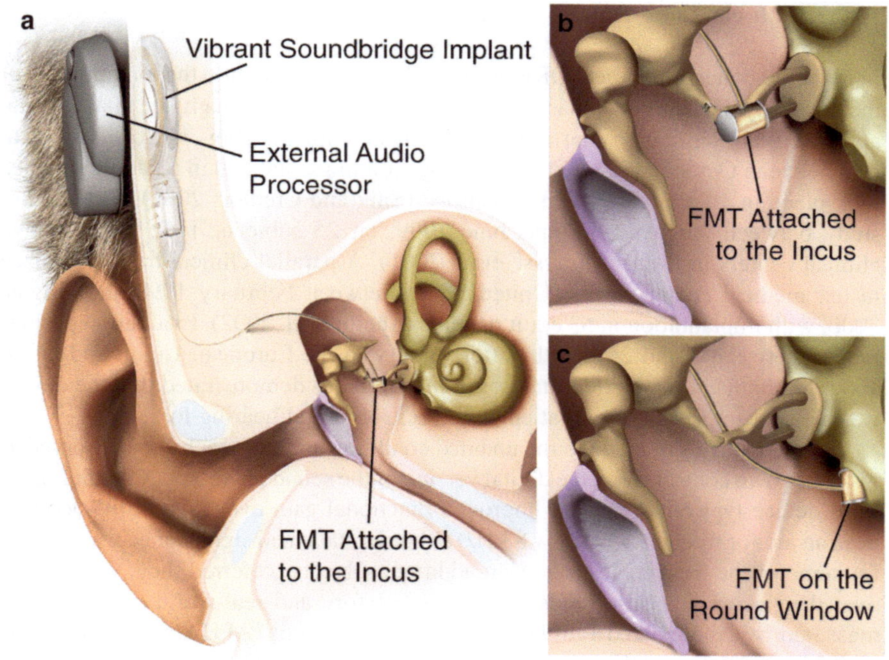

Fig. 10.2 (**a**) The partially implantable (*PI*) Vibrant Soundbridge middle ear hearing device (*MEHD*). A floating-mass transducer (*FMT*) in the one-connection-point (*OCP*) category is shown attached to the incus (**a** and **b**) and to the round window (*RW*) (**c**) (Images reproduced with permission from MED-EL Corporation)

10.7.2 The Vibrant Soundbridge

If the RDE is the device with the record for the longest usage period, then the device with the record for implantation in the largest number of subjects—by a long shot—is the Vibrant Soundbridge. More than 1,100 patients were already implanted by 2005 (Labassi and Beliaeff 2005), and by now there have been more than 10,000 devices implanted in the European Union, United States, and other parts of the world. The FMT itself was and continues to be marketed as a vibrating ossicular prosthesis (VORP), while the system incorporating all of the subcomponents (Fig. 10.2a) is referred to as the Vibrant Soundbridge (VSB). A search for the words "Vibrant Soundbridge" in PubMed under titles and abstracts shows that there have been nearly 110 publications on the device since its initial clinical studies began (Tjellstrom et al. 1997; Lenarz et al. 1998; Snik and Cremers 1999).

10.7.2.1 Incus Long-Process Attachment

In its initial conception, the VORP of the VSB was crimped onto the long process of the incus by the surgeon via a facial-recess approach through the mastoid, an approach very similar to that used for cochlear implants (Fig. 10.2b). A clinical trial between October 1998 and April 1999, with implantations in 63 adult patients, took place in 10 clinical centers in Europe (Snik and Cremers 2001); and another study between August 1997 and June 1999, on 25 subjects, took place in five clinical centers in France (Fraysse et al. 2001). A parallel clinical trial took place in ten medical centers in the United States between February 1998 and August 2000, with implantation in 53 adult patients (Luetje et al. 2002). Evaluation periods ranged between 3 and 25 months after surgery in the European Union and lasted about 5 months in the United States. All three studies demonstrated that the device was safe for implantation in patients with sensorineural hearing loss.

Comparisons were made to the unaided condition and thus each subject served as his or her own control. With the digital version of the sound processor (Vibrant D), the VSB device typically provided improved functional gain over a “best-fit” acoustic hearing aid for frequencies below 8 kHz, although statistical significance was lacking for some of the frequencies tested. In addition, subjects preferred their VSB device over their acoustic hearing aids in many laboratory and real-world conditions, as measured by Abbreviated Profile of Hearing Aid Benefit (APHAB) scores (Luetje et al. 2002). Subsequent long-term follow-up studies suggest that the device was stable 5–8 years postimplantation (Schmuziger et al. 2006; Mosnier et al. 2008).

Despite user satisfaction with the system, the company was not commercially successful. This was likely due to multiple factors including cost, lack of insurance reimbursement, and slow adoption rates by otologists.

In June of 2003, MED-EL (Innsbruck, Austria) acquired the VSB technology, its manufacturing equipment, and all patent rights. The subsidiary Vibrant Hearing Technology GmbH was established to commercialize the VSB system. Judging by the number of devices implanted since moving to Europe, the VSB technology appears to be on its way to becoming successful in the marketplace. A more detailed and personal history of the design and development of this device is nicely chronicled in Geoffrey Ball’s autobiography (2011) *No More Laughing at the Deaf Boy*.

10.7.2.2 Implantation on the Round-Window Membrane

With initial inclusion criteria requiring normal outer and middle ear function to qualify for a VORP, an area of recent excitement has been the extension of the inclusion criteria to include subjects that have a conductive hearing impairment in addition to a sensorineural hearing impairment, and with that to allow VORP attachment to anatomical sites other than the incus. This patient group is referred to as having “mixed hearing loss.”

One of the first "off-label" reports of this kind was on subjects that had ossicular-chain pathologies, and involved implanting the VORP on the RW membrane (Fig. 10.2c), thus bypassing the middle ear altogether (Colletti et al. 2006). Another was on subjects who had congenital malformations of the auricle, atresia of the outer ear canal, or pathologies of the ossicles. In a two-stage procedure, the auricle was reconstructed during stage 1 and the VORP implanted on the RW during stage 2 (Kiefer et al. 2006; Wollenberg et al. 2007). With the VORP implanted on the RW, the VSB was able to make up for the conductive impairment and provide additional gain for part of the sensorineural impairment, as has been demonstrated by other follow-up studies (Beltrame et al. 2009; Colletti et al. 2009; Bernardeschi et al. 2011). This has proven to be a pivotal finding in that it has opened up the new idea that subjects with mixed hearing impairment could also be treated with the VSB device. There is now a concerted effort to understand the acousto-mechanical advantages and disadvantages of placing the VSB on the RW using temporal-bone studies (Nakajima et al. 2010).

10.7.2.3 Implantation in the Oval-Window Niche

Although the RW was shown to be a safe and reliable implantation site, it required a surgical widening of the entrance of the round window niche in order to optimally couple the FMT to the membrane, and not all surgeons are comfortable with this procedure. Could the footplate of the oval window (OW) niche be an alternative site for implantation that would overcome this limitation? A small study was conducted in patients with the stapes superstructure already missing but for whom the footplate was still mobile. The FMT was coupled to the footplate with a piece of cartilage and partially wrapped in perichondrium within the OW niche (Zehlicke et al. 2010). The post-operative results indicated that the FMT coupled to the footplate appears to be as good a coupling site for the FMT as the RW. An advantage of this approach over RW implantation is that otologists are typically more experienced with OW surgeries than with RW surgeries.

10.7.3 The Otologics MET Systems

Another method of stimulation is through mechanical vibration of the incus. In the Otologics Middle Ear Transducer (MET) Ossicular Stimulator, implantation of the output transducer is performed through a postauricular incision to expose the body of the incus and the head of the malleus. This is a TMLCP-type transducer with one end attached to the incus and the other end attached to the bony wall of the MEC. To allow insertion of the probe tip, a 1-mm hole is made in the incus using a laser. A flexible fibrous connection forms between the probe and the incus tip during the post-operative healing process. The microphone and sound processor are similar to those for other partially implantable systems, although specific details differ.

A total of 282 patients were implanted with this device in 22 countries and at more than 100 centers (Jenkins et al. 2004). The MET system progressed into the Fully Implantable Ossicular Stimulator (FIMOS) and has subsequently been marketed as the Carina system, discussed later.

10.8 Active Ossicular-Replacement Prostheses for Mixed-Hearing-Impairment Devices

There are a number of passive ossicular-replacement prostheses (ORPs) used in ossiculoplasty surgical procedures that attempt to close the air-bone gap (ABG) by repairing ossicular-chain pathologies (Merchant and Rosowski, Chap. 9). A common type of passive prosthesis is the partial ORP (PORP), used in cases where the incus has eroded or is missing altogether. The PORP is interposed between the eardrum or the malleus and the stapes head. Yet another prosthesis is the total ORP (TORP), for cases in which both the incus and the stapes superstructure have eroded or are missing. A TORP is interposed between the eardrum or the malleus and the stapes footplate. These prostheses are quite successful in repairing the conductive pathology, as has been shown in many clinical and temporal-bone studies (e.g., Murugasu et al. 2005; Beutner and Hüttenbrink 2009).

Patients who undergo ossiculoplasty to repair their conductive pathology often also have a concomitant sensorineural hearing loss, such that after the surgery they would be sent to obtain an acoustic hearing aid for their sensorineural hearing loss. But there are patients (e.g., with blocked canals or no eardrum or ossicles) for whom acoustic hearing aids can not help them due to inadequate gain, particularly at frequencies above 2–3 kHz. To devise methods to both close the ABG and provide active amplification in a single device and procedure, surgeons and engineers got together and combined an FMT with an ORP. The use of this combined device in a surgical procedure is known as an ORP-Vibroplasty.

One of the first designs of this type was the mechanical attachment of an FMT along the long axis of a Bell Tübingen prosthesis to produce a combined device called a "Bell-Vibroplasty", and this device was first tested in human-cadaver temporal bones (Huber et al. 2006). The Bell Tübingen prosthesis is a lightweight titanium PORP (Heinz Kurz GmbH Medizintechnik, Dusslingen, Germany) with an optimized design for coupling to the head of the stapes (Schmerber et al. 2006). In the passive mode, the Bell-Vibroplasty was shown to be as effective as the Bell PORP. In the active mode, the Bell-Vibroplasty could provide an average maximum equivalent pressure output (MEPO) of between 105 and 125 dB SPL, in the 0.5–8 kHz frequency range.

In a subsequent design, a clip holder was built to connect to the FMT along the same axis, and the assembly was interposed between the eardrum (natural or artificial) and the stapes footplate (Hüttenbrink et al. 2010). This device is known as a TORP-FMT assembly (Fig. 10.3a). Direct implantation on the footplate does require the stapes superstructure to be missing, however, and because no surgeon

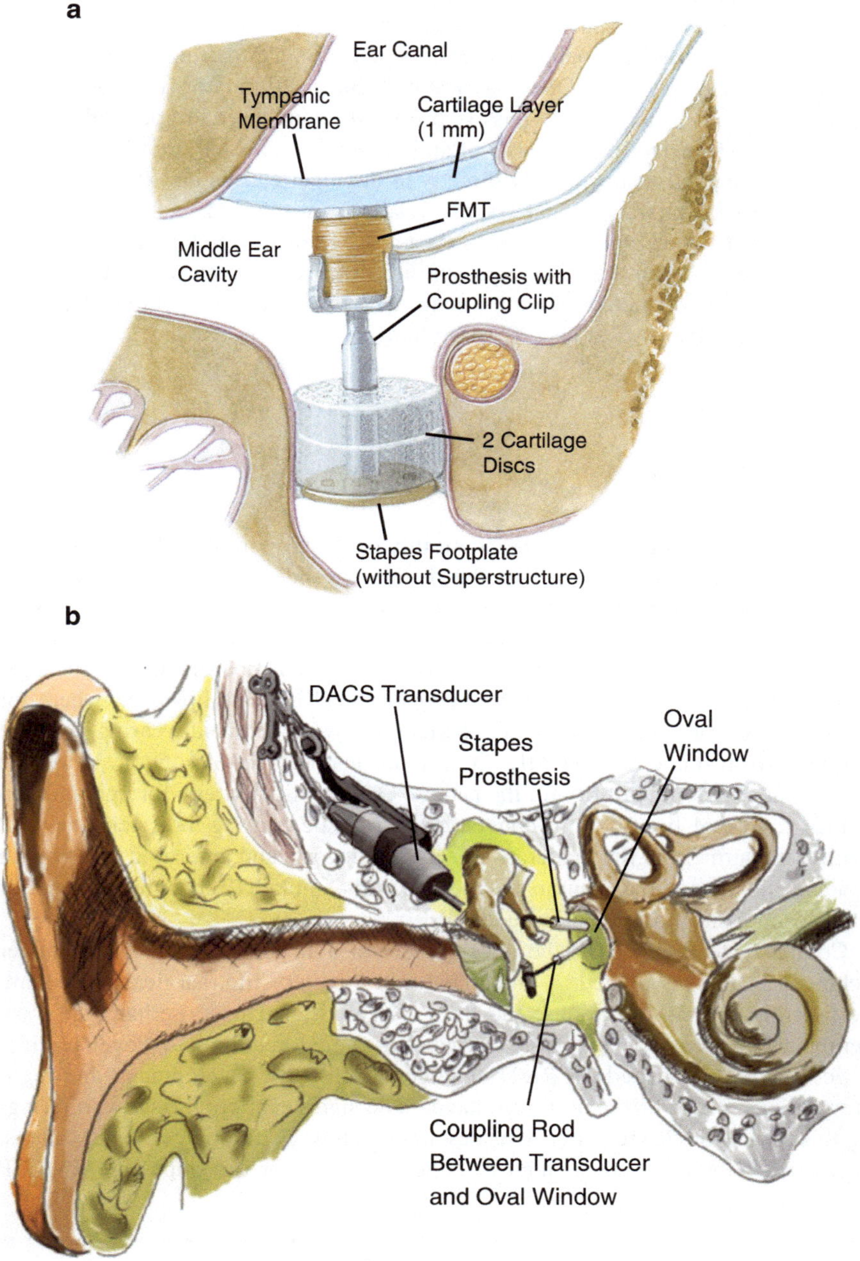

Fig. 10.3 Active total/partial ossicular-replacement prostheses (*T/PORPs*) for mixed hearing loss. (**a**) An FMT clipped to a TORP assembly is placed between the eardrum and the oval window using pieces of cartilage. (Image reproduced with permission from MED-EL Corporation.) (**b**) A passive stapes PORP for otosclerosis is combined in parallel with an active implant in the Direct

would intentionally want to remove the stapes superstructure if it were already intact, the number of candidates for this approach is limited.

A system intended for patients with mixed hearing impairment arising from advanced otosclerosis is the Direct Acoustic Cochlear Stimulator (DACS) system. It consists of both an active and a passive prosthesis, configured in a parallel fashion (Fig. 10.3b). The active part consists of an implantable TMLCP transducer, which is said to couple "acoustic" energy directly to the inner ear via an externally worn audio processor. A conventional stapes prosthesis is attached to the mechanical transducer and placed in the OW after stapes-crus removal, which permits coupling to the inner ear. A parallel passive prosthesis, attached between the incus and the OW, repairs the conductive hearing impairment as is done in a routine stapedectomy procedure (Häusler et al. 2008; Bernhard et al. 2011).

10.9 TI MEHDs

One of the promises of IHDs is the possibility of a totally implantable device with no external components, such that the device is rendered completely invisible. Significant advantages of these systems are the near elimination of the occlusion effect due to the lack of anything blocking the ear canal, and a significant reduction in feedback problems. Cosmetic concerns and convenience are likely the most significant reason that a patient would consider a TI MEHD, however, where for social and/or professional reasons the patient may wish to avoid wearing a visible hearing aid. In this case a TI MEHD may not just be a viable option, but could become an overriding priority for the patient. Qualification criteria for implantation include a lack of benefit from conventional acoustic hearing aids, sensorineural hearing impairment in the moderate-to-severe category, and a mastoid region with an adequate volume to accommodate the size of the implant (Zenner and Leysieffer 2001; Zenner and Rodriguez Jorge 2010).

Compared to PI MEHDs, the development of a TI MEHD faces even greater technical challenges related to biocompatibility, battery life, surgical placement, the potential for complications, and the relatively higher cost. Perhaps the most significant challenge is the large amount of capital required to develop these devices, which is needed for R&D, intellectual property development, and regulatory filings. Three different groups have made significant progress toward bringing this type of device to patients, as described below. In all three of their systems, mechanical vibrations are applied to either the incus or the stapes via TMLCP

Fig. 10.3 (continued) Acoustic Cochlear Stimulator (*DACS*). The active implant transmits its motions directly to the oval window (*OW*) via a connecting rod (Image reproduced with permission from Dr. Christof Stieger)

output transducers. Another common feature in all three systems is the use of a remote control required to either alter the program settings or change the volume of the output.

10.9.1 The Implex TICA System

The Totally Implantable Communication Assistant (TICA) system from Implex Corporation is the earliest example of a TI MEHD (Fig. 10.4a). A signal from a subcutaneous microphone, located in the external auditory meatus near the eardrum, is amplified and vibrations are delivered to the head of the incus (Zenner et al. 1998). The system was designed to have a bandwidth of about 10 kHz, which provided the possibility for sound localization cues due to pinna diffraction and thus the possibility for better hearing in noisy situations. However, the close proximity of the microphone to the eardrum limited its functional gain. The device was implantated in 20 subjects in Europe (Zenner and Leysieffer 2001), but no trials were conducted in the United States. Cochlear Corporation acquired Implex in 2004, but has not reported any further progress in the literature.

10.9.2 The Envoy Esteem System

The Esteem from Envoy Medical Corporation is the first TI MEHD approved in the United States by the FDA. It consists of three interconnected implanted components (Fig. 10.4b). Because of its clever input- and output-transducer designs, there are no microphones or loudspeakers in the system. A TMLCP input transducer senses the input sound with one of its ends attached to the malleus head. Similarly, a TMLCP output transducer is affixed to the stapes head at one end. The other ends of both transducers are attached to the bony wall of the mastoid. The incudostapedial joint (ISJ) is separated and the lenticular process removed to allow fixation of the output transducer to the stapes head; the ISJ separation is also required to eliminate mechanical feedback to the input transducer. A disadvantage of this approach is that when the device is off, there is a loss of hearing in comparison to the preimplantation condition. However, owing to its high efficiency, the system can remain on all of the time. One of the design choices made for efficiency reasons is the use of linear amplification for sound processing rather than a more standard multiband compressor. The implanted battery is expected to last about 5 years before requiring replacement, but battery replacement does require an additional surgery (this is similar to what is done for pacemakers).

In a Phase I U.S. clinical trial, the feasibility of the Esteem was shown in five patients (Chen et al. 2004). Two-month postactivation results indicated mean functional gain equal to a best-fit conventional hearing aid up to 1 kHz; at 2 and 3 kHz, the gain was below that of a conventional aid. Patients reported a high level of satisfaction in the APHAB test at 2 and 4 months postactivation (Chen et al. 2004).

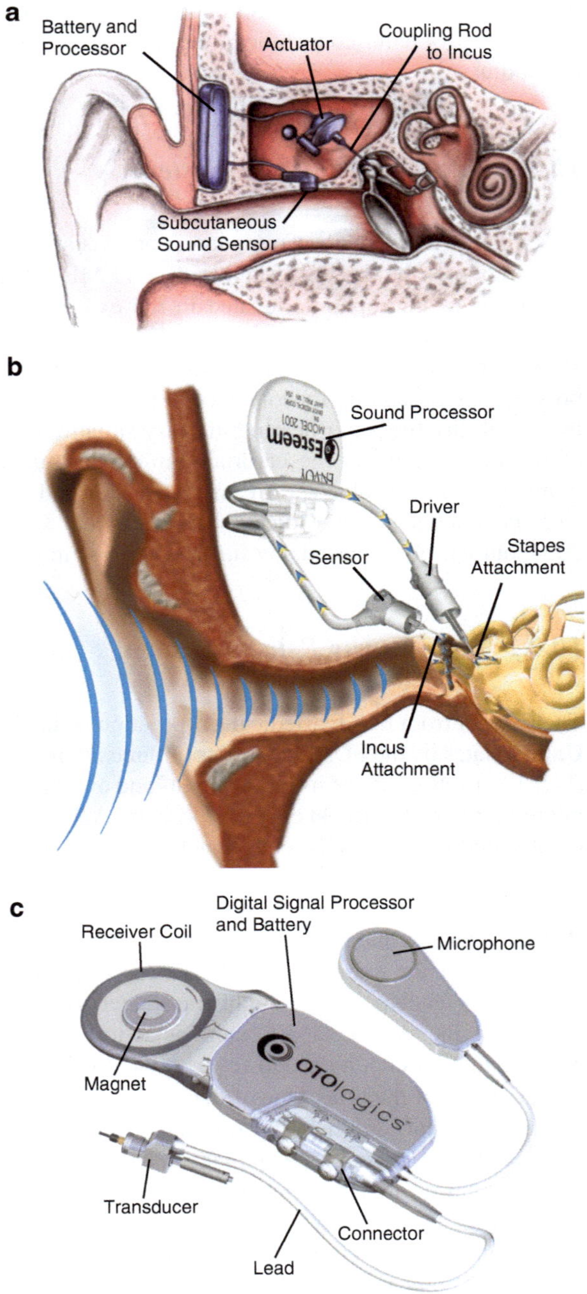

Fig. 10.4 Totally implantable (*TI*) MEHDs. All three systems use a TMLCP output transducer. (**a**) An Implex TICA system with the microphone under the skin of the ear-canal wall near the eardrum and the output transducer contacting the head of the incus. (Reproduced with permission from Cochlear Corporation.) (**b**) An Envoy Esteem system with the sensor attached to the malleus and the output transducer attached to the stapes. The long process of the incus is disarticulated from the stapes to minimize feedback (Kraus et al. 2011). (Reproduced with permission from Envoy Medical.) (**c**) Components of the Otologics MET FIMOS and Carina systems, which are similar to the TICA system in that the output transducer also vibrates the head of the incus, although the microphone is now placed retroauricularly instead of within the ear-canal wall (Reproduced with permission from Otologics, LLC)

A Phase II pivotal trial was conducted at three clinical sites with implantation in a total of 57 patients (Kraus et al. 2011; Shohet et al. 2011). Comparisons were made between the unaided condition, a best-fit acoustic hearing aid, and the Esteem system. This time, gain with the Esteem was superior to the acoustic hearing aids below 4 kHz, while no comparison data with acoustic hearing aids was reported above 4 kHz. There was overall improvement in the speech-in-quiet test condition with the Esteem when compared to acoustic hearing aids. Using the QuickSIN test, hearing in noise was not statistically improved with the Esteem when compared to acoustic hearing aids. However, there was some preference for the Esteem in the background-noise category of the APHAB questionnaire.

The device received approval from the FDA in March of 2010. Patients with the device include celebrities such as the actor Lou Ferrigno, best known for his role as the Incredible Hulk, who has had one ear implanted with the device and has said he is looking forward to having his second ear implanted. Another patient, with bilaterally implanted devices, is the otologist and medical advisor to Envoy Dr. Michael Glasscock.

10.9.3 The Otologics MET FIMOS and Carina Systems

The design of the Otologics TI systems, initially called the MET FIMOS III system and subsequently marketed as the Carina system (Fig. 10.4c), improve on the TICA system by moving the input microphone out of the ear canal in favor of a retroauricular subcutaneously implanted microphone, which improves the feedback gain margin in comparison to the TICA microphone location (Zenner and Rodriguez Jorge 2010). Owing to the improved gain margin, the ossicles do not have to be disrupted—unlike the Esteem system. The output transducer is essentially the same as the PI Otologics MET system described earlier (Jenkins et al. 2007).

The implanted battery lasts for up to 36 h per charge; it is typically recharged daily for 1–1.5 h with an external charging system (Bruschini et al. 2009). With this configuration, the battery will not need to be replaced for about 15 years, which is nearly 3 times longer than that of the Esteem system, but with the Esteem system the battery does not need to be recharged.

The system was approved in Europe after a clinical trial completed on eight patients (Bruschini et al. 2009, 2010). The U.S. Phase I clinical trial of the MET FIMOS III was conducted in parallel on 20 patients at multiple sites, under IDE approval from the FDA (Jenkins et al. 2007, 2008). Both studies indicate that pure-tone average and monaural recognition-of-words scores were slightly better for the acoustic hearing aid (the subject's own aid, fitted as it was when they first entered the study) 3 months postimplantation. Patient-benefit scales favored the implanted device. However, device complications were reported that resulted in deterioration of speech perception at 6 months postimplantation. After refitting, device performance improved at 12 months and onwards. The results from the ongoing Phase II U.S. trial have not yet been published.

10.10 Nonsurgical Nonimplantable MEHDs

The foregoing summaries suggest that amplification and speech-understanding results with either the partially or totally implanted amplification systems are somewhat mixed in comparison to conventional acoustic hearing aids. However, one consistent positive outcome has been that patients prefer the sound quality of implanted devices and do perceive benefit from them in some real-world situations, as measured using questionnaires such as the APHAB (Tysome et al. 2010). It is not clear if patients are willing to pay the significantly higher cost and go through an elective surgical procedure to attain this benefit, however.

Arguably one of the primary advantages of implanted versus acoustic devices is that, in the implanted systems, the middle ear is mechanically driven rather than acoustically driven. An alternative approach on the horizon for achieving mechanical drive without requiring implantation is to move the transducer to the lateral surface of the eardrum and provide direct mechanical stimulation from there without ever penetrating into the middle ear. This would have many of the benefits that come from mechanically driving the middle ear, but without the downsides of surgery and the accompanying high cost. The first examples of such an approach involved using an electromagnetic field to generate forces on a magnet attached to the tympanic membrane (TM) (Rutschmann 1959; Wilska 1959; Goode 1970). These early attempts were impractical from a commercialization point of view, however, because the magnets were glued to the TM and would soon be displaced due to the migratory forces of the eardrum epithelium. The attachment problem was later solved with the development of a silicone platform with an embedded magnet in the center, with the entire assembly floating on top of the eardrum rather than being attached to it (Perkins and Shennib 1993; Perkins 1996). A drop of mineral oil between the silicone platform and the eardrum was used to allow the eardrum epithelial layer to freely migrate without displacing the platform. In this early design, the magnetic-field-generating coil was worn around the neck, which proved to be inefficient and required that a large battery be worn around the neck (ReSound Corporation). This was a type of TMUCP transducer.

In subsequent devices, the field-generating coil was moved inside the ear canal, which was significantly better in that a much smaller rechargeable battery could be worn in a unit placed behind the ear (Perkins et al. 2007, 2010; Puria and Perkins 2010). It provided sufficient output to reach audibility for subjects with up to 60 dB HL of hearing impairment, at frequencies up to 8 kHz in most of the study population and up to 11.2 kHz in half of the study population. However, there was little headroom left to provide gain, and subject variability was high. An important outcome was that directly driving the eardrum did provide an average gain margin of 30 dB, with a microphone at the canal entrance, except in the limited regions of 3 (± 1) kHz and 9 (± 1) kHz, where the mean gain margins dipped down to 12 dB and 23 dB, respectively, as a result of ear-canal resonances.

Recently, the same group in Redwood City, CA (EarLens Corporation) has taken a very different approach to overcome these limitations. Balanced-armature transducers used in hearing-aid receivers are very power efficient. So, an especially efficient approach is to couple the mechanical output of a balanced-armature transducer directly to the umbo (Puria et al. 2012). The transducer also has to provide a bias force to maintain contact with the eardrum. In practice, the motor and biasing mechanisms were integrated onto a ring-like platform residing on the skin of the anterior sulcus and the peritympanic canal epithelium, with the fixed end of the transducer attached to the platform such that its mass did not load the eardrum (Fig. 10.5).

The anatomy of the TM, including its cone depth and diameter, varies for each individual; thus the assembly has to be customized to each individual's ear anatomy, which is currently obtained by taking impressions of the subject's eardrums and ear canals but could in the future be obtainable using some form of scanning technology. In the transducer classification scheme presented here, this design is a TMLCP output transducer. For the TMLCP transducers used in the TI MEHDs discussed previously, the surgeon made the two attachments in situ during surgery. In this new design, the manufacturer makes the TMLCP transducers on the bench before placement, based on the highly precise anatomical information obtained from the impressions. The challenge of this class of transducers is in supplying them with power and signal: a wireless solution is required because the traditional approach of using wires is not feasible in this case, as motions of the wires by the subject would cause displacement of the transducer.

10.10.1 Photonic Approaches

The many different wireless energy-transmission possibilities include light, RF signals, and ultrasound. However, analysis suggests that light is likely the most viable and energy-efficient approach. There are several proposals in the prior art that use optical energy to transmit audio signals for nonsurgical approaches to amplification (Fay et al. 2009; Puria and Perkins 2010; Pluvinage and Perkins 2011).

In Taiwan, a different group (National Taiwan University) was developing a hybrid approach that combined light transmission with a magnet on the eardrum (Lee et al. 2008). A platform with a magnet was placed on the eardrum and two coils were placed in the sulcus surrounding it: one for pushing the magnet and the other for pulling the magnet. Two photodetectors received and converted optical energy to drive their respective coil, and the combined efforts of the coils in turn drove the magnet on the TM. Light was transmitted by two LEDs, each emitting a different wavelength. In this approach, the transducer mass of 115 mg loaded the eardrum. Initial results on human cadavers showed movement of the eardrum, but it was not clear if significant amplification could be obtained with a reasonable battery using such an approach; likely the distance between the magnet and drive coils would lead to inefficiencies. There have been no reports that the system was used for subject testing.

a

b

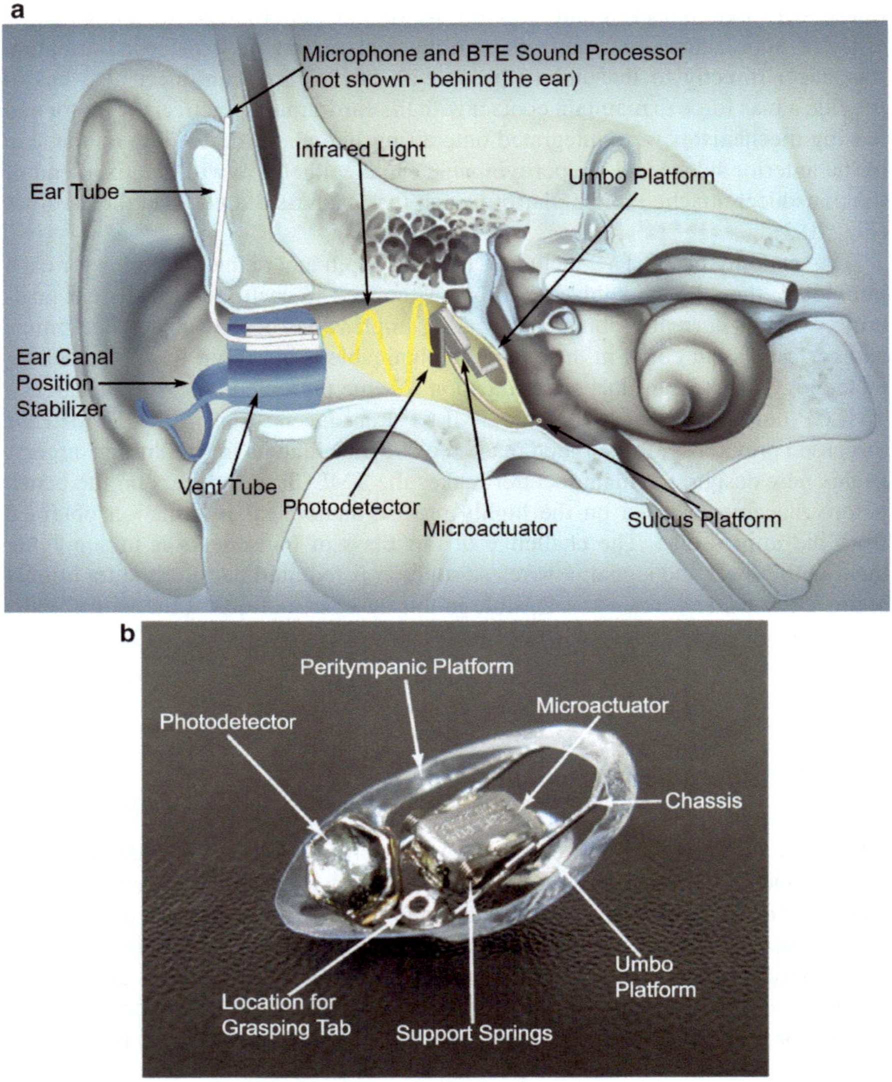

Fig. 10.5 Schematic illustration of the (**a**) Light-based Contact Hearing Device (*CHD*) with the light-activated Tympanic Contact Actuator (*TCA*) on the eardrum. The light-generating behind-the-ear (*BTE*) sound processor (including the microphone, sound processor, laser diode, and battery [not shown]) and the positional stabilizer aim light down the canal, where it is picked up by the TCA and converted into direct drive of the eardrum. The vent tube in the stabilizer minimizes occlusion and allows natural sounds to reach the eardrum. (**b**) A close-up view of the TCA, which is customized to the anatomy of each ear (Images reproduced with permission from EarLens Corporation)

10.10.2 The Light-Based Contact Hearing Device (CHD)

Meanwhile, the balanced-armature approach coupled to a photodetector led to the development of the light-activated Tympanic Contact Actuator (TCA), which, when customized for each subject's ear canal and TM anatomy, could be inserted and removed by an otologist within minutes (Fig. 10.5b). This is similar to wearing a contact lens, except that it improves hearing rather than vision. The TCA, along with an accompanying light-generating BTE prototype containing a laser diode, drive electronics, and rechargeable battery, together form the light-based Contact Hearing Device (CHD).

An IDE-approved FDA study of the CHD on 13 subjects was completed in February of 2012. Measurements indicated that the TCA was safe and stable for the duration of the 4-month study and that it provided significant output and gain margin up through 10 kHz (Fay et al. 2013). The measurements also indicated that there was significant aided benefit with the CHD for speech in quiet and in some situations with spatially separated speech maskers. Also, the APHAB results showed that the subjects preferred the CHD in comparison to the unaided condition.

10.11 Functional Comparisons Across Devices

This review of middle ear hearing devices and systems would not be complete without a comparison of some sort of clinical outcome across devices. Speech results are difficult to compare across studies because of the different tests that are used. The effects of the devices on hearing thresholds—that is, the functional gain—are the most basic measurements since they at least provide a glimpse of the low- and high-frequency audibility of speech, music, and other sounds. The one caveat here is that functional gain and functional improvement, although likely correlated, are not the same. As mentioned previously, bandwidth is thought to be important for both speech understanding in noisy places and perceived sound quality.

Figure 10.6 shows the audiometric hearing-level (HL) inclusion criteria used in seven studies where adequate data were available for functional-gain analysis. Of the seven devices reported, one is the nonsurgical CHD from EarLens; three devices are from Otologics, consisting of one PI device (MET) and two TI devices (FIMOS and Carina); two are PI devices from MED-EL, the Vibrant Soundbridge with the FMT attached to the incus long process (FMT) and the Vibrant Soundbridge with the FMT placed in contact with the round window (RW FMT); and one TI device is from Envoy Medical (Esteem).

The HL inclusion ranges (Fig. 10.6a and b) indicate the acceptable range of hearing impairment at each frequency that a subject can have in order to qualify for each type of device. Generally, the overall system noise floor determines the lower-HL specification (plotted as the top of each range), and the maximum equivalent pressure output (MEPO) determines the upper-HL specification (plotted as the

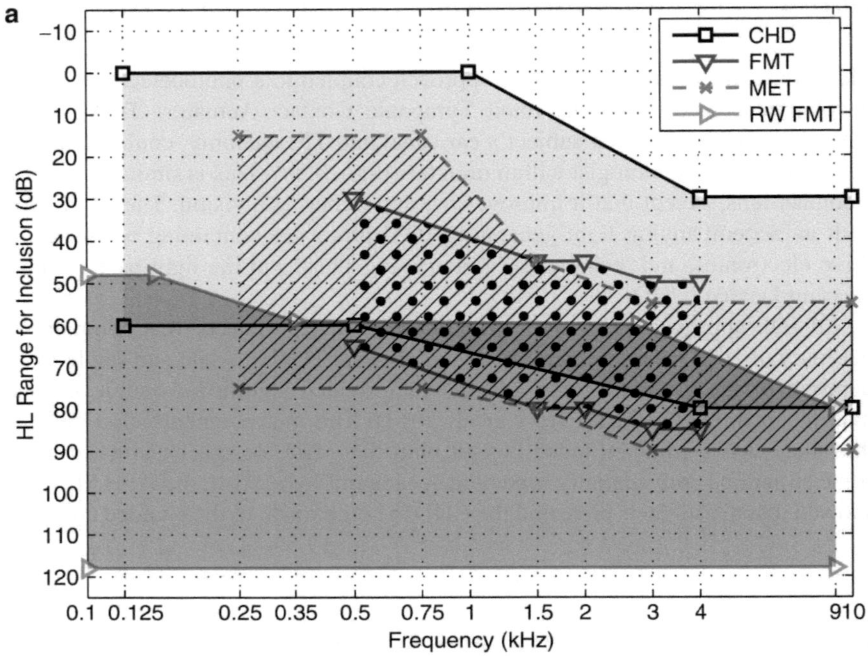

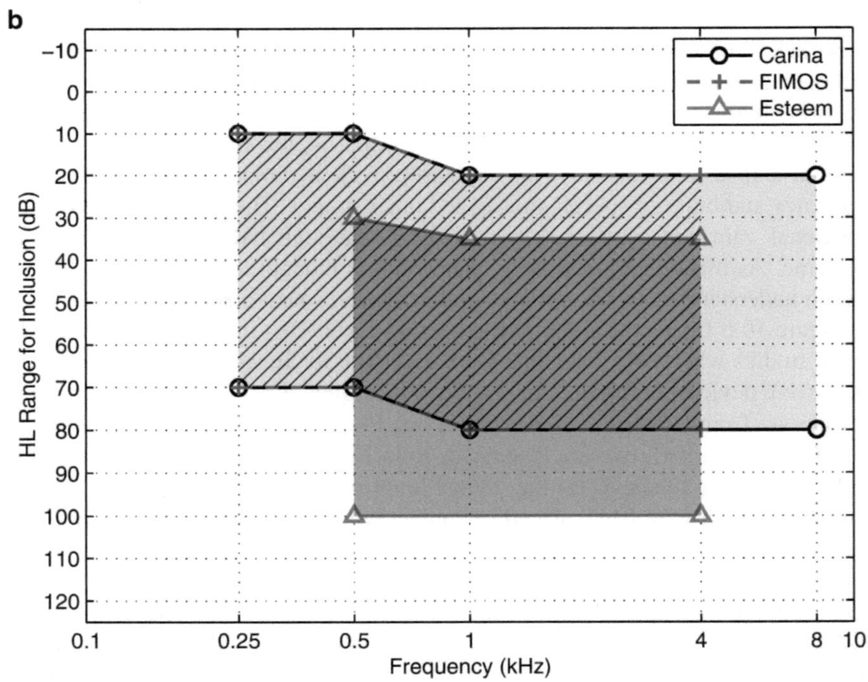

Fig. 10.6 Hearing-level (*HL*) inclusion criteria for clinical studies involving each type of middle ear device. (**a**) CHD refers to the nonimplantable (*NI*) EarLens Contact Hearing Device (Fay et al. 2013); FMT refers to the partially implantable (*PI*) incus-attached Symphonix Vibrant

bottom of each range). Although the CHD is designed to provide amplification over the widest frequency range of the seven devices (0.125–10 kHz), its worst-case HL of 80 dB is also on the lower end. The other three devices grouped in Fig. 10.6a are the PI MEHDs. Of these, the device allowing the most-severe HL (just above 115 dB) and widest bandwidth (0.1–9 kHz) is the FMT on the RW (Beltrame et al. 2009). The 0.5–4 kHz frequency range for the FMT attached to the incus is less inclusive than when it is attached to the RW (Luetje et al. 2002). The Otologics MET transducer includes a similar range of HLs as the FMT on the incus, but covers a wider range of frequencies (0.25–10 kHz).

The HL inclusion criteria for the three TI MEHDs are grouped together in Fig. 10.6b. The frequency range of 0.25–8 kHz for the Carina system is the widest of the TI systems (Zenner and Rodriguez Jorge 2010). The Carina's earlier incarnation, the FIMOS, only extended to 4 kHz (Jenkins et al. 2007). The worst-case HL of 80 dB for both of these Otologics systems is the same as that of the non-implantable CHD shown in Fig. 10.6a. On the other hand, the Esteem's specified bandwidth of 0.5–4 kHz is one of the narrowest, but its maximum allowable HL goes all the way to 100 dB (Kraus et al. 2011), which is second only to the FMT on the RW.

10.11.1 Hearing Levels in the Study Populations

While Fig. 10.6 shows the inclusion criteria specified by the different manufacturers of the devices, Fig. 10.7 shows the average HLs of the subjects recruited for each study. The subject populations generally fall into one of three regions. The CHD subject population has an average hearing impairment of 15–20 dB below 1 kHz and around 65 dB in the 8–10 kHz range, which is the least severe among the study populations and extends into the "mild" category of hearing loss. With the exception of the subjects with the FMT on the RW, the subjects for the other IHDs cluster together in the vicinity of 40 dB HL at 0.25 kHz and 70 dB HL in the 4–8 kHz range, which fall into the "moderate" to "severe" categories of hearing loss. The subject population with the FMT on the RW has

Fig. 10.6 (continued) Soundbridge FMT system (Luetje et al. 2002); MET refers to the PI Otologics Middle Ear Transducer Ossicular Stimulator system as reported in Jenkins et al. (2004); and RW FMT refers to the PI Vibrant Soundbridge FMT system with coupling to the round window instead of to the incus (Beltrame et al. 2009). (**b**) Carina refers to the totally implantable (*TI*) Otologics Carina system (Zenner and Rodriguez Jorge 2010) while FIMOS refers to the TI Otologics MET Fully Implantable Ossicular System III (Jenkins et al. 2007), both of which have an output transducer similar to the PI MET system shown in (**a**); and Esteem refers to the TI Envoy Esteem system (Kraus et al. 2011)

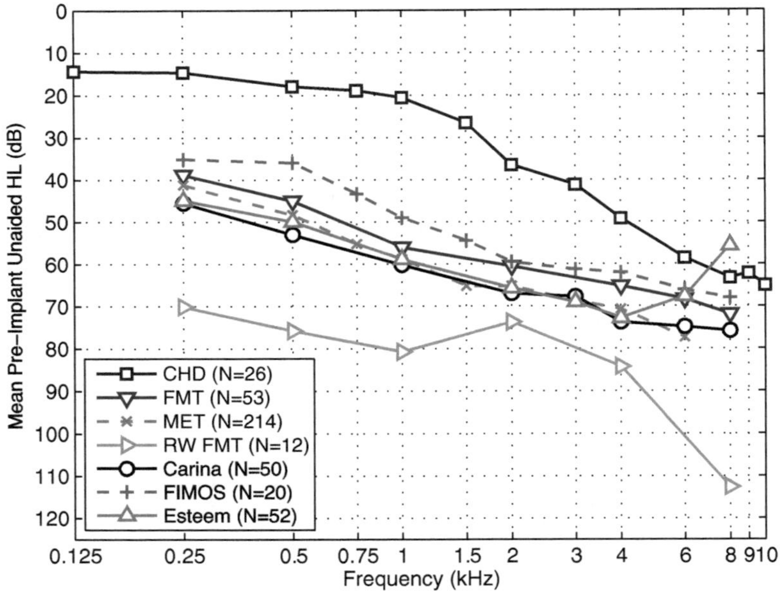

Fig. 10.7 Average HL before implantation for each device's clinical-trial subject population (see references in the caption to Fig. 10.6)

significantly more severe hearing impairments of 70–112 dB HL, depending on the frequency, which extends into the "profound" category of hearing loss.

10.11.2 Functional Gains Across Studies

One of the most important ways to evaluate an MEHD is by the amount of functional gain that it can provide as a function of frequency. This gain is the difference between the audiometric threshold obtained before implantation (or placement) and that obtained after the device has been turned on postimplantation (or postplacement). Generally speaking, the measured gain depends on the amount of gain prescribed by the fitting algorithm. A complication here is that the hearing devices typically employ multiband compression algorithms in which the gain depends on the input level. The gain is highest for low input levels, and above the compression knee point it decreases until it approaches unity at the highest input levels. Therefore, measurements of the functional gain near the threshold of hearing (in which the effects of compression are minimal) can be considered to represent the maximum gain for a given HL.

Figure 10.8a shows the mean functional gains corresponding to the HLs shown in Fig. 10.7. The two FMT-based PI MEHDs attain large peak gains of about 38–43 dB in the 1–4 kHz range. At frequencies above 4 kHz, their gain either begins to decrease (FMT) or is not reported (RW FMT). A large peak gain of just over 35 dB at 2 kHz is attained by the Esteem, but it decreases for frequencies above this, with 6 dB of gain remaining at 8 kHz. The three Otologics systems (MET, Carina, and FIMOS) follow similar gain contours in the 0.25–6 kHz range, but appear to be approximately shifted vertically from one another. The MET also has a large peak gain of 38 dB at 1.5 kHz and still offers about 27 dB of gain at 6 kHz. When the MET transducer is used in the TI FIMOS system, the peak gain was only about 14 dB at 2 kHz (Jenkins et al. 2007). Changes in the subsequent Carina system result in a higher peak of about 24 dB at 2 kHz (Zenner and Rodriguez Jorge 2010).

The NI CHD reaches its peak gain of about 22 dB in the 6–10 kHz range. The (highly repeatable) dip in the gain at 8 kHz is due to the use of a fixed KEMAR transfer function in the calibration procedure. Because the average HL of the subjects in the CHD study is in the normal range at 0.5 kHz and below (Fig. 10.7), no gain was prescribed in this range; this is reflected in the measured gain values for the CHD being close to zero at 0.5 kHz and below (Fig. 10.8a). The CHD has the widest reported range of 0.125–10 kHz, with substantial measurable gain at frequencies above 4 kHz.

10.11.3 Functional Gains Normalized by HL

As was shown in Fig. 10.7, the initial HLs of the different study populations are not the same, and the amount of programmed gain for a given device is a function of the amount of hearing impairment at a given frequency. Thus it is not surprising that the gains of the various devices in Fig. 10.8a differ from one another and depend on frequency. Furthermore, a reported peak gain is not necessarily the maximum gain that a device could provide if it were programmed with the highest device HLs shown in Fig. 10.6. One way to normalize the measured gains is to subtract out the programmed gain. However, the programmed gains used in the different studies are generally not reported. One common fitting prescription is to use the "1/2-gain rule," for which the prescribed gain is simply 1/2 times the subject's HL. So, if a subject has a 60 dB hearing impairment at 6 kHz, then by the 1/2-gain rule the fitting algorithm would prescribe a gain at threshold of 30 dB for 6 kHz. Another fitting prescription is the "1/3-gain rule," in which case the prescribed gain in this example would be a less aggressive 20 dB.

Figure 10.8b shows the measured gain of Fig. 10.8a for each device, but with a corresponding gain prescription, calculated using the 1/3-gain rule based on the average HLs shown in Fig. 10.7, subtracted from it. These curves will be referred to as the "normalized gain." Results below 0 dB indicate that the actual gain prescribed for the device is typically less than that from the 1/3-gain rule, and those above 0 dB indicate that the actual gain prescribed for the device is typically higher than that from the 1/3-gain rule.

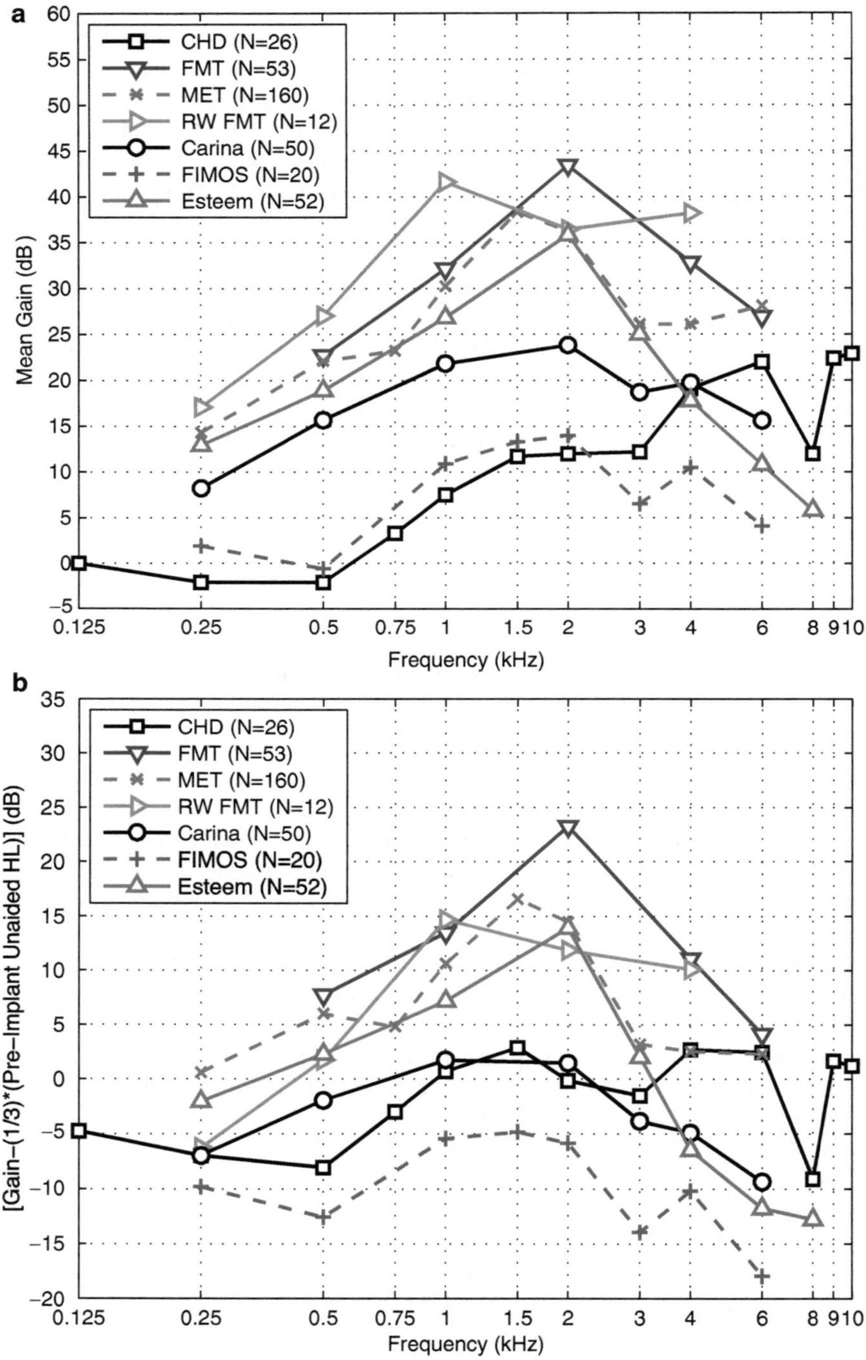

Fig. 10.8 (**a**) Mean functional gain from the clinical trial for each device type (see references in the Fig. 10.6 caption). The prescribed gain depends on the HL of each subject, which contributes to

Generally the FMT on the incus has the highest normalized gain of 15–25 dB above 0 dB (FMT), and the normalized gain also ends up being fairly high when the FMT is placed on the RW (RW FMT). In the 1–2 kHz range the MET results lie 10–15 dB above 0 dB. These systems appear to follow a prescription closer to the 1/2-gain rule, at least in their mid-frequencies. The Esteem also prescribes a gain higher than that from the 1/3-gain rule at 2 kHz, but it falls off below 0 dB above 3 kHz.

The CHD has a lower measured gain below 1 kHz (Fig. 10.8a) because the HL at those frequencies was already in the normal hearing range (Fig. 10.7), and thus no gain was prescribed. In the 6–10 kHz range, its normalized gain is typically higher than those reported for the TI systems.

10.12 Summary

Middle ear and cochlear pathologies lead to various hearing impairments. Although a variety of available acoustic hearing aids have been the standard of care for mild to severe cochlear hearing impairment for decades, these solutions do not work for all patients due to individual variations in ear anatomy as well as the degree and spectrum of hearing impairment. In addition, lifestyle preferences can also dictate the need for alternate, less conspicuous, and more convenient solutions. Creating devices to meet all of these demands has been the quest of engineers and clinical practitioners for some time, but the technical and regulatory challenges have been many.

Decades of progress have led to the development of middle ear hearing devices (MEHDs), and these choices are now finally becoming available to ear surgeons and audiologists for their patients to consider. Many of these devices are now possible because of vast technological improvements, but they have also come about due to improved scientific understanding of the structure–function relationships of the middle ear.

As with many new technologies, there are pros and cons for MEHDs relative to conventional acoustic hearing aids. One of the biggest drawbacks of acoustic hearing aids is that in noisy situations they typically do not work well, and many people that have them tend to take them out and put them in their purse or pocket during these times. One demonstrated way to alleviate this has been to extend the upper end of their frequency range from the current 4–5 kHz in acoustic hearing

Fig. 10.8 (continued) the variability seen across the different studies. (**b**) Mean normalized functional gains, in which an amplification prescription based on a 1/3-gain rule has been subtracted from each of the measured gains in order to compensate for some of the variability across subject populations. The 1/3-gain rule is the prescribed gain at low levels equal to 1/3 times the unaided pre-implantation HL. Deviations from 0 dB here indicate the degree by which a device varies from the 1/3-gain target

aids up to about 10 kHz. MEHD output transducers have been designed de novo for mechanical vibrational output rather than acoustic output, and have the potential to achieve wider bandwidths than acoustic hearing aids. Devices that mechanically stimulate the middle ear are the totally implantable (TI) types, the partially implantable (PI) types, and systems being developed that do not require surgical implantation (NI). These MEHDs typically leave the ear canal open or widely vented, yet they are still capable of delivering amplification at low frequencies—unlike open-canal acoustic hearing aids. These MEHDs have been shown to be preferred over acoustic hearing aids in some real-world situations, as measured using self-reporting questionnaire-type evaluations.

Several decades ago the field of ophthalmology began to give patients options other than eyeglasses. Now the treatment of hearing loss is beginning to enter such an era, as novel and differentiated solutions start to become increasingly available. Over the coming decades, hearing scientists, otologists, engineers, and audiologists will continue to work together to change the standard of hearing health care and bring a wider variety of treatment options to those with hearing impairment.

Acknowledgments The author thanks Kevin N. O'Connor for generating figures, providing significant editorial assistance, and providing general help in putting this book chapter together. The author also thanks Geoffrey R. Ball, Suzanne Carr Levy, Rodney Perkins, Jason Shelton, and Richard L. Goode for critical comments and suggestions on draft versions of this chapter. This work was supported by grants R01 DC 005960 and R448499 from the National Institute on Deafness and Other Communication Disorders (NIDCD) of the National Institutes of Health.

Disclosure Dr Puria declares that he has a financial interest in the EarLens Corporation which is in the process of developing the CHD discussed in this chapter.

References

Ball, G. R. (1996). Implantable electromagnetic hearing transducer. USA Patent Pub. No. 5,554,096. USPTO.

Ball, G. R. (2011). *No more laughing at the deaf boy: A technological adventure from Silicon Valley and the Alps*. Innsbruck: Haymon Verlag.

Ball, G. R., Huber, A., & Goode, R. L. (1997). Scanning laser Doppler vibrometry of the middle ear ossicles. *Ear, Nose, & Throat Journal*, 76(4), 213–218, 220, 222.

Ball, G. R., Robertson, W., III, & Julian, C. A. (1999). Two stage implantable microphone. USA Patent Pub. No. 5,859,916. USPTO.

Beltrame, A. M., Martini, A., Prosser, S., Giarbini, N., & Streitberger, C. (2009). Coupling the Vibrant Soundbridge to cochlea round window: Auditory results in patients with mixed hearing loss. *Otology & Neurotology*, 30(2), 194–201.

Bernardeschi, D., Hoffman, C., Benchaa, T., Labassi, S., Beliaeff, M., Sterkers, O., & Grayeli, A. B. (2011). Functional results of Vibrant Soundbridge middle ear implants in conductive and mixed hearing losses. *Audiology and Neuro-Otology*, 16(6), 381–387.

Bernhard, H., Stieger, C., & Perriard, Y. (2011). Design of a semi-implantable hearing device for direct acoustic cochlear stimulation. *IEEE Transactions on Biomedical Engineering*, 58(2), 420–428.

Beutner, D., & Hüttenbrink, K. B. (2009). Passive and active middle ear implants. *GMS Current Topics in Otorhinolaryngology, Head and Neck Surgery*, 8, Doc09.

Bruschini, L., Forli, F., Santoro, A., Bruschini, P., & Berrettini, S. (2009). Fully implantable Otologics MET Carina device for the treatment of sensorineural hearing loss. Preliminary surgical and clinical results. *Acta Otorhinolaryngologica Italica*, 29(2), 79–85.

Bruschini, L., Forli, F., Passetti, S., Bruschini, P., & Berrettini, S. (2010). Fully implantable Otologics MET Carina(™) device for the treatment of sensorineural and mixed hearing loss: Audio-otological results. *Acta Oto-Laryngologica*, 130(10), 1147–1153.

Carlile, S., & Schonstein, D. (2006). Frequency bandwidth and multi-talker environments. *Proceedings of the 120th Convention of the Audio Engineering Society*, 118(1), 353–363.

Chen, D. A., Backous, D. D., Arriaga, M. A., Garvin, R., Kobylek, D., Littman, T., Walgren, S., & Lura, D. (2004). Phase 1 clinical trial results of the Envoy System: a totally implantable middle ear device for sensorineural hearing loss. *Otolaryngology – Head and Neck Surgery*, 131(6), 904–916.

Cho, J. H. (2004). Middle ear hearing aid transducer. USA Patent Pub. No. 6,735,318. USPTO.

Colletti, V., Soli, S. D., Carner, M., & Colletti, L. (2006). Treatment of mixed hearing losses via implantation of a vibratory transducer on the round window. *International Journal of Audiology*, 45(10), 600–608.

Colletti, V., Carner, M., & Colletti, L. (2009). TORP vs round window implant for hearing restoration of patients with extensive ossicular chain defect. *Acta Oto-Laryngologica*, 129(4), 449–452.

Donahue, A., Dubno, J. R., & Beck, L. (2010). Guest editorial: Accessible and affordable hearing health care for adults with mild to moderate hearing loss. *Ear and Hearing*, 31(1), 2–6.

Fay, J. P., Puria, S., Rucker, P., Winstead, J. H., & Perkins, R. C. (2009). Energy delivery and microphone placement methods for improved comfort in an open ear canal hearing aid. USA Patent Pub. No. US 2009/0092271. USPTO.

Fay, J. P., Puria, S., Felsenstein, L., Stone, J., & Pluvinage, V. (2010). Optomechanical electromechanical hearing devices with combined power and signal architecture. USA Patent Pub. No. 2010/0034409 A1. USPTO.

Fay, J. P., Perkins, R., Levy, S. C., Nilsson, M., & Puria, S. (2013). Preliminary evaluation of a light based Contact Hearing Device for the hearing impaired. *Otology & Neurotology* (in press).

Fraysse, B., Lavieille, J. P., Schmerber, S., Enée, V., Truy, E., Vincent, C., Vaneecloo, F. M., & Sterkers, O. (2001). A multicenter study of the Vibrant Soundbridge middle ear implant: Early clinical results and experience. *Otology & Neurotology*, 22(6), 952–961.

Goode, R. L. (1970). An implantable hearing aid. State of the art. *Transactions – American Academy of Ophthalmology & Otolaryngology*, 74(1), 128–139.

Häusler, R., Stieger, C., Bernhard, H., & Kompis, M. (2008). A novel implantable hearing system with direct acoustic cochlear stimulation. *Audiology and Neuro-Otology*, 13(4), 247–256.

Huber, A. M., Ball, G. R., Veraguth, D., Dillier, N., Bodmer, D., & Sequeira, D. (2006). A new implantable middle ear hearing device for mixed hearing loss: A feasibility study in human temporal bones. *Otology & Neurotology*, 27(8), 1104–1109.

Hüttenbrink, K. B., Beutner, D., & Zahnert, T. (2010). Clinical results with an active middle ear implant in the oval window. *Advances in Oto-Rhino-Laryngology*, 69, 27–31.

Jenkins, H. A., Niparko, J. K., Slattery, W. H., Neely, J. G., & Fredrickson, J. M. (2004). Otologics Middle Ear Transducer Ossicular Stimulator: Performance results with varying degrees of sensorineural hearing loss. *Acta Oto-Laryngologica*, 124(4), 391–394.

Jenkins, H. A., Atkins, J. S., Horlbeck, D., Hoffer, M. E., Balough, B., Arigo, J. V., Alexiades, G., & Garvis, W. (2007). U.S. Phase I preliminary results of use of the Otologics MET Fully-Implantable Ossicular Stimulator. *Otolaryngology – Head and Neck Surgery*, 137(2), 206–212.

Jenkins, H. A., Atkins, J. S., Horlbeck, D., Hoffer, M. E., Balough, B., Alexiades, G., & Garvis, W. (2008). Otologics fully implantable hearing system: Phase I trial 1-year results. *Otology & Neurotology*, 29(4), 534–541.

Jung, E. S., Seong, K. W., Lim, H. G., Lee, J. H., & Cho, J. H. (2011). Implantable microphone with acoustic tube for fully implantable hearing devices. *IEICE Electronics Express*, 8(4), 215–219.

Kiefer, J., Arnold, W., & Staudenmaier, R. (2006). Round window stimulation with an implantable hearing aid (Soundbridge) combined with autogenous reconstruction of the auricle—a new approach. *ORL; Journal for Oto-Rhino-Laryngology and its Related Specialties*, 68(6), 378–385.

Kim, M. W., Kim, M. K., Yoon, Y. H., Kim, S. H., Park, I. Y., & Cho, J. H. (2007). Electrical model and analysis of vibration characteristic of differential floating mass transducer for fully-implantable middle ear hearing devices. *World Congress on Medical Physics and Biomedical Engineering 2006. IFMBE Proceedings*, Vol. 14, 3208–3211.

Kochkin, S. (2002). MarkeTrak VI: Hearing aid industry market tracking survey 1984–2000. Retrieved from http://www.knowles.com/search/ppt/MarkeTrak6.ppt. Accessed August 20, 2012.

Kochkin, S. (2009). MarkeTrak VIII: 25-Year trends in the hearing health market. *Hearing Review*, 16(11), 12–31.

Komori, M., Yanagihara, N., Hinohira, Y., Hato, N., & Gyo, K. (2010). Long-term results with the Rion E-type semi-implantable hearing aid. *Otolaryngology – Head and Neck Surgery*, 143(3), 422–428.

Komori, M., Yanagihara, N., Hinohira, Y., Hato, N., & Gyo, K. (2012). Re-implantation of the Rion E-type semi-implantable hearing aid: Status of long-term use and hearing outcomes in eight patients. *Auris, Nasus, Larynx*, 39(6), 572–576.

Kraus, E. M., Shohet, J. A., & Catalano, P. J. (2011). Envoy Esteem totally implantable hearing system: Phase 2 trial, 1-year hearing results. *Otolaryngology – Head and Neck Surgery*, 145(1), 100–109.

Labassi, S., & Beliaeff, M. (2005). Retrospective of 1000 patients implanted with a vibrant Soundbridge middle-ear implant. *Cochlear Implants International*, 6 (Suppl 1), 74–77.

Lee, C. F., Shih, C. H., Yu, J. F., Chen, J. H., Chou, Y. F., & Liu, T. C. (2008). A novel opto-electromagnetic actuator coupled to the tympanic membrane. *Journal of Biomechanics*, 41(16), 3515–3518.

Lenarz, T., Weber, B. P., Mack, K. F., Battmer, R. D., & Gnadeberg, D. (1998). [The Vibrant Soundbridge System: A new kind of hearing aid for sensorineural hearing loss. 1: Function and initial clinical experiences]. *Laryngorhinootologie*, 77(5), 247–255.

Lesinski, S. G., & Neukermans, A. P. (1998). Implantable hearing aid. USA Patent Pub. No. 5,772,575. USPTO.

Luetje, C. M., Brackman, D., Balkany, T. J., Maw, J., Baker, R. S., Kelsall, D., Backous, D., Miyamoto, R., Parisier, S., & Arts, A. (2002). Phase III clinical trial results with the Vibrant Soundbridge implantable middle ear hearing device: A prospective controlled multi-center study. *Otolaryngology – Head and Neck Surgery*, 126(2), 97–107.

Moore, B. C., & Tan, C. T. (2003). Perceived naturalness of spectrally distorted speech and music. *The Journal of the Acoustical Society of America*, 114(1), 408–419.

Moore, B. C., Stone, M. A., & Alcántara, J. I. (2001). Comparison of the electroacoustic characteristics of five hearing aids. *British Journal of Audiology*, 35(5), 307–325.

Moore, B. C., Füllgrabe, C., & Stone, M. A. (2010). Effect of spatial separation, extended bandwidth, and compression speed on intelligibility in a competing-speech task. *The Journal of the Acoustical Society of America*, 128(1), 360–371.

Mosnier, I., Sterkers, O., Bouccara, D., Labassi, S., Bebear, J.-P., Bordure, P., Dubreuil, C., Dumon, T., Frachet, B., Fraysse, B., Lavieille, J.-P., Magnan, J., Martin, C., Meyer, B., Mondain, M., Portmann, D., Robier, A., Schmerber, S., Thomassin, J.-M., Truy, E., Uziel, A., Vanecloo, F.-M., Vincent, C., & Ferrary, E. (2008). Benefit of the Vibrant Soundbridge device in patients implanted for 5 to 8 years. *Ear and Hearing*, 29(2), 281–284.

Murugasu, E., Puria, S., & Roberson, J. B., Jr. (2005). Malleus-to-footplate versus malleus-to-stapes-head ossicular reconstruction prostheses: Temporal bone pressure gain measurements and clinical audiological data. *Otology & Neurotology*, 26(4), 572–582.

Nakajima, H. H., Dong, W., Olson, E. S., Rosowski, J. J., Ravicz, M. E., & Merchant, S. N. (2010). Evaluation of round window stimulation using the floating mass transducer by intracochlear sound pressure measurements in human temporal bones. *Otology & Neurotology*, 31(3), 506–511.

O'Connor, K. N., & Puria, S. (2006). Middle ear cavity and ear canal pressure-driven stapes velocity responses in human cadaveric temporal bones. *The Journal of the Acoustical Society of America*, 120(3), 1517–1528.

Park, I. Y., Shimizu, Y., O'Connor, K. N., Puria, S., & Cho, J. H. (2011). Comparisons of electromagnetic and piezoelectric floating-mass transducers in human cadaveric temporal bones. *Hearing Research*, 272(1–2), 187–192.

Park, W. T., O'Connor, K. N., Chen, K. L., Mallon, J. R., Jr., Maetani, T., Dalal, P., Candler, R. N., Ayanoor-Vitikkate, V., Roberson, J. B., Jr., Puria, S., & Kenny, T. W. (2007). Ultraminiature encapsulated accelerometers as a fully implantable sensor for implantable hearing aids. *Biomedical Microdevices*, 9(6), 939–949.

Perkins, R. (1996). Earlens tympanic contact transducer: A new method of sound transduction to the human ear. *Otolaryngology – Head and Neck Surgery*, 114(6), 720–728.

Perkins, R. C., & Shennib, A. A. (1993). Contact transducer for hearing devices. USA Patent Pub. No. 5,259,032. USPTO.

Perkins, R., Puria, S., Fay, J., & Winstead, J. H. (2007). Output transducers for hearing systems. USA Patent Pub. No. 2007/0100197. USPTO.

Perkins, R., Fay, J. P., Rucker, P., Rosen, M., Olson, L., & Puria, S. (2010). The EarLens system: New sound transduction methods. *Hearing Research*, 263(1–2), 104–113

Pluvinage, V., & Perkins, R. C. (2011). Systems and methods for photo-mechanical hearing transduction. USA Patent Pub. No. 7,867,160. USPTO.

Puria, S. (2003). Measurements of human middle ear forward and reverse acoustics: implications for otoacoustic emissions. *The Journal of the Acoustical Society of America*, 113(5), 2773–2789.

Puria, S., & Perkins, R. C. (2003a). Flextensional output actuators for surgically implantable hearing aids. USA Patent Pub. No. US 6,629,922. USPTO.

Puria, S., & Perkins, R. C. (2003b). Flextensional microphones for implantable hearing devices. USA Patent Pub. No. 6,554,761. USPTO.

Puria, S., Vermiglio, A. J., Sr., Fay, J. P., & Soli, S. D. (2008). Hearing restoration: Better multitalker speech understanding. Presented at the Combined Otolarygological Spring Meeting (COSM) in Orlando, FL, May 1–4.

Puria, S., & Perkins, R. C. (2010). Hearing system having an open chamber for housing components and reducing the occlusion effect. USA Patent Pub. No. 7,668,325. USPTO.

Puria, S., Perkins, R. C., & Rucker, P. (2010a). Optically coupled acoustic middle ear implant systems and methods. USA Patent Pub. No. 2010/0312040 A1

Puria, S., Fay, J. P., Felsenstein, L., Stone, J., Killion, M., & Pluvinage, V. (2010b). Optomechanical electro-mechanical hearing devices with separate power and signal components. USA Patent Pub. No. 2010/0034409 A1. USPTO.

Puria, S., Rosen, M., Fay, J. P., Rucker, P., & Stone, J. (2012). Balanced armature devices and methods for hearing. USA Patent Pub. No. 13/069,262. USPTO.

Rutschmann, J. (1959). Magnetic audition—Auditory stimulation by means of alternating magnetic fields acting on a permanent magnet fixed to the eardrum. *IRE Transactions on Medical Electronics*, *ME*-6(1), 22–23.

Schmerber, S., Troussier, J., Dumas, G., Lavieille, J. P., & Nguyen, D. Q. (2006). Hearing results with the titanium ossicular replacement prostheses. *European Archives of Oto-Rhino-Laryngology*, 263(4), 347–354.

Schmuziger, N., Schimmann, F., Wengen, D., Patscheke, J., & Probst, R. (2006). Long-term assessment after implantation of the Vibrant Soundbridge device. *Otology & Neurotology*, 27(2), 183–188.

Shohet, J. A., Kraus, E. M., & Catalano, P. J. (2011). Profound high-frequency sensorineural hearing loss treatment with a totally implantable hearing system. *Otology & Neurotology*, 32(9), 1428–1431.

Silverstein, H., Atkins, J., Thompson, J. H., Jr., & Gilman, N. (2005). Experience with the SOUNDTEC implantable hearing aid. *Otology & Neurotology*, 26(2), 211–217.

Snik, A. F., & Cremers, C. W. (1999). First audiometric results with the Vibrant soundbridge, a semi-implantable hearing device for sensorineural hearing loss. *Audiology*, 38(6), 335–338.

Snik, A. F., & Cremers, C. W. (2001). Vibrant semi-implantable hearing device with digital sound processing: Effective gain and speech perception. *Archives of Otolaryngology – Head & Neck Surgery*, 127(12), 1433–1437.

Song, B. S., Park, J. H., Yoon, Y. H., Kim, M. N., Park, S. K., Lee, S. H., & Cho, J. H. (2000). Differential floating mass type vibration transducer for MEI system. *Engineering in Medicine and Biology, 2000. Proceedings of the 22nd Annual International Conference of the IEEE*, Vol. 4, 2575–2578.

Tjellstrom, A., Luetje, C. M., Hough, J. V., Arthur, B., Hertzmann, P., Katz, B., & Wallace, P. (1997). Acute human trial of the floating mass transducer. *Ear, Nose, & Throat Journal*, 76(4), 204–206, 209–210.

Tysome, J. R., Moorthy, R., Lee, A., Jiang, D., & O'Connor, A. F. (2010). Systematic review of middle ear implants: do they improve hearing as much as conventional hearing AIDS? *Otology & Neurotology*, 31(9), 1369–1375.

Valente, M. (2002). *Hearing aids: Standards, options, and limitations* (2nd ed.). New York: Thieme.

Wilska, A. (1959). A direct method for determining threshold amplitudes of the eardrum at various frequencies. In H. G. Kobrak (Ed.), *The middle ear* (pp. 76–79). Chicago: University of Chicago Press.

Wollenberg, B., Beltrame, M., Schönweiler, R., Gehrking, E., Nitsch, S., Steffen, A., & Frenzel, H. (2007). [Integration of the active middle ear implant Vibrant Soundbridge in total auricular reconstruction]. *HNO*, 55(5), 349–356.

Yanagihara, N., Sato, H., Hinohira, Y., Gyo, K., & Hori, K. (2001). Long-term results using a piezoelectric semi-implantable middle ear hearing device: The Rion Device E-type. *Otolaryngologic Clinics of North America*, 34(2), 389–400.

Zehlicke, T., Dahl, R., Just, T., & Pau, H. W. (2010). Vibroplasty involving direct coupling of the floating mass transducer to the oval window niche. *The Journal of Laryngology and Otology*, 124(7), 716–719.

Zenner, H. P., & Leysieffer, H. (2001). Total implantation of the Implex TICA hearing amplifier implant for high-frequency sensorineural hearing loss—The Tubingen University experience. *Otolaryngologic Clinics of North America*, 34(2), 417–446.

Zenner, H. P., & Rodriguez Jorge, J. (2010). Totally implantable active middle ear implants: Ten years' experience at the University of Tubingen. *Advances in Oto-Rhino-Laryngology*, 69, 72–84.

Zenner, H. P., Maassen, M. M., Plinkert, P. K., Zimmermann, R., Baumann, J. W., Reischl, G., & Leysieffer, H. (1998). [First implantation of a totally implantable electronic hearing aid in patients with inner ear hearing loss]. *HNO*, 46(10), 844–852.

Zenner, H. P., Limberger, A., Baumann, J. W., Reischl, G., Zalaman, I. M., Mauz, P. S., Sweetow, R. W., Plinkert, P. K., Zimmermann, R., Baumann, I., De Maddalena, H., Leysieffer, H., & Maassen, M. M. (2004). Phase III results with a totally implantable piezoelectric middle ear implant: Speech audiometry, spatial hearing and psychosocial adjustment. *Acta Oto-Laryngologica*, 124(2), 155–164.

MIX
Papier aus verantwortungsvollen Quellen
Paper from responsible sources
FSC® C105338

If you have any concerns about our products,
you can contact us on
ProductSafety@springernature.com

In case Publisher is established outside the EU,
the EU authorized representative is:
Springer Nature Customer Service Center GmbH
Europaplatz 3, 69115 Heidelberg, Germany

Printed by Libri Plureos GmbH
in Hamburg, Germany